ADVANCED GRAVITATIONAL WAVE DETECTORS

After decades of research, physicists now know how to detect Einstein's gravitational waves. Advanced gravitational wave detectors, the most sensitive instruments ever created, will be almost certain to detect the births of black holes throughout the Universe. This book describes the physics of gravitational waves and their detectors.

The book begins by introducing the physics of gravitational wave detection and the likely sources of detectable waves. Case studies on the first generation of large-scale gravitational wave detectors introduce the technology and set the scene for a review of the experimental issues in creating advanced detectors in which the instrument's sensitivity is limited by Heisenberg's Uncertainty Principle. The book covers lasers, thermal noise, vibration isolation, interferometer control and stabilisation against opto-acoustic instabilities. This is a valuable reference for graduate student and researchers in physics and astrophysics entering this field.

D. G. BLAIR is Director of the Australian International Gravitational Research Centre (AIGRC), The University of Western Australia.

E. J. HOWELL is a Research Fellow at the The University of Western Australia.

L. JU is an Associate Professor at the Australian International Gravitational Research Centre, The University of Western Australia.

C. ZHAO is Research Director of Gingin High Optical Power Facility (HOPF) and Associate Professor at the Australian International Gravitational Research Centre, The University of Western Australia.

"This book is not only a monograph on advanced gravitational wave detectors and the astrophysical phenomena they will explore, it also contains a pedagogically fine introduction to the field of gravitational wave science. I recommend it to any budding or mature scientist or engineer who wants an overview of this exciting field and where it is going."
  Kip. S. Thorne, Feynman Professor of Theoretical Physics, Emeritus, Caltech

"Almost 100 years after Einstein introduced his Theory of General Relativity, we are finally on the threshold of making direct detections of gravitational waves … *Advanced Gravitational Wave Detectors* gives us an up-to-date view of the science and techniques for making the first detections and then developing yet more sensitive future detectors … This comprehensive review, written by experts in gravitational waves physics, covers these topics in depth and will serve as a very good introduction for students, while at the same time, being a valuable resource for practitioners in the field."
  Barry C. Barish, Linde Professor of Physics, Emeritus, Caltech

ADVANCED GRAVITATIONAL WAVE DETECTORS

Edited by

DAVID G. BLAIR
University of Western Australia, Perth

ERIC J. HOWELL
University of Western Australia, Perth

LI JU
University of Western Australia, Perth

CHUNNONG ZHAO
University of Western Australia, Perth

CAMBRIDGE UNIVERSITY PRESS
Cambridge, New York, Melbourne, Madrid, Cape Town,
Singapore, São Paulo, Delhi, Mexico City

Cambridge University Press
The Edinburgh Building, Cambridge CB2 8RU, UK

Published in the United States of America by Cambridge University Press, New York

www.cambridge.org
Information on this title: www.cambridge.org/9780521874298

© Cambridge University Press 2012

First published 2012

Printed in the United Kingdom at the University Press, Cambridge

A catalogue record for this publication is available from the British Library

Library of Congress Cataloguing in Publication data
Advanced gravitational wave detectors / David G. Blair ... [et al.].
p.   cm.
Includes bibliographical references and index.
ISBN 978-0-521-87429-8 (hardback)
1. Astronomical instruments.   2. Gravitational waves–Measurement–Instruments.
3. Laser interferometers.   4. Gravimeters (Geophysical instruments)
I. Blair, David G.
QB117.A38   2012
522$'$.68–dc23   2011052116

ISBN 978-0-521-87429-8 Hardback

In memory of Stefano Braccini, our co-author and respected colleague.

Contents

Contributors

David G. Blair
School of Physics, University of Western Australia, Crawley, WA 6009, Australia

Pablo Barriga
School of Physics, University of Western Australia, Crawley, WA 6009, Australia

Stefano Braccini
INFN, Sezione di Pisa, I-56127 Pisa, Italy

Yanbei Chen
Theoretical Astrophysics 130-33, California Institute of Technology, Pasadena, CA 91125, USA

Jerome Degallaix
GEO – Albert Einstein Institute, Hannover, Germany,
Leibniz Universität, Hannover, Germany

Jean-Charles Dumas
School of Physics, University of Western Australia, Crawley, WA 6009, Australia

Maik Frede
Laser Zentrum Hannover, Germany

Peter Fritschel
LIGO Laboratory, Massachusetts Institute of Technology,
Cambridge, MA 02139, USA

Slawomir Gras
School of Physics, University of Western Australia, Crawley, WA 6009, Australia

Hartmut Grote
GEO – Albert-Einstein-Institute, Hannover, Germany,
Leibniz Universität Hannover, Germany

Gregg Harry
Massachusetts Institute of Technology, NW22-257, 175 Albany Street, Cambridge, MA 02139, USA

Eric J. Howell
School of Physics, University of Western Australia, Crawley, WA 6009, Australia

Li Ju
School of Physics, University of Western Australia, Crawley, WA 6009, Australia

Brian Lantz
Ginzton Laboratory, Stanford University, Stanford CA 94305, USA

Ben Lee
School of Physics, University of Western Australia, Crawley, WA 6009, Australia

Harald Lück
GEO – Albert Einstein Institute, Hannover, Germany,
Leibniz Universität Hannover, Germany

Haixing Miao
Theoretical Astrophysics 130-33, California Institute of Technology, Pasadena, CA 91125,
USA,
School of Physics, University of Western Australia, Crawley, WA 6009, Australia

Michele Punturo
Istituto Nazionale di Fisica Nucleare (INFN), Perugia, Italy,
European Gravitational Observatory (EGO), Italy

Bangalore S. Sathyaprakash
School of Physics and Astronomy, Cardiff University, Cardiff, UK

Bernard F. Schutz
Max Planck Institute for Gravitational Physics (Albert Einstein Institute) Potsdam-Golm,
Germany,
School of Physics and Astronomy, Cardiff University, Cardiff, UK

Linqing Wen
Australian International Gravitational Research Centre, School of Physics, University of
Western Australia,Crawley, WA 6009, Australia,
International Centre for Radio Astronomy Research, University of Western Australia,
Crawley, WA 6009, Australia,
California Institute of Technology, Pasadena, CA 91125, USA

Benno Willke
GEO – Albert Einstein Institute, Hannover, Germany, Leibniz Universität, Hannover,
Germany

Chunnong Zhao
School of Physics, University of Western Australia, Crawley, WA 6009, Australia

Foreword

In 1905, Albert Einstein published a series of papers that revolutionised physics. They demonstrated the existence of molecules as physical entities, started the thinking that led to quantum mechanics, and laid the foundations of Special Relativity. Einstein then spent the next decade developing his Theory of General Relativity – a work that arguably was his greatest achievement. A key feature of general relativity was the prediction of the existence of gravitational waves. More generally, this new theory of gravity has come to be universally recognised as giving our best description of the Universe.

Now, almost 100 years after Einstein introduced his theory, we are finally on the threshold of making direct detections of gravitational waves. This greatly anticipated achievement will enable us to make rigorous tests of general relativity, as well as give us a completely new way to view the Universe.

Advanced Gravitational Wave Detectors, gives us an up-to-date view of the science and techniques for making the first detections and then developing yet more sensitive future detectors. There are many ingenious ideas and advanced technologies incorporated into the large-scale instruments that are poised to detect gravitational waves. The detections will come from neutron stars, black holes or other such objects that were unknown in Einstein's time. The techniques will involve lasers, photodiodes and digital data acquisition systems, also unknown at that time. This comprehensive review, written by experts in gravitational waves physics, covers these topics in depth and will serve as a very good introduction for students, while at the same time, being a valuable resource for practitioners in the field.

Barry C. Barish
Linde Professor of Physics, Emeritus,
California Institute of Technology

Preface

The detection of gravitational waves is sometimes described as the Holy Grail of Modern Physics. This is somewhat of a misnomer. Like the search for the holy grail, the search has appeared endless and fruitless, especially to non-scientific observers who cannot believe that it could take so long to make a detector, test it and come up with a firm answer. But unlike the search for the holy grail, physicists *know* that gravitational waves exist, not only from the beauty and elegance of Einstein's General Theory which predicts their existence, but also from the observations of binary pulsar systems which lose energy exactly in accordance with the theoretical predictions. This work by Joseph Taylor was rewarded with the 1993 Nobel Prize in physics.

The saga of gravitational wave detection goes back a long way: Einstein believed they existed but thought they were not physically detectable. Eddington queried their existence: he suggested that 'they travel at the speed of thought'. But in the 1950's Pirani, Feynman, Bondi and later Isaacson proved their physical reality, and in about 1960 Joseph Weber began to develop his famous resonant mass detectors. One now resides in the Smithsonian museum and another at one of LIGO's gravitational wave observatories. About 1970 his claims of detection (which turned out to be false) fired up a whole community. Astronomers were the most skeptical of Weber's results because they implied that thousands of stellar masses of matter were being turned into gravitational waves in the Milky Way every year. Weber's claims alerted physicists to the challenge and the possibility of detecting gravitational waves. A concentrated effort began, both to repeat his results, and to design vastly better detectors.

By 1975 there was a consensus that Weber's results were false. By this time a programme to build cryogenic versions of Weber's detectors was well underway, in the USA and Italy and a few years later in Australia. Cryogenic techniques and superconducting sensors offered at least a million-fold improvement in energy sensitivity to sharp bursts of gravity waves emanating from supernovae and neutron star births. At the time it appeared relatively straightforward to lower the detector temperatures to a degree or so above absolute zero, enabling the use of new superconducting vibration transducers, and simultaneously reducing thermal acoustic noise until vibrations as small as 10^{-20} metres could be detected, corresponding to gravitational wave strains in the 10^{-21} range.

The difficulty in creating cryogenic detectors was seriously underestimated. A whole range of new technologies had to be integrated, from ultra-low acoustic loss materials to new superconducting vibration sensors and amplifiers, to vibration isolation at a level never before encountered. There were numerous setbacks, including the vacuum implosion of one cryostat and the earthquake destruction of the detector at Stanford.

Only in the 1990s did five cryogenic detectors come into long-term operation. They set definitive limits on the strength and rate of gravitational wave bursts, and firmly disproved earlier claims. But even at vastly improved sensitivity, no signals were detected. This was not really a surprise since by this time likely sources were better understood and the bursts to which these detectors were most sensitive to were likely to be very rare – say once every century.

The detector builder's optimism had been sustained by the hope that there may have been a class of unknown sources waiting to be detected. The hope was dashed! Still, the resonant mass detectors set important upper limits which were not to be broken until an entirely new technology became operational in the first decade of the 21st century.

The detection of gravitational waves with laser interferometers was first considered in the 1960s, with experiments beginning in Europe and the USA in the 1970s and 1980s. There was a long period of setbacks and innovation as entirely new technologies were developed. Finally in the 1990s huge laser interferometer instruments began construction in USA (the LIGO project), in Italy (the French–Italian Virgo project), in Germany (the British–German GEO project) and in Japan (the TAMA project). These detectors have broad bandwidth, and are particularly designed to detect the final stages of the coalescence of neutron stars. In their last seconds such systems create rapidly rising chirp signals as the stars sweep around each other at up to 500 times per second.

Once built, the detectors took a massive effort as physicists learnt how to bring them into sensitive operation, but by 2009 the detectors had operated for long periods of time, and greatly exceeded all previous limits. Again, no signals were detected, but this time the limits began to place significant constraints on the astrophysics of sources.

The next step in the saga will take us well into the second decade of the 21st century, and will bring about, for the first time, detectors capable of detecting known gravitational wave sources at a frequent rate. Many years before they demonstrated that their detectors worked in accordance with predictions, the growing band of gravitational wave physicists had embarked on a worldwide collaboration to develop designs for detectors that would be able to reach reach into the Universe at least 10 times further than the first laser interferometers. Such detectors are known as 'advanced' detectors. The keys to the advances required are extremely high laser power, massive mirrors and new concepts in interferometer design. The physics of the detectors is presented here. At the frontier of knowledge there is room for new difficulties, but as I write there is a feeling that gravitational wave detection in imminent.

It is impossible to guess when the saga of gravity wave detection will end. While advanced detectors are constructed the next generation of third generation detectors is being planned. The field remains fascinating and intensely innovative. The next detectors will exceed the standard quantum measurement limit, and avoid the effects of classical environmental gravity gradient forces.

There appears no doubt that gravitational astronomy will give us a new sense. It will allow humanity to explore the beginning of time in the Big Bang, and the end of time in black holes. It will allow us to observe the most energetic events in the Universe through the detection of vibrations so tiny that 50 years ago they were beyond our imagination. Gravitational wave detectors will listen in to most of the visible Universe, making a census of the dark side of the Universe: black hole births, and stellar deaths. This book, written by experts in the field, introduces you to the physics of this exciting new field.

David G. Blair

Introduction

By late 2010 five large-scale laser interferometer gravitational wave detectors had been operating for several years at unprecedented sensitivity. They were searching for gravitational wave signals created by matter in its most extreme and exotic form – neutron stars, black holes and the Big Bang itself. The detectors were the most sensitive instruments ever created, able to detect fractional changes in spacetime geometry at the level of parts in 10^{23}, corresponding to the measurement of energy changes of less than 10^{-31} joules per hertz of bandwidth. Despite this extraordinary achievement, the sensitivity was about 10 times below the level where we could be confident of detecting predicted signals. For example, the mean time between detectable chirp signals from the coalescence of pairs of neutron stars was likely to be once every 50 years, so that in a year of operation the chance of detection was only about 2%.

Despite this pessimistic prognosis, many of the 1000 physicists in the worldwide collaborations involved with the above detectors remained optimistic that nature might to kind enough to provide a first signal. Optimism was high enough that a system had been put in place to alert optical telescopes to slew to the part of the sky corresponding to the arrival times of any significant event.

On 16 September 2010 a coincident signal appeared in LIGO detectors spaced 2000 kilometres apart in the USA. It was immediately recognised as a significant event, especially after it was also identified in the data of the Virgo detector in Italy. By triangulation of the arrival times, it was determined to have come from the direction of the constellation Canis Major, the Big Dog. Within minutes telescopes in Australia, France and Chile and in orbit had been automatically alerted and many images of the region of the sky were taken. The signal appeared to have come from a coalescing pair of black holes at a distance of order 100 million light years. The absence of an optical signal was not surprising.

Members of the collaborations were soon alerted, but we were all required to keep the event secret until two things had happened. First, the data needed to have been fully analysed and all possible ways that the signal could be a false positive needed to be considered. Second, we had to wait until the *blind injection* envelope was opened. We were reminded that we had agreed to the process of blind injections. In this process rare events *might* be injected into the detectors to test the ability of the detectors to distinguish between real signals and accidental glitches in the data.

It took many months to complete the data analysis. More than 100 scientists were involved in the data analysis. Groups pored over data channels searching for possible electrical, optical or acoustic interference. Different search algorithms were tested and compared. Eventually it was determined that the probability that the event was accidental was one in 7000 years. A paper announcing the first discovery of gravitational waves was written. Finally, on 14 March 2011, the collaboration met in Arcadia, California, where the blind injection envelope was to be opened.

A large lecture hall equipped with six data projectors was packed and more than 100 scientists all over the planet were present via internet connection. First, a series of presentations presented the scientific case for discovery of gravitational waves from a coalescing pair of black holes. People took bets on the event being real. Most people agreed that it was 99% certain to be an injection, and yet the tension and the suspense was palpable. A leading member of the collaboration declared that he was sure the event was real. Champagne glasses were brought out, and filled, and a thumb drive was handed to Jay Marx, director of LIGO. He plugged it in. A presentation appeared on all six screens entitled 'The Envelope'. One click and we had the answer: it was a blind injection! Still, we drank our champagne because this was really a triumph. The effort had proved that these enormously complex instruments could detect single rare events, and determine their nature. Rather than toasting to the discovery of gravitational waves, we toasted to Advanced LIGO, the detector that would mark the beginning of gravitational wave astronomy in about 2015, along with Advanced Virgo, LCGT in Japan, and hopefully an Australian detector.

As we write, gravitational waves have still not been detected. The LIGO detectors are being rebuilt to create the first *advanced* detectors. Later the Virgo detector will shut down for its upgrade. The GEO high frequency detector in Germany and two resonant mass gravitational wave detectors in Italy will continue to listen for a rare and strong event that could occur any time in our galaxy.

This book is designed to help train the young scientists who will be the explorers of the gravitational wave spectrum. The first chapters are intended for a general scientific audience, and for undergraduate level students. Here we discuss the breadth of gravitational wave spectrum, and proposed methods for exploring the spectrum, as well as the exquisite new technologies that make the exploration possible. Later chapters specialise on the different technical topics that combine to cover the entire field of ground-based interferometric gravitational wave detectors. This part is designed for advanced students and researchers in the field. Many of the chapters contain new and previously unpublished results.

All the authors of this book are members of the large international collaborations mentioned above. All the chapters were peer reviewed through the LIGO Scientific Collaboration. We are all grateful to the collaboration for this process which helped to ensure the quality and the accuracy of all the chapters. Additionally, we would like to thank and acknowledge Luciano Rezzolla and Anthony Mezzacappa for providing valuable insight on state-of-the-art simulations of gravitational wave burst sources. On behalf of all the authors we thank all our colleagues who helped and contributed to the work presented in this book.

The editors wish to acknowledge financial support from the West Australian Government Centres of Excellence scheme, the Australian Research Council and the University of Western Australia.

David G. Blair, Eric J. Howell, Li Ju and Chunnong Zhao
Australian International Gravitational Research Centre,
University of Western Australia
Perth, June 2011

Part 1

An introduction to gravitational wave astronomy
and detectors

1

Gravitational waves

D. G. Blair, L. Ju, C. Zhao and E. J. Howell

This chapter describes the theory of gravitational waves. We first introduce gravitational waves and describe how they are generated and propagate through space. We then show how the luminosity, frequency and amplitude of a gravitational wave source can be defined. A brief mathematical summary of how gravitational waves are a natural consequence of Einstein's general theory of relativity is then provided. To finish, we summarise some important quantities that are used to describe gravitational wave signal strengths and the response of detectors to different types of signal.

1.1 Listening to the Universe

Our sense of the Universe is provided predominantly by electromagnetic waves. During the 20th century the opening of the electromagnetic spectrum successively brought dramatic revelations. For instance, optical astronomy gave us the Hubble law expansion of the Universe. Radio astronomy gave us the cosmic background radiation, the giant radio jets powered by black holes in galactic nuclei, and neutron stars in the form of radio pulsars. X-ray astronomy gave us interacting neutron stars and black holes. Infrared astronomy gave us evidence for a massive black hole in the nucleus of our own galaxy.

Gravitational waves offer us a new sense with which to understand our Universe. If electromagnetic astronomy gives us eyes with which we can see the Universe, then gravitational wave astronomy offers us ears with which to hear it. We are presently deaf to the myriad gravitational wave sounds of the Universe. Imagine you are in a forest: you see a steep hillside, massive trees and small shrubs, bright flowers and colourful birds flitting between the trees. But there is much more to a forest: the sound of the wind in the treetops, the occasional crash of a falling branch, the thump thump of a fleeing kangaroo, the pulse of cicadas, the whistles of parrots and honking of bell frogs. The sense of hearing dramatically enriches our experience.

The gravitational wave universe is likely to be rich with 'sounds' across a frequency range from less than one cycle per year (the nanohertz band) up to tens of kilohertz. Sources emitting in the audio frequency band are detectable using Earthbased detectors. Observations in the microhertz to millihertz range require space-based detectors, while radio pulsar timing can detect in the nanohertz band. At even lower frequencies a gravitational wave signature from the Big Bang is expected to be frozen in the cosmic microwave background radiation.

Gravitational waves are produced whenever there is a non-spherical acceleration of mass–energy distributions. The nanohertz to millihertz frequencies consist of highly redshifted signals from the very early Universe, the slow interactions of very massive black holes and a weak background from binary star systems. Signal frequencies often scale inversely with the mass of the relevant systems. Black holes below 100 solar masses and neutron stars will produce gravitational waves in the audio frequency range: nearly monochromatic whistles from millisecond pulsars, short bursts from their formation, and chirps as binary systems spiral together and finally coalesce. Past experience tells us that our imagination and ability to predict is often limited. The sources we predict today may be just a fraction of what we will hear when advanced detectors (under construction at the time of writing) and future third generation detectors are operating at sufficient sensitivity.

Gravitational waves are waves of tidal force. They are vibrations of spacetime which propagate through space at the speed of light. They are registered as tiny vibrations in the relative spacings of carefully isolated masses. Their detection is primarily an experimental science, consisting of the development of the necessary ultra-sensitive measurement techniques. While gravitational waves can be considered as classical waves, the measurement systems must be treated quantum mechanically since the expected signals generally approach the limits set by the uncertainty principle.

The binary pulsar system PSR 1913+16 has played a key role in the unfolding story of gravitational waves. This system has proved Einstein's theory of general relativity to high precision, including the quadrupole formula which states that the total emitted gravitational wave power from any system is proportional to the square of the third time derivative of the system's quadrupole moment. The system loses energy exactly as predicted by this formula (Weisberg and Taylor, 1984; Weisberg and Taylor, 2005). Figure 1.1 shows the impressive fit of the measured values with the relativistically predicted accumulated shift in periastron (point of closest approach) due to orbital decay. Hulse and Taylor, who discovered the system in 1974 (Hulse and Taylor, 1975), were rewarded by a Nobel prize almost 20 years later. By this time careful monitoring had shown a gravitational wave energy loss from the system in agreement with theory to better than 1%.

1.2 Gravitational waves in stiff-elastic spacetime

In Newtonian physics spacetime is an infinitely rigid conceptual grid. Gravitational waves cannot exist in this theory. They would have infinite velocity and infinite energy density, because in Newtonian gravitation the metrical stiffness of space is infinite. Conversely, general relativity introduces a finite coupling coefficient between curvature of spacetime, described by the Einstein curvature tensor, and the stress energy tensor which describes the mass–energy which gives rise to the curvature. This coupling is expressed by the Einstein equation

$$\mathbf{T} = \frac{c^4}{8\pi G} \mathbf{G}, \qquad (1.1)$$

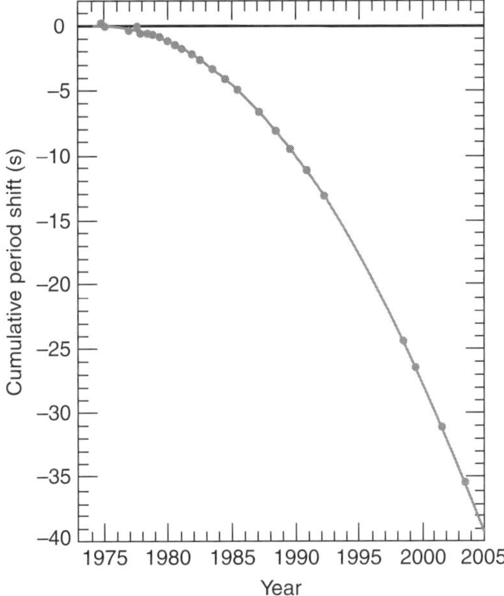

Figure 1.1 The curve indicates the general relativistically predicted accumulated shift in periastron for PSR 1913 + 16 due to the orbital decay by gravitational radiation. The data points indicate the measured values of the epoch of periastron (courtesy of Wikimedia Commons; data from Weisberg and Taylor, 2005).

Here $\mathbf{T}$ is the stress energy tensor and $\mathbf{G}$ is the Einstein curvature tensor, c is the speed of light and G is Newton's gravitational constant. The coupling coefficient, $c^4/8\pi G$, is an enormous number of order 10^{43}. This expresses the extremely high stiffness of space, which is the reason that the Newtonian law of gravitation is an excellent approximation in normal circumstances, and why gravitational waves have a small amplitude, even when their energy density is very high.

The existence of gravitational waves is intuitively obvious as soon as one recognises that spacetime is an elastic medium. The basic properties of gravity waves can be easily deduced from our knowledge of Newtonian gravity, combined with knowledge of the fact that curvature is a consequence of mass distributions.

First, consider how gravitational waves might be generated. Electromagnetic waves are generated when charges accelerate. Because a negative charge accelerating to the left is equivalent to a positive charge accelerating to the right, it is impossible to create a time-varying electric monopole. The process of varying the charge on one electrode always creates a time-varying dipole moment. Hence it follows that electromagnetic waves are generated by time-varying dipole moments. In contrast to electromagnetism, gravity has only one charge: there is no such thing as negative mass! Hence it is not possible to create an oscillating mass dipole. Action equals reaction. That is, momentum is conserved and the acceleration of one mass to the left creates an equal and opposite reaction to the right. For two equal masses, their spacing can change but the centre of mass is never altered.

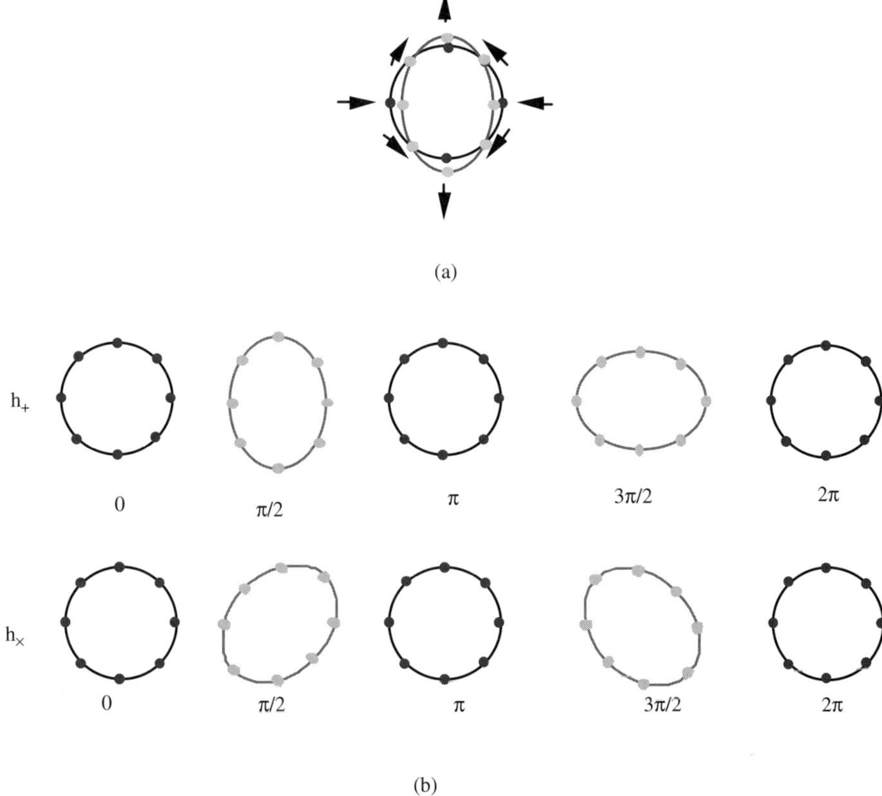

(a)

(b)

Figure 1.2 (a). The lowest order non-spherical deformation of a ring: the diagonal masses are not moved. (b) The deformation of a ring of test particles in one cycle of a gravitational wave field.

This means that there is a time-varying quadrupole moment, but there is no variation in monopole or dipole moment.

To be certain of the quadrupole nature of gravitational waves, think of a system which collapses under its own gravity. First think of a spherically symmetrical array of masses that collapse gravitationally towards a point. At a distance there is no difference between the gravitational field of a point mass and that of the same mass distributed in a uniform spherical distribution (this is a consequence of the inverse square law, and is also true for electric fields). Hence the process of gravitational collapse of a spherical distribution creates no variation in the external gravitational field, and hence no gravitational waves. Clearly gravitational waves must be created by non-spherical motions of masses. Consider a ring of eight test masses, such as that illustrated in Figure 1.2.

The simplest non-spherical motion is one in which the edge masses move inwards and the top and bottom masses move apart, as shown in Figure 1.2(a). Such a quadrupole motion does vary the external field and does create gravitational waves. For a small amount of vertical stretching, and an equal horizontal shrinking, it is obvious that the diagonally placed masses have no radial motion. There is clearly a second polarisation 45 degrees

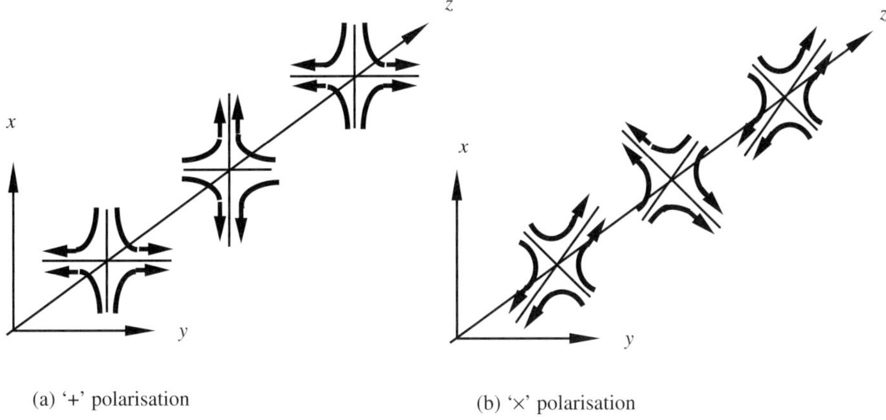

(a) '+' polarisation (b) '×' polarisation

Figure 1.3 Gravitational wave field force lines. (a) '+' polarisation; (b) '×' polarisation.

rotated from the first in which the diagonal masses move radially, and the top, bottom and edge masses have no radial motion. Unlike electromagnetic waves, gravitational wave polarisations are just 45 degrees apart.

Gravitational wave detection can be easily understood from the symmetry between sources and detectors – time reversal invariance. A gravitational wave will distort a ring of test masses in exactly the same way that the distortion of a ring of test masses creates gravitational waves. The non-spherical deformation pattern we just observed is exactly like the tidal deformation of the earth created by the gravity gradient due to the moon. A gravitational wave is indeed a wave of time varying gravity gradient. The amplitude of a gravitational wave is measured by the relative change in spacing between masses. That is, the wave amplitude, usually denoted h, is given by $\Delta L/L$, where L is the equilibrium spacing and ΔL is the change of spacing of two test masses. Whereas electromagnetic luminosity depends on the square of the second time derivative of the electric dipole moment, the gravitational wave luminosity is proportional to the square of the third time derivative of the mass quadrupole moment. The extra derivative arises because gravitational wave generation is associated with the differential acceleration of masses.

The above deformation patterns also apply to solid or fluid bodies. The rigidity of normal matter is so low compared with that of spacetime that the stiffness of the matter is utterly negligible. Considering the deformations of Figure 1.2(a) applied to a solid sphere, such as the Earth, it also follows that the $45°$ points must involve circumferential motions since the deformation shown acts to transfer matter from the 'equator' to the poles in the same way that the lunar tides act on the Earth.

The gravitational wave has an effective force field determined by the displacement vectors of the test masses. The force field is discussed further below, and is shown in Figure 1.3. The force field indicates that detectors can be designed to couple to gravity waves in several different ways. They may detect straight linear strains, orthogonal strains, or circumferential strains.

1.3 The luminosity of gravitational waves

The weak coupling of gravitational waves to matter is a consequence of the enormous elastic stiffness of spacetime. If the elastic stiffness of spacetime were infinite (Newtonian physics) the coupling would be zero. In general relativity, the generation of gravitational waves is given quantitatively by combining the third time derivative of the quadrupole moment, D, described previously, with the appropriate coupling constant. The latter can only depend on the constants G and c (for classical waves), and by dimensional analysis this constant must have the form G/c^5. The gravitational wave luminosity of a source is given by

$$L_{\mathrm{G}} \sim \frac{G}{c^5} \left(\frac{\mathrm{d}^3 D}{\mathrm{d}l^3} \right)^2 . \tag{1.2}$$

Except for a numerical factor, this is the Einstein quadrupole formula (Einstein, 1916). There are two useful formulae one can derive from equation (1.2). The first is the formula for a hypothetical terrestrial source or binary star system. The second is for an interacting black hole system. The terrestrial source might be a pair of oscillating masses joined with a spring. Ideally the spacing of the masses should change from zero to L. This is achieved in the edge-on view of a rotating dumbbell or binary star system in a circular orbit, as shown in Figure 1.4. Viewed edge-on, the masses appear to move in and out periodically twice per rotation cycle. The quadrupole moment for two masses a distance x apart is Mx^2. If the motion is sinusoidal at an angular frequency of ω, the square of the third time derivative is $\sim M^2 L^4 \omega^6$. Thus the gravitational wave luminosity of such a system is

$$L_{\mathrm{G}} \sim \frac{G}{c^5} M^2 L^4 \omega^6 . \tag{1.3}$$

This equation applied to any natural or artificial source in our Solar System gives a depressingly small luminosity. This is a consequence of the extraordinarily small value of G/c^5. However, the situation is different in an astrophysical context.

Suppose that the system is a similar binary system, except that it consists of a pair of gravitationally bound masses, of size such that their escape velocity approaches c and each has a radius near to the Schwarzschild radius: that is, a pair of black holes. In this case, using the Schwarzschild radius, $r_S = 2GM/c^2$, the luminosity can be expressed in relativistic units:

$$L_{\mathrm{G}} \sim \frac{c^5}{G} \left(\frac{v}{c} \right)^6 \left(\frac{r_S}{r} \right)^2 . \tag{1.4}$$

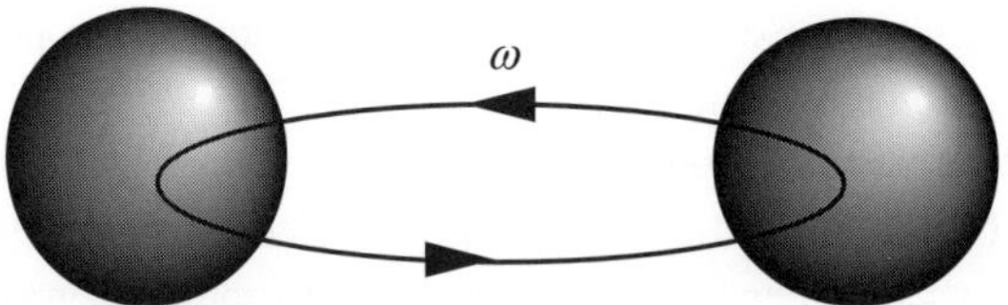

Figure 1.4 A rotating dumbbell or a binary star system, viewed edge-on, has a maximal variation of quadrupole moment.

The remarkable difference between equation (1.3) and equation (1.4) expresses the difference between the physics of normal matter and that of black holes. Equation (1.3) is scaled by the tiny factor G/c^5, while equation 1.4 is scaled by its enormous reciprocal. Normal matter in our Solar System creates negligible curvature of spacetime. A black hole creates an extreme distortion of spacetime. Hence normal matter sources are intrinsically extremely weak, while very large amplitude waves are created in events such as the coalescence of a pair of black holes (for which we would expect $v \sim c$ when $r_S \sim r$). The factor c^5/G is roughly the total electromagnetic luminosity of the Universe. This is the upper limit to the gravitational wave luminosity of black hole systems. In reality equation (1.4) does not take into account the gravitational redshift effects and other spacetime curvature effects which act to reduce the maximum luminosity. However, to an order of magnitude, equation (1.4) indicates the extreme luminosity of gravitational waves that can be expected in short bursts when gravitationally collapsed systems with strong gravity, such as black holes (escape velocity $= c$) and neutron stars (escape velocity $\sim 0.1c$), are involved.

Any source can be characterised by an amplitude h and flux F detected at the Earth or by a luminosity L_G, which characterises the total rate of energy loss from the system. The energy flux (in $\mathrm{W\,m^{-2}}$) in terms of the amplitude h is given by:

$$F = \frac{\pi}{4}\frac{c^3}{G} f^2 h^2 . \tag{1.5}$$

In general h is the amplitude for two polarisations $h^2 = h_+^2 + h_\times^2$. Numerically, we can write

$$F = 30 \, W \, m^{-2} \left(\frac{f}{1\,kHz}\right)^2 \frac{h^2}{10^{-20}} . \tag{1.6}$$

This value represents a considerable energy flux, 3% of the solar intensity on Earth, although such high flux densities can only be sustained in short bursts. Hence, the energy of a gravitational wave is extremely high for a very small amplitude.

1.4 The amplitude and frequency of gravitational wave sources

As we saw in the previous section, a gravitational wave is a wave of gravity gradient which causes relative displacements, or strains, between test masses. The detection of gravitational waves requires the detection of small strain amplitudes. We should now consider the typical size of such strain amplitudes. One can very crudely estimate this by scaling the amplitude of the gravitational wave relative to the extreme amplitude at the point of coalescence of two masses to form a black hole. At the point of black hole formation spacetime curvature is very large. For example, the deflection of light for a light beam passing near to the event horizon can approach a complete orbit of a black hole. At the point of generation the dynamic curvature of space that will become the outgoing gravitational wave is unlikely to be able to exceed the static curvature represented by the maximal deflections of light past a black hole. The strain $\Delta L/L$ represented by such deflections can be estimated from the difference in light travel time for the deflected path around the black hole (say half an orbit) and the direct path between the same points in the absence of the black hole. For a half orbit (in Euclidean geometry) the circular path is $\pi/2$ longer than the direct path, so roughly $\Delta L \sim L$, and the maximum possible strain amplitude is $\sim$ unity. But by the inverse square law, the amplitude of the wave reduces as $1/r$. (The energy density, which is proportional to the square of the amplitude, reduces as $1/r^2$.) So for such a black hole source we can give the strain amplitude at distance r as simply $h \sim r_S/r$. For more realistic sources only a fraction of the total energy will participate in quadrupole motion. Thus it is more reasonable to include an efficiency factor ϵ, which characterises the fraction of the total system rest mass which can convert to gravitational waves. In this case we can write

$$h \sim \epsilon^{1/2} \frac{r_S}{r} \; . \tag{1.7}$$

Since the Schwarzschild radius of a solar mass is a few kilometres, the maximum strain amplitude that can be expected from any stellar mass source is numerically equal to the reciprocal of its distance in kilometres. Because r_S is linearly proportional to the mass, gravitational wave amplitudes from very high mass sources, such as colliding blackholes of 10^9 solar mass in galactic nuclei, will be of correspondingly larger amplitude. Putting in numbers, equation (1.7) gives $h \sim 10^{-16}$ for 10 solar masses and 100% efficiency at the galactic centre, and $h \sim 10^{-13}$ for 3 billion solar masses at 3 Gpc (towards the edge of the visible Universe).

Clearly these maximal numbers are very small. It might seem that the supermassive black hole sources might be much more detectable than the stellar mass source. The strain amplitude in this case corresponds to the detection of a motion equal to the size of an atomic nucleus on a one metre baseline, or one metre between here and Neptune. In fact the detection of such small strains on Earth is probably impossible. This is because the frequency of the waves from supermassive black hole sources must always be very low. The peak frequency, or its reciprocal, the burst duration, can be estimated from the time the binary black hole system takes to complete its final orbit before coalescence. Its value is about 10 kHz for one solar mass, reducing inversely as the mass. Thus the peak frequency will be about 1 kHz for the above galactic centre source, and 3×10^{-6} Hz for the distant massive black holes. The latter frequency will be reduced towards 10^{-6} Hz by cosmological redshifts. At such low frequencies, environmental effects, in particular gravity gradients associated with tides and weather variations in the surrounding environment, create perturbations which greatly exceed the desired signal.

There are two known ways to get around this obstacle. One is by using drag-free satellite technology in which a spacecraft is servo controlled by thrusters to follow a central protected freely floating test mass. In this way it is possible to create very stable free floating masses in space. Laser interferometry between the spacecraft and the test mass can then measure the gravitational wave strains. In this case detection does look relatively straightforward, though expensive, since it requires several widely separated spacecraft.

For frequencies even lower than 10^{-6} Hz, radio pulsars can replace man-made spacecraft in detection systems. The pulsar ideally provides a perfect monochromatic timing signal. The radio beams from the pulsar are traversed by incoming gravitational waves. If several pulsars are observed in the same part of the sky, a gravitational wave signals would appear as correlated arrival time variations of pulses from pulsars in different directions. In this case it is more convenient to consider the gravitational wave acting not on the pulsar, but on the Earth. By timing many pulsars and looking for pulse arrival time errors, a random gravitational wave background gives a unique statistical signature. Because pulsar signals are very weak, long integrations are required to obtain sufficient time resolution. Adequate sensitivity is not achieved unless the integration time is more than 1 year.

Terrestrial detection appears to be possible only for frequencies above 1 Hz. Although there are plenty of noise sources in this frequency band, they have been beaten down by first generation detectors to allow gravitational strain measurements of $h \sim 10^{-21}$ at $\sim 100\,\mathrm{Hz}$. Improvements beyond this level have been mapped out; this book focuses on the various approaches, which involve high power laser metrology, vibration isolation, thermal noise reduction and clever quantum measurement schemes.

1.5 Gravitational waves in general relativity

A detailed understanding of general relativity is not essential for experimentalists constructing detectors. However, for completeness we give a brief summary of the mathematical basis for gravitational waves. For further details on the theory of gravitational waves, readers are referred to: Weinberg (1972), Misner *et al.* (1973), Damour (1986), Thorne (1987), Giazotto (1989), Kenyon (1991), Saulson (1994), and Schutz (2001).

The geometry of spacetime can be expressed by the metric tensor $g_{\alpha\beta}$ which connects the space-time coordinate $dx^{\alpha\beta}$ ($\alpha, \beta = 1, 2, 3, 4$) to the spacetime interval ds by way of the relation

$$ds^2 = g_{\alpha\beta}dx^\alpha dx^\beta. \tag{1.8}$$

As we have seen, gravitational waves in the vicinity of the Earth will always be very weak. The background curvature can be ignored and the background metric can be approximated as that of the flat Minkowski metric of special relativity η. An approximation of the gravitational wave field can then be expressed in the form (Misner *et al.*, 1973)

$$g_{\alpha\beta} = \eta_{\alpha\beta} + h_{\alpha\beta}, \tag{1.9}$$

where $\eta_{\alpha\beta}$ is the metric of the flat background, and $|h| << 1$ is the perturbation on this background. If there is no stress-energy source term in Einstein's field equation, i.e. $\mathbf{T} = 0$ in equation (1.1), we are left with the weak field vacuum approximation to the Einstein equations. To obtain an explicit statement of the metric perturbations h it is necessary to make a gauge choice. The most useful gauge is the transverse-traceless gauge in which the coordinates are defined by the geodesics of freely falling test bodies. In this gauge, and in the weak field limit discussed above, the equations of general relativity become a system of linear equations; specifically a system of wave equations (Misner *et al.*, 1973)

$$\left(-\frac{1}{c^2}\frac{\partial^2}{\partial t^2} + \nabla^2\right)h_{\alpha\beta} = 0. \tag{1.10}$$

Equation (1.10) is a three-dimensional wave equation, telling us that gravitational waves travel at the speed of light c. The gravitational wave curvature tensor h can be considered as the gravitational wave field. The wave field is transverse and traceless, and for waves travelling in the z-direction may be expressed as follows:

$$h_{\alpha\beta} = \begin{pmatrix} 0 & 0 & 0 & 0 \\ 0 & h_{xx} & h_{xy} & 0 \\ 0 & h_{yx} & h_{yy} & 0 \\ 0 & 0 & 0 & 0 \end{pmatrix}; \quad h_{xx} = h_+; \; h_{xy} = h_\times. \tag{1.11}$$

There is no z-component due to the transverse nature of the waves. Because the wave field is traceless, h satisfies

$$h_{xx} = -h_{yy} . \tag{1.12}$$

Because the Riemann tensor is symmetric, h also satisfies

$$h_{xy} = h_{yx} . \tag{1.13}$$

The symmetry of h means that there are just two possible independent polarisation states, which are usually denoted h_+ and $h_\times$. This was deduced qualitatively in Section 1.2. In the case of sinusoidal gravitational waves we can express these polarisations as

$$h_+ = h_{xx} = \Re[A_+ e^{-i\omega(t-z/c)}] ,$$
$$h_\times = h_{xy} = \Re[A_\times e^{-i\omega(t-z/c)}] . \tag{1.14}$$

Here A_+ and $A_\times$ are the strain amplitudes of each polarisation.

We have already seen that a gravitational wave field moves masses in the same way that an electromagnetic wave sets charged particles in motion. Each wave field exerts tidal forces on objects through which it passes. The corresponding lines of force have a quadrupole pattern as shown in Figure 1.3, where (a) shows the force lines of the '+' polarisation and (b) shows the '$\times$' polarisation, which is rotated 45° with respect to the '+' state. These time-varying tidal forces can deform an elastic body or change the distance between mass points in free space. A ring of particles placed perpendicular to the wave propagation direction will be distorted, as we already saw in Figure 1.2.

Einstein's famous quadrupole formula describes the gravitational wave amplitude from a source. Einstein derived his formula in a slow motion weak field approximation. However, Thorne (1987) emphasises that the result is accurate as long as the reduced wavelength of the gravitational wave $\lambda/2\pi$ exceeds the source size. This condition applies to all but the most compact sources, such as forming or coalescing black holes. The latter are potentially the strongest and most detectable sources. It is unfortunate that these are just the ones where the non-linearity of general relativity, and in particular the gravitational redshift of the outgoing gravitational waves due to the gravitational energy of the spacetime curvature itself, makes the gravitational radiation amplitude extremely difficult to estimate.

The quadrupole formula states that the gravitational wave amplitude h at a distance R from a source is proportional to the second time derivative of the transverse traceless projection of the quadrupole moment evaluated at the retarded time $(t - R/c)$. That is

$$h_{jk} = \frac{2}{r}\frac{G}{c^4}\frac{\partial^2}{\partial t^2}[D_{jk}(t - R/c)]^{\mathrm{TT}}, \quad (j,k = 1,2,3), \tag{1.15}$$

where $[D_{jk}(t - R/c)]^{\mathrm{TT}}$ is the above mentioned transverse traceless projection of the quadrupole moment. The transverse traceless requirement relates to the transverse nature of gravitational waves, and the lack of wave generation from spherically symmetrical motions. For weak fields, for which gravitational self energy is small (see Damour, 1987), D is given by the second moment of the source mass density ρ:

$$D_{jk} = \int \rho(t)[x^j x^k - x^2 \delta^{jk}/3]d^3x , \tag{1.16}$$

where the term with the Kronecker delta ensures that D is trace free.

As described in Section 1.3, the total gravitational wave power is proportional to the square of the third time derivative of the mass quadrupole moment (Einstein, 1918). In general, the total energy radiation rate, L_{G}, is given by sum of the squares of all the projections of the quadrupole moment

$$L_{\mathrm{G}} = \frac{1}{5}\frac{G}{c^5}\sum_{j,k}\left|\frac{\mathrm{d}^3 D_{jk}}{\mathrm{d}t^3}\right|^2 . \tag{1.17}$$

The very small universal constant, which we derived earlier from dimensional analysis (except for the numerical factor) as

$$\frac{G}{5c^5} = 5.49 \times 10^{-54}\, s\, J^{-1} , \tag{1.18}$$

sets the characteristic gravitational radiation power output. Unless $\dddot{D}_{jk}$ involves energy of astronomical proportions, the gravitational wave power will be extremely small. Thus it appears to be impossible to generate detectable gravitational waves on the laboratory scale, even at extreme limits of known technology. However, we express caution about this statement – it may be the failure of our imagination.

It is useful to consider a laboratory source, simply as an application of equations (1.16) and (1.17). Suppose the source consists of a pair of masses a distance L apart, and joined by a spring to allow sinusoidal oscillation of their spacing at angular frequency ω. From equation (1.16), $D = ML^2$, and if $L = L_0 + a\,\sin\,\omega t$, it follows that

$$D = ML_0^2 + 2ML_0 a\,\sin\,\omega t + Ma^2\,\sin^2\,\omega t. \tag{1.19}$$

From equation (1.19), taking the third time derivative, it follows that this source will produce gravitational waves at the frequencies of ω and 2ω. If the amplitude a is small compared with L_0, the 2ω term is small and the gravitational wave luminosity is given by

$$L_{\mathrm{G}} = \frac{G}{5c^5}4M^2 L_0^2\, a^2\omega^6 . \tag{1.20}$$

For any practical harmonic oscillator on Earth, L_{G} is infinitesimal. However such mass quadrupole oscillators have been created as sources of near field dynamic gravity gradients (not waves) for the purpose of calibrating gravitational wave detectors. Such systems were successfully used for low frequency detectors tuned to the Crab pulsar (Suzuki *et al.*, 1981) and also by the Rome group to calibrate resonant bar detectors and measure the inverse square law of gravitation (Astone *et al.*, 1991).

1.6 Gravitational wave detector response and signal strength

Before we move on to explore gravitational wave sources, it will be useful first to define some of the quantities used throughout this book to describe the response of a gravitational wave detector to different kinds of signal. The brief definitions described here will be given a more comprehensive treatment in Chapter 4, when we will discuss issues of data analysis. The output from a single gravitational wave detector will consist of a time series, $s(t)$, composed of the detector response to a gravitational wave signal, $h(t)$, described in terms of a linear combination of the two orthogonal transverse polarisations, $h_{+,\times}$, and the detector noise $n(t)$:

$$s(t) = F_+(t,\,\theta,\,\phi,\,\psi)\,h_+(t) + F_\times(t,\,\theta,\,\phi,\,\psi)\,h_\times(t) + n(t)\,. \qquad (1.21)$$

$F_{+,\times}$ are the dimensionless detector antenna pattern functions for the two polarisations (Shultz and Tinto, 1987; Tinto, 1987; Jaranowski *et al.*, 1998). These describe the detector sensitivity to radiation of different polarisations, incident from different directions. The angles θ and ϕ represent the direction to the source and ψ is the polarisation angle of the wave.

As a gravitational wave detector can follow the phase of a gravitational wave signal, the time series, $s(t)$, is often represented in the frequency domain in terms of the power spectral density, $S_h(f) = \tilde{s}^*(f)\tilde{s}(f)$, where $\tilde{s}(f)$ is the Fourier transform of the time series

$$\tilde{s}(f) = \int_{-\infty}^{+\infty} \mathrm{e}^{-2\pi i f t} s(t) \mathrm{d}t \qquad (1.22)$$

The information contained in $s(t)$ can therefore be represented by a *strain amplitude spectrum*, which takes the form:

$$\tilde{h}(f) = \sqrt{S_h(f)}\,. \qquad (1.23)$$

This quantity has dimensions of $Hz^{-1/2}$ (Thorne, 1987). Another useful definition is the dimensionless *characteristic strain*:

$$h_c(f) = \sqrt{f}\tilde{h}(f) = \sqrt{f S_h(f)}\,, \qquad (1.24)$$

which defines the root-mean-square signal in bandwidth $\Delta f = f$, around a frequency, f. The definition of h_c often includes additional factors to represent the increase in signal-to-noise ratio obtained by basing a detection strategy around prior knowledge of the emission characteristics of a source. For example, if the signals are well predicted, as is the case for monochromatic sources or compact binary inspiral sources, the optimal signal detection method of *matched filtering* can be used. This method enhances the value of h_c by a factor $\sqrt{n}$, where n is the number of cycles used in the integration. For monochromatic sources, the value of n will increase with integration time. For compact binary inspirals, the value of n increases with the compactness of the system as it approaches merger. Matched filtering is an important technique in gravitational wave data analysis and as such, will be given a detailed treatment in Chapter 4.

In the next chapter we will focus on astrophysical sources of gravitational waves. We will see that detectable gravitational waves can be emitted by sources at both extremes of size and distance – from neutron stars within our own Galaxy to supermassive black holes at the edge of the Universe.

Acknowledgements

The authors wish to acknowledge financial support from the West Australian Government Centres of Excellence scheme, the Australian Research Council and the University of Western Australia.

References

Astone, P., *et al.* 1991. *Zeitschrift-fur-Physik-C-(Particles-and-Fields)*, **50**, 21.

Damour, T. 1986. *An introduction to the theory of gravitational radiation, in Gravitation in Astrophysics*. Plenum Press, New York.

Damour, T. 1987. *In Gravitation in Astrophysics*. Plenum, New York.

Einstein, A. 1916. *Preuss. Akad. Wiss. Berlin, Sitzungsberichte der Physikalisch-mathematischen Klasse, Preuss. Akad. Wiss. Berlin, Sitzungsberichte der Physikalisch-mathematischen Klasse*.

Einstein, A. 1918. *Sitzungsber. K. Preuss. Akad. Wiss., Sitzungsber. K. Preuss. Akad. Wiss.*

Giazotto, A. 1989. *Phys. Report*, **182**, 365.

Hulse, R. A. and Taylor, J. H. 1975. *Astrophys. J.*, **195**, L51.

Jaranowski, P., Krolak, A. and Schutz, B. F. 1998. *Phys. Rev. D*, **58**, 063001.

Kenyon, I. R. 1991. *General Relativity*. Oxford: Oxford University Press.

Misner, C. W., Thorne, K. S., and Wheeler, J. A. 1973. *Gravitation*. San Francisco: W. H. Freeman and Company.

Saulson, P. R. 1994. *Fundamentals of Interferometric Gravitational Wave Detectors*. World Scientific Pub Co Inc.

Schutz, B. F. 2001. *A First Course in General Relativity*. Cambridge, UK: Cambridge University Press.

Schutz, B. and Tinto, M. 1987. *MNRAS*, **224**, 131.

Suzuki, T., *et al.* 1981. *Jap. J. Appl. Phys.*, **20**, L498.

Thorne, K. S. 1987. *Gravitational Radiation*. 1st edn. Cambridge, UK: Cambridge University Press.

Tinto, M. 1987. *MNRAS*, **226**, 829.

Weinberg, S. 1972. *Gravitation and Cosmology: Principles and Applications of the General Theory of Relativity*. Wiley, New York.

Weisberg, J. M. and Taylor, J. H. 1984. *Phys. Rev. Lett*, **52**, 1348.

Weisberg, J. M. and Taylor, J. H. 2005. Page 25 of: F. A. Rasio & I. H. Stairs (ed), *Binary Radio Pulsars*. Astronomical Society of the Pacific Conference Series, vol. 328.

<h1 style="text-align:center">2</h1>

Sources of gravitational waves

D. G. Blair and E. J. Howell

This chapter introduces the different classes of gravitational wave sources targeted by terrestrial and space-based detectors. The possibility and implications of gravitational wave emissions from supernovae and coalescing binary systems of neutron stars and/or black holes are discussed, as well as the possible connection between gravitational wave sources and gamma-ray bursts. The chapter also discusses continuous gravitational wave sources and describes how a stochastic gravitational wave background could be produced from astrophysical sources or from events in the early Universe.

2.1 Introduction

Astrophysics provides us with a variety of candidate systems which should be observable in the spectrum of gravitational waves. However, it is important to remember that our powers of prediction of new phenomena are limited, so any list of sources is almost certain to be incomplete.

Amongst stellar mass systems we expect detectable gravitational radiation from the formation of black holes and neutron stars (Fryer *et al.*, 2002), and from the coalescence of binary neutron stars and final collapse of such binaries to form a black hole (Phinney, 1991). We would expect not only discrete sources, but also continuous stochastic backgrounds created from large numbers of discrete sources. In our Galaxy the very large populations of binary stars create a stochastic background in the 10^{-3} to 10^{-5} Hz range (Hils *et al.*, 1990; Cutler, 1998; Nelemans *et al.*, 2001).

In the Universe as a whole, stellar mass black holes and neutron stars are born at a rate of $\sim 10^2$ per second (Ferrari *et al.*, 1999a; Howell *et al.*, 2004). All these create an almost continuous background in the audio frequency part of the spectrum. This particular background provides an exciting opportunity to observe the earliest epochs of star formation in the first galaxies. Gravitational waves should also allow us to map the birth and growth of the massive black holes that appear to reside in the nuclei of most galaxies (Merritt, 2005). We may also be able to observe gravitational waves from the Big Bang, amplified during the inflationary era, and possible signatures of cosmological phase transitions and topological defects such as cosmic strings (Grishchuk, 1975; Maggiore, 2000). These very earliest sources in the Universe would constitute a probe of physics at energy scales far beyond those accessible in particle accelerators and hence represent the best opportunity we have to obtain experimental data from the era of inflation.

Back in our own Galaxy we would also expect to find many quasi-monochromatic sources of gravitational waves such as individual binary star systems, including binary neutron stars as they evolve towards coalescence, and various rotating neutron star systems such as millisecond and X-ray pulsars (Sathyaprakash and Schutz, 2009).

As we will see throughout this chapter, the gravitational wave spectrum spreads across at least 13 decades from below 10^{-9} Hz to $> 10^3$ Hz. Even lower frequencies are likely to exist: so low that they will appear as a frozen signature in the cosmic microwave background. The spectrum conveniently divides into a terrestrial detection band, above 1 Hz (generally within the audio frequency band), mainly associated with stellar mass compact objects; a space detection band, from 10^{-4} to 10^{-1} Hz, where laser interferometers in space can detect sources including both binary star systems in our Galaxy, and cosmological sources associated with massive black hole interactions; and the pulsar timing band, which is sensitive between 10^{-9} and 10^{-8} Hz, able to detect both stochastic gravitational waves and signals created from the coalescence of massive black holes.

Classification of sources

All the above sources can be naturally divided into three distinct classes, according to the methods of data processing and signal extraction. The first class consists of catastrophic *burst sources* such as the final coalescence of compact binary star systems, or the formation of neutron stars and black holes in core-collapse supernova events. The binary coalescence events can consist of binary neutron stars, binary black holes, or neutron star–black hole binaries. The burst signal consists of a very short single event, consisting of one or very few cycles, and hence is characterised by a broad bandwidth, roughly determined by the reciprocal of the event duration.

The second class consists of *narrow band sources*. These include the rotation of single non-axisymmetric stars, particularly pulsars and accreting neutron stars, as well as binary star systems far from coalescence. All such systems are *quasi-periodic* because gravitational wave energy loss must cause period evolution, and in general they are also periodically Doppler shifted by binary motion and Earth orbital motion. Such sources are generally weaker than the burst sources, but in principle they are always amenable to long-term integration to extract signals from the noise. This requires accurate knowledge of the frequency modulations to maintain a coherent integration. Assuming a white noise background and perfect knowledge of the frequency evolution, the signal-to-noise ratio increases as $N^{1/2}$, where N is the number of cycles. For narrow band sources it may be possible to integrate for 10^8 seconds, compared with a burst source which could have a duration as small as ~ 10 ms. Thus, at 100 Hz N can be 10^{10}, allowing a 10^5-fold improvement in signal-to-noise ratio.

The third class of sources are the *stochastic backgrounds* produced from the integrated effects of many weak periodic sources in our Galaxy, or from a large population of burst sources at very large distances, as well as the above-mentioned cosmological processes in the early Universe. Stochastic backgrounds are difficult to detect in a single detector because they are practically indistinguishable from instrument noise. If the source was not isotropically distributed (such as a population of binary stars towards the centre of our Galaxy), it might be detectable from the variation of observed instrument noise as the detector orientation varies on the rotating Earth. However, a much better way of detecting stochastic backgrounds is by cross-correlating two nearby detectors. The correlated stochastic signal will integrate up in relation to the uncorrelated instrument noise (assuming both detectors to be truly independent). In this case the signal-to-noise ratio increases as $N^{1/4}$, where N is the effective number of cycles, determined by the observation frequency. This technique allows a roughly 300-fold improvement in signal-to-noise ratio in 10^8 seconds of integration (compared, again, with a $\sim$10 ms burst source).

Binary neutron star systems

Binary neutron star systems can produce gravitational waves in all the three classes. First, a large population of binary neutron stars in our Galaxy, with orbital periods in the range from days to minutes, can produce a stochastic background of individually unresolvable sources in our Galaxy in the frequency band 10^{-5}–10^{-2} Hz (Regimbau, 2007). Nearby individual systems which are far from binary coalescence could produce detectable nearly monochromatic waves at any frequency up to 0.1 Hz. In addition to the binary orbit, the individual rotation of the stars themselves (if they are non-axisymmetric), will also give rise to quasi-periodic gravitational waves. For example, the spin-down of a millisecond pulsar can be entirely due to gravitational wave emission if mass equal to just 10^{-7} solar mass is located in a non-axisymmetric configuration on the star. Finally, as a binary evolves and radiates away gravitational potential energy, it will gradually spiral inwards. As a result, the frequency of the gravitational wave signal will *increase* with time. In addition, any periodic waves from the rotation of the individual stars will cause loss of rotational kinetic energy, so that this frequency will *decrease* with time. Eventually the stars will coalesce, resulting in a short, intense burst of gravitational waves.

 The time evolution of the frequency of two 1.4 solar mass neutron stars in a binary system is shown in Figure 2.1. Over a period of about 1000 seconds the frequency rises from about 10 Hz to 1 kHz as the neutron stars spiral together. This part of the merger begins when the stars are within about 1000 km of each other. The orbital velocity is $\sim 0.1c$. Signal detection can make use of exactly the same principles used to extract narrow band signals due to the fact that the time evolution of the signal frequency and phase is predictable. Matched filtering, based on the existence of a family of accurately predictable waveforms, can allow integration over all of the observed signal cycles. A terrestrial detector may be able to observe more than 1000 gravitational wave cycles from a neutron star binary. The total number of observable cycles increases strongly as the lower cut of frequency is reduced. This provides a strong incentive for creating detectors at the lowest possible frequency. For 1000 observable cycles, the signal-to-noise ratio is improved by the square root of this number, or about 30.

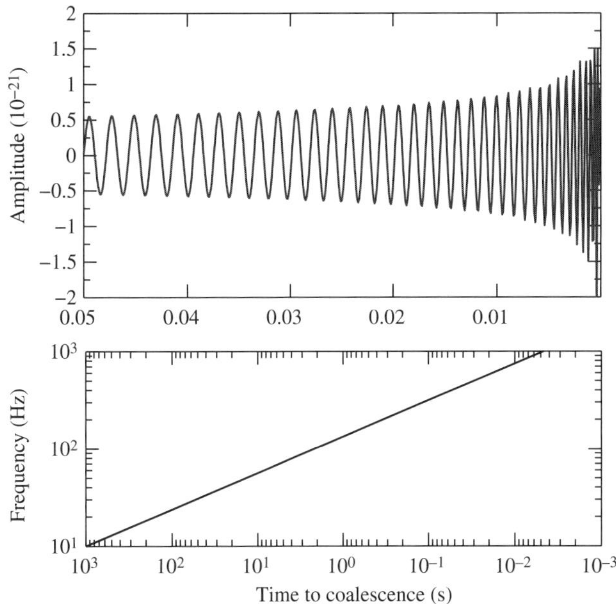

Figure 2.1 The top panel shows the predicted gravitational waveform produced by the inspiral of a binary made of two neutron stars. During the final minutes of a coalescing binary the waveform is highly distinctive. The lower panel shows the frequency evolution of the system for the last 1000 s before merger as it sweeps across the detection band of ground-based interferometers.

Binary black hole systems

Exactly the same concepts as described in the last section may be applied to both black hole systems of tens of solar masses and to supermassive black hole binaries. As the signal frequency decreases inversely with the black hole mass, a pair of 10^8 solar mass black holes would produce a chirp of gravitational waves rising from one cycle per year to a cycle per day over a period of 10^{11} seconds! This time-scale is much too long for the entire event to be observed, so in reality one could only expect to observe a very few cycles. However, since there appear to be a large population of quasars and galaxies containing massive black holes, as well as a large population of interacting galaxies, such events may not be uncommon, and may give rise to numerous strongly detectable sources at very low frequencies that would be a target for space-based interferometric detectors.

For many years the precise waveforms for the gravitational waves from black hole binary systems were uncertain because of difficulties in obtaining stable numerical solutions in full general relativity. This problem was very important because the signal-to-noise ratio in signal detection depends critically on the quality of the waveform match between the real signal and a theoretical template. The phase errors of an inaccurate template can wash out the signal in just the region where the waveform has the greatest amplitude.

In a most significant breakthrough, Pretorius (2005), Baker *et al.* (2006) and Campanelli *et al.* (2006) developed numerical techniques that enable precise calculation of such systems.

There have been several very important insights from this work. First has been the general confirmation that the gravitational wave luminosity L_G does indeed reach the enormous values predicted by the naive application of the quadrupole formula discussed in Chapter 1: it is found that $L_G \sim 10^{23} L_\odot$. Second has been the discovery of very large kick velocities that can be applied to newborn black holes as a result of the non-spherically symmetric emission from such a coalescing binary. The magnitude depends on the spins (their magnitude and direction) of the initial black holes. Thus we now know that black holes powered by gravitational wave radiation reaction can in theory have sufficient velocity to be expelled from a galaxy: a sort of gravitational wave-powered black hole rocket! Third, the numerical results show that the transition from inspiral to merger is a simple smooth process: there is not an exceptionally complex waveform. There is a smooth transition from inspiral to strongly damped quasi-normal mode oscillation of the final black hole as the system settles down into a stationary Kerr state.

In parallel with this work, analytical calculations have been developed known as Effective One Body (EOB) methods (Pretorius, 2009). This analytical approach is a post-Newtonian scheme, yielding expansions of the gravitational binding energy and gravitational wave luminosity as a function of single parameter – the velocity of the reduced mass. Thus, a two body problem is effectively reduced to a one body problem. As will be shown later in Chapter 4, this analytical treatment complements the numerical approaches. Using numerical calculations to calibrate the EOB expansions, codes have been developed that allow rapid analytical estimations for black hole coalescence. Efforts are underway to generalise them to the coalescence of arbitrary spinning black holes.

2.2 Rough guide to signal amplitudes

It is useful to have some formulae with which to make rough estimates of signal amplitudes. The simplest estimate of the gravitational wave amplitude from a source can be obtained by recognising that the maximum possible value of the strain amplitude h is ~ 1, and that this would occur near the event horizon of a black hole source at distance $r = r_S$, where r_S is the Schwarzschild radius. Its value will scale inversely as the square root of the gravitational wave energy conversion efficiency ϵ, and like all waves, the amplitude will scale inversely with distance. Thus $h \sim r_S / r \epsilon^{1/2}$.

For a continuous gravitational wave of frequency f_g, the strain amplitude h is related to the power density w through the relation (Hawking, 1971; Press and Thorne, 1972)

$$w \approx \frac{\pi c^3}{4G} f_g^2 \langle h^2 \rangle = 3.18 \times 10^{35} f_g^2 \langle h^2 \rangle \, \mathrm{W m^{-2}}, \tag{2.1}$$

where $h^2 = h_+^2 + h_\times^2$. Because of the large numerical constant in equation (2.1), the amplitude h is extremely small even for a fairly large power density. For a gravitational wave with strain amplitude of $h \sim 10^{-21}$ (typical of possible signals from the Virgo cluster) at a frequency of 1 kHz, the flux would be $0.3 \, \mathrm{W \, m^{-2}}$, which is about 10^{20} times bigger than typical radio source energy fluxes. The strain amplitude can be written as

$$h = 4\pi R^2 w \sim \left(\frac{G}{\pi^2 c^3} \right)^{1/2} \frac{L^{1/2}}{f_g R} , \tag{2.2}$$

$$h \sim 1.7 \times 10^{-22} \left(\frac{1\,\mathrm{kHz}}{f_g} \right) \left(\frac{10\,\mathrm{Mpc}}{R} \right) \left(\frac{L}{10^{46}\,\mathrm{W}} \right)^{1/2} . \tag{2.3}$$

As we saw above, the modelling of gravitational waveforms in gravitational collapse is extremely difficult. However, for a gravitational wave burst event, the characteristic time-scale of the event τ_g, and the total gravitational energy released E_g, provide a reasonable basis for estimating source parameters. The energy radiation rate L is related to τ_g and E_g by

$$L \sim \frac{E_g}{\tau_g} . \tag{2.4}$$

Burst sources naturally have a broad band spectral distribution. The characteristic frequency of a burst of duration τ_g is roughly

$$f_g = \frac{1}{2\pi \tau_g} . \tag{2.5}$$

This frequency roughly defines the peak frequency in the spectrum. For a roughly Gaussian burst, the width of the spectrum Δf is of the same order of f_g. The strain amplitude can then be written as

$$h \sim \left(\frac{G}{\pi^2 c^3} \right)^{1/2} \frac{(E_g/\tau_g)^{1/2}}{f_g R} , \tag{2.6}$$

$$h \sim 5.8 \times 10^{-20} \left(\frac{E_g}{M_\odot c^2} \right)^{1/2} \left(\frac{1\,\mathrm{kHz}}{f_g} \right)^{1/2} \left(\frac{10\,\mathrm{Mpc}}{R} \right) . \tag{2.7}$$

If a gravitational collapse forms a black hole, we can be more specific in estimating the event duration. Defining a characteristic time τ to be the time for the gravitational wave to travel across the region of strong gravitation d_s, which is assumed to be about twice the gravitational radius $2GM/c^2$, and assuming τ_g is approximately the same as the characteristic time τ, we have

$$\tau = \frac{1}{2\pi f_g} \sim \frac{d_s}{c} \sim \frac{4GM}{c^3} . \tag{2.8}$$

For a system of several solar masses, this corresponds to a frequency of a few kHz. Putting equation (2.8) into equation (2.6) and using $E_g = \epsilon M c^2$, where ϵ is the efficiency factor, the strain amplitude of a burst event then becomes

$$h \sim \frac{1}{2\pi} \frac{c}{f_g R} \epsilon^{1/2} , \tag{2.9}$$

$$h \sim 5 \times 10^{-21} \left(\frac{1\,\mathrm{kHz}}{f_g} \right) \left(\frac{10\,\mathrm{Mpc}}{R} \right) \left(\frac{\epsilon}{10^{-3}} \right)^{1/2} . \tag{2.10}$$

2.3 Supernovae

Supernovae have long been considered a primary source of gravitational wave bursts. Unfortunately, the modelling of supernovae is very complex, involving a combination of general relativistic hydrodynamics, magnetic fields, rotation, neutrino transport and nuclear physics (Ott, 2009). Until recently supernova models were unable to account for the supernova explosion itself. Models show extremely complex and chaotic behaviour that includes shock formation and turbulence, which create highly complex waveforms that include multiple sharp bursts over a duration ~ 1 second. The numerical intensity of supernova modelling means that, even with huge supercomputer resources, very few models can be run in full, and important aspects such as the rotation of the progenitor and magnetic processes are often overlooked. To date core collapse models predict that only $\sim 10^{-8}$ of a solar rest mass is radiated by the collapse itself. Researchers in the field hope eventually to be able to create models that can accurately account for the full range of core-collapse scenarios, which could include supernovae and the hypernovae and collapsars that produce most of the observed gamma-ray bursts (Woosley, 1993). Undoubtedly these models will require experimental verification that can best be done by detecting the gravitational wave emission.

Strong gravitational wave emission is only possible for core-collapse to a neutron star or black hole. The efficiency of gravitational wave radiation emission can be described in terms of the energy available for emission or explosion, given by the binding energy of the neutron star that is formed by the collapse (for black hole formation the efficiency can be assumed to be of the same order, as no additional radiation can escape). To date many core-collapse calculations have predicted rather low efficiency, of order 10^{-6} to 10^{-10}, while the post-collapse evolution calculations have predicted higher efficiencies, of order 10^{-3} to 10^{-7}. We will discuss some of these results below. However, it must be emphasised that all the models used so far suffer from uncertainties in the equation of state, the viscosity and the difficulty in constructing a full three-dimensional numerical general relativistic hydrodynamical code which must also include magnetic and neutrino phenomena. In particular, the stochastic dynamics mean that it will be impossible to predict the exact waveform. However, numerical simulations will be valuable in setting constraints on properties such as the characteristic amplitudes and frequencies, the durations and the gravitational wave energy emissions. However, almost all models show a few common features: the efficiency of gravitational wave production depends on the angular momentum of the progenitor star, the frequency is in the 100 Hz – few kHz range, and the duration of the event is short.

Although many studies have focused on the gravitational wave emissions occurring around the time of core bounce, recent two-dimensional, multi-physics core-collapse supernova simulations by Marek *et al.* (2009) and Yakunin *et al.* (2010) suggest that the most significant emissions may occur hundreds of milliseconds post-bounce owing to turbulent flows induced by the standing accretion shock instability (SASI) (Blondin *et al.*, 2003).

The SASI is an instability of the supernova shock wave now believed to be central to the supernova explosion mechanism itself. In the two-dimensional simulations mentioned above, the SASI leads to gross deformations of the shock and, in turn, to significant accretion funnels beneath it. These accretion funnels collide with the proto-neutron star (PNS) surface and induce violent mass motions there, as well as perturbing deeper PNS regions. These mass motions dominate the production of the post-bounce gravitational wave emission above $\sim$200 Hz. On much longer time-scales for shocks that remain in a stalled state, simulations by Ott *et al.* (2006) suggest that the SASI-induced accretion funnels may also excite vibrational modes of the PNS, which in turn would generate pressure waves. These waves would accumulate behind the shock and may eventually drive the shock out if an explosion were not generated sooner through neutrino heating (Burrows *et al.*, 2006). These acoustically driven pulsations may continue until accretion subsides and may produce gravitational wave emissions an order of magnitude greater than through core bounce, for several hundred milliseconds. Additionally, it has been shown in three-dimensional studies of the SASI by Blondin and Mezzacappa (2007) that the SASI can produce the rapid initial rotations observed in newly born neutron stars. In three dimensions, the SASI induces spiralling flows, in addition to the sloshing flows seen in two-dimensional simulations, that accrete onto the PNS and deposit significant angular momentum. The gravitational wave implications of these new SASI-induced modes of the post-shock flow must be explored in future three-dimensional multi-physics simulations. Although these post-collapse signals produce an enhanced gravitational wave signal, the emissions could be greatly exceeded, should dynamical instabilities occur after the collapse. Studies of the time evolution of a newly born rapidly rotating neutron stars have shown that the star can be driven into a non-axisymmetric configuration due to a bar-mode instability (Chandrasekhar, 1969; Houser *et al.*, 1994; Lai and Shapiro, 1995; Shibata and Karino, 2004; Baiotti *et al.*, 2007; Dimmelmeier *et al.*, 2008). These instabilities derive their name from the 'bar-like' deformation they induce, transforming a spheroidal body into an elongated bar that tumbles end-over-end. The resulting highly non-axisymmetric structure resulting from a compact astrophysical object encountering this instability makes it a potentially strong source of gravitational radiation (New *et al.*, 2000; Shibata *et al.*, 2000). An approximation of the amplitude of the emitted gravitational waves can be obtained through the Newtonian quadrupole approximation (Ott, 2009)

$$h = 5 \times 10^{-21} \left(\frac{\psi}{0.2} \right)^{1/2} \left(\frac{f}{1\,\text{kHz}} \right)^2 \left(\frac{10\,\text{kpc}}{D} \right) \left(\frac{M}{1.4 M_\odot} \right) \left(\frac{R}{10\,\text{km}} \right)^2, \qquad (2.11)$$

where D, M and R are the source distance, the mass and radius of the star respectively, ψ is the degree of ellipticity and f is the gravitational wave frequency (twice the rotational frequency).

A system susceptible to bar-mode formation is parameterised by the stability parameter $\beta = T/|W|$ where T is the rotational kinetic energy and W is the gravitational potential energy. There exist two different time-scales and mechanisms for bar-mode instabilities. Newtonian studies reveal that uniformly rotating, incompressible stars are *secularly* unstable if $\beta \geq 0.14$, and have a growth time that is determined by the time-scale of dissipative processes in the system (such as viscosity or gravitational radiation) – this is usually much longer than the dynamical time-scale of the system (New *et al.*, 2000; Saijo *et al.*, 2001). In contrast, a *dynamical* instability sets in when $\beta \geq 0.27$, and has a growth time of the order of the rotation period of the object (New *et al.*, 2000). This is expected to be the fastest growing mode. In addition, recent work in full general relativity has revealed that a stronger gravitational field tends in general to favour the development of the dynamical instability, which can develop for $\beta \geq 0.255$ (Baiotti *et al.*, 2007) and for even smaller values if the stars are very compact (Manca *et al.*, 2007). Because bar-mode instabilities are a potentially important source of gravitational radiation, they have been the subject of many numerical studies. Some studies have indicated that gravitational radiation efficiencies of order $\sim 10^{-4}$ can be produced (Shibata and Sekiguchi, 2005; Baiotti *et al.*, 2007). Studies by Dimmelmeier *et al.* (2008), however, suggested that the post-bounce core cannot reach sufficiently rapid rotation to become unstable to the classical high-β dynamical bar-mode instability. However, they found that many of their post-bounce core models had sufficiently rapid rotation to become subject to the low-β dynamical instability first by seen by Centrella *et al.* (2001). The potential for enhancements in the gravitational wave emissions by dynamical instabilities at low-β is encouraging and has been demonstrated in a number of other studies (Shibata *et al.*, 2002; Ott *et al.*, 2005; Scheidegger *et al.*, 2008), the latter of which noted a conversion efficiency of $\epsilon \sim 10^{-7}$ in one of their models. Although the low-β instability has so far been found mostly in stellar models that have large degrees of differential rotation, further work has revealed that the low-β instability is the manifestation of a more generic class of instabilities, i.e. a *shear instability* (Watts *et al.*, 2005) and that it can therefore develop in a larger space of parameters, in particular for larger values of β (Corvino *et al.*, 2010).

The above examples seem to indicate that supernovae which produce rapidly rotating neutron stars may be reasonably efficient sources of gravitational radiation. Large amounts of angular momentum may be radiated away in gravitational waves, but if the duration and frequency evolution are uncertain this presents an additional complication when it comes to trying to dig a signal out of the detector noise.

The fraction of supernovae for which high gravitational wave emission occurs is unknown. In the following discussion we shall adopt an optimistic efficiency of 0.1% where we need to use a numerical value. However, the uncertainty of this number must be recalled. To start with, it is useful to note that a supernova of 0.1% efficiency produces a characteristic strain amplitude of $\sim 10^{-18}$ at 10 kpc (within our Galaxy), and 10^{-21} at 10 Mpc (half-way to the Virgo cluster of galaxies).

The chance of detecting gravitational wave bursts obviously depends strongly on the rate of the burst events. Due to the isolation of the Milky Way galaxy, and the large distances required to increase substantially the size of the target population, the amplitude distribution of bursts is extremely non-uniform. Strong events from our Galaxy are almost certainly rare, and to increase the event rate substantially one needs to be able to detect events in the Virgo cluster. Thus, considering our optimistic estimate of the gravitational wave efficiency, to have a chance of detecting several events per year the detectors will be required to measure characteristic amplitudes very much less than 10^{-21}.

2.4 Neutron star coalescence

The modelling of gravitational wave emission from neutron star coalescence has been studied extensively. For most of their evolution, the neutron stars can be considered as point masses, and much of their evolution is well described by the quadrupole formula equations (1.16) and (1.17) of Chapter 1. The waveform as shown in Figure 2.1 is quite distinctive and amenable to the method of matched filtering for signal detection. A numerical template is used, defined by the set of possible waveforms. When this is cross-correlated with the data and correctly matched in phase, it will produce a large positive correlation. The signal-to-noise ratio is substantially enhanced by this means.

The apparent signal enhancement achievable is expressed in terms of the characteristic amplitude h_c. The characteristic amplitude represents the effective amplitude detected after optimal filtering of the waveform. Roughly, h_c includes an enhancement of the signal by the square root of the number of cycles within the spectral band of interest. The enhancement is around a factor of 30 for a neutron star coalescence detected by a terrestrial laser interferometer detector, although this increases strongly if the waveform is detectable at much lower frequencies where the frequency evolution is slow. For example, the number of observable cycles increases almost 50 times if the detector is able to observe down to 10 Hz instead of 100 Hz. This means that we can only roughly estimate the size of the detectable signal, as it depends on the detailed frequency response of the detector.

Thorne (1992) gives the characteristic strain amplitude of the waves from inspiralling binaries as

$$h \sim 0.237 \, \frac{\mu^{1/2} M^{1/3}}{r_0 f_c^{1/6}}, \tag{2.12}$$

$$h \sim 4 \times 10^{-22} \left(\frac{\mu}{M_\odot} \right) \left(\frac{M}{M_\odot} \right)^{1/3} \left(\frac{100 \, \text{Mpc}}{R} \right) \left(\frac{100 \, \text{Hz}}{f_c} \right)^{1/6}. \tag{2.13}$$

Here M and μ are the total and reduced masses: $M = M_1 + M_2$, $\mu = M_1 M_2 / M$, and f_c is roughly the frequency of maximum detector sensitivity.

The first numerical simulations of binary neutron star coalescence used Newtonian physics and simple polytropic equations of state (Oohara and Nakamura, 1989; Rasio and Shapiro, 1992; Shibata *et al.*, 1992). These studies suggested that the binary system merged into a single massive, rapidly rotating star. However, to determine if the final object collapses to a black hole requires general relativistic simulations. Intense efforts were required to construct reliable code for three-dimensional hydrodynamic simulations in full general relativity (Font *et al.*, 2000; Miller *et al.*, 2001). The first fully general relativistic simulations were performed by Shibata *et al.* (1992), after which efforts were made to also include realistic microphysics and equations of state, as well as magnetic field evolution.

Recent work on numerical modelling of the coalescence of neutron stars shows that small changes in the initial conditions of neutron star binaries – specifically differences in magnetic field and mass ratio – led to a rich variety of phenomena. Rezzolla and collaborators have recently shown that coalescence waveforms after merger depend very strongly on the equation of state and on the mass ratio of the binary stars. A small change in mass ratio and also a change in the equation of state strongly alters the waveform, while much smaller changes are introduced by the degree of magnetisation, unless the latter is enormous (Baiotti *et al.*, 2008; Rezzolla *et al.*, 2010; Giacomazzo *et al.*, 2011).

For the case of equal-mass binaries, the merger leads to a hypermassive neutron star (i.e. a star that has a mass much larger than the maximum that can be supported by uniform rotation), which develops a bar deformation and which can be sustained for quite long periods of time, greatly increasing the total post-merger gravitational wave emission.

The lifetime of the hypermassive neutron star depends sensitively on a number of factors such as: the mass of the neutron stars, the equation of state and the degree of magnetisation. As a result, the merger can either lead to a prompt black hole (if, for instance, the neutron stars have masses $\gtrsim 1.6$ solar masses) or to a hypermassive neutron star that can survive up to a fraction of a second (if, for instance, the neutron stars have masses $\lesssim 1.4$ solar masses) (Rezzolla *et al.*, 2010). For hypermassive neutron stars that do not collapse promptly to a black hole, neutrinos can represent a source of cooling and yield, through their extraction of energy, to a rapid collapse to form a black hole. In either case, not all of the matter will fall immediately into the black hole. Rather, a small fraction of the total mass of the system will remain outside the black hole because of its large angular momentum and produce a hot torus of high-density matter orbiting at fraction of the speed of light around the black hole. The torus will progressively accrete onto the black hole as it loses energy and angular momentum, in a process that will depend on the torus mass and magnetisation and which could last 0.1–1 s.

For unequal-mass binaries and with moderate mass ratios (e.g. 1.6 solar masses and 1.4 solar masses), the merger and subsequent collapse lead to a much more massive and extended torus, which accretes onto the rapidly rotating black hole (Rezzolla *et al.*, 2010). Additionally, some of the matter can be swung to large distances from the black hole while remaining gravitationally bound. This process leads to the possibility of further fall-back and enhanced accretion episodes later in time; a similar phenomenology is observed in some short gamma-ray bursts.

It should also be noted that although the black hole formation usually marks the end of a strong gravitational signal, the black hole will still ring down and emit gravitational waves as a result of the accreting material, although the amplitude of this latter signal is considerably smaller. Additionally, the numerical models of Giacomazzo *et al.* (2011) for the coalescence of magnetised neutron stars show that the twisting and shearing motions in the torus can lead to substantial growth of the magnetic field, with amplifications that grow exponentially as a result of magneto hydrodynamic instabilities.

The progress that has been made in this area is shown by the eye-catching illustration on front cover of this book. This image shows a single frame from the simulations of Baiotti *et al.* (2008) – the first complete and accurate calculation of the inspiral and merger of a system of equal-mass binary neutron stars in 3D and full general relativity. The different colours represent different degrees of density – green indicates high density and orange low density. After merger, a hypermassive neutron star will form which, although hot, will be unable to resist gravity and will collapse to produce a black hole. This transition will take place within a fraction of a second. The low-density material (orange in colour) will produce a massive disc or torus that will orbit the black hole. As discussed earlier, energy that is is extracted, either from the disc or the black hole's spin, could be used to power the ultra-relativistic outflows of short-hard gamma-ray bursts. A first indication that this paradigm could be the correct one has been obtained in a recent breakthrough study, which has revealed that the merger of magnetised neutron stars does ultimately lead to the formation

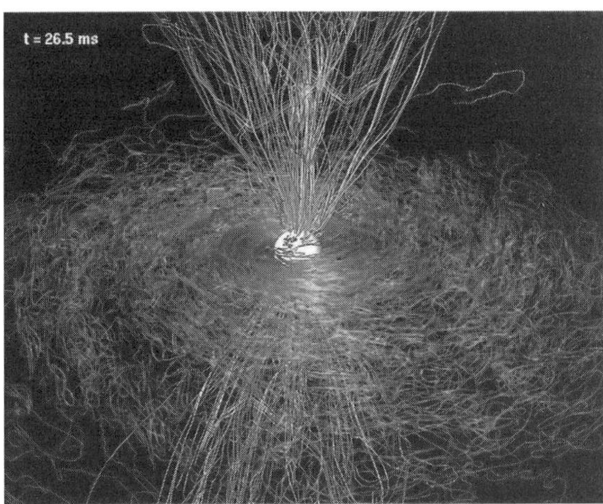

Figure 2.2 The magnetic-field structure in a hypermassive neutron star after the collapse to a black hole. The lighter lines are the magnetic-field lines inside the torus and on the equatorial plane, while the white lines are magnetic-field lines outside the torus and near the axis (reproduced from Rezzolla *et al.* 2011).

of a very strong magnetic field with a jet-like structure (Rezzolla *et al.*, 2011), as shown in Figure 2.2.

2.5 Rates of coalescing compact binaries

As coalescing compact binaries are expected to emit strongly in gravitational radiation and are well modelled, they are excellent candidate sources for ground-based interferometric detectors. Thus, estimations of detection rates have received great attention. Unfortunately, making such predictions is difficult – estimates typically spread over two orders of magnitude. There are generally two methods used in estimating coalescing compact binary rates. First, for binary neutron star systems, one can extrapolate from the observed population of binary pulsar systems. Unfortunately this population is very small. Uncertainties also arise from selection effects in pulsar surveys. Second, predictions can be made using population synthesis in which the population evolution of progenitor systems is numerically simulated. This approach suffers from poor constraints on the astrophysical parameters. For example, the supernova birth kick velocities demonstrated by pulsar proper motion studies may cause binary neutron star systems to be disrupted. Modelling also depends critically or the component mass distributions, which are uncertain.

At the time of writing, radio observations have provided evidence of only four neutron star binary systems in our Galaxy with merger times less than the Hubble time, $\tau_H = 1/H_0$. One of these resides in a globular cluster. Small number statistics dictate that any newly discovered system can lead to a rapid revision of estimates. This was the case with the discovery of the system, PSR J0737−3039, which prompted an order-of-magnitude increase in estimates (Burgay *et al.*, 2003). This is because this system will merge 3.5 times sooner than the Hulse–Taylor pulsar PSR 1913+16. Estimates are heavily dependent on the shortest

Table 2.1. *Coalescing systems of neutron star (NS) and/or black hole (BH) binaries. The table shows the estimated plausible upper, realistic and plausible lower detection rate estimates (R_{pl}, R_{re}, R_{low}) for future networks of advanced interferometric detectors (Abadie et al., 2010)*

	NS/NS	NS/BH	BH/BH
R_{pl}, yr^{-1}	400	300	1000
R_{re}, yr^{-1}	40	10	20
R_{low}, yr^{-1}	0.4	0.2	0.4

estimated coalescence time. The relative close proximity and low luminosity of PSR J0737–3039 provided additional grounds for optimism. For coalescing compact binary systems containing black holes, there are no observations. As a result, rate estimates rely solely on population synthesis techniques. Research indicates that the merger rates of systems that include at least one neutron star could be strongly dominated by field populations (which do not belong to a larger cluster). Coalescing black hole systems are formed much more effectively through dynamical interactions in dense stellar environments such as globular clusters (Sadowski *et al.*, 2007). Due to significant uncertainties in this latter scenario, most predictions consider only the field populations. Therefore, the detection rates of inspiralling systems that include black holes, in spite of their greater luminosities, are below those of neutron star/neutron star mergers.

Table 2.1 presents the rate estimates for initial and advanced LIGO–Virgo networks. These estimates are calculated by scaling to the average volume within which a given detector can detect coalescing compact objects above a signal-to-noise ratio of 8. The rates assume that detector noise is Gaussian and stationary and two or more detectors are operating in coincidence. These estimates suggest that the detection of coalescing compact objects is inevitable by Advanced LIGO.

The discovery of black hole + Wolf–Rayet star systems provides further grounds for optimism in regards to coalescing systems of black holes. At the time of writing there are two known systems: NGC300 X1 lies at a distance of 1.8 Mpc and is composed of a ~ 15 solar mass black hole and a Wolf–Rayet star of ~ 26 solar masses (Crowther *et al.*, 2010); IC10 X-1 contains of a ~ 23 solar mass black hole (Prestwich *et al.*, 2007). As discussed earlier, Wolf–Rayet stars are the progenitors of Type 1b/c supernovae. If such systems survive the supernova explosion, binary black hole systems could form and eventually coalesce within a time-scale of Gyrs (Bulik *et al.*, 2011).

Results from the Sloan Digital Sky Survey indicate that half of recent star formation involved galaxies with low metallicity (Panter *et al.*, 2008). This could have a profound effect on the rates of compact object systems containing black holes. Survival of these systems is highly dependent on whether they can overcome two key obstacles in their evolution. First, postnatal supernova kicks can disrupt a significant proportion of systems. Second, orbital shrinkage during the common envelope phase, during which the larger star transfers mass to its smaller companion, can cause the stars to merge before they can become compact objects.

Belczynski *et al.* (2010b) has shown that a lower metallicity environment helps to suppress these two effects. First, observational evidence suggests that larger black hole masses, which are produced at low metallicity – are born with lower supernova kick velocities (Mirabel and Rodrigues, 2003; Belczynski *et al.*, 2010a). Second, in lower metallicity environments, slower radial expansion occurs during the common envelope phase, thus increasing binary retention. The greater fraction of systems that can survive, in combination with a greater detection range from more massive and hence luminous systems, is highly encouraging. Belczynski states, 'We are therefore truly on the cusp of seeing gravitational waves, and the first source ever to be seen is likely to be a black hole binary'. It will require gravitational wave detection to prove whether this optimism is justified.

2.6 Gravitational wave standard sirens

During the inspiral phase, the rate at which the system chirps, $\mathrm{d}f/\mathrm{d}t$, is determined solely through the system's chirp mass, $\mathcal{M} = (m_1 m_2)^{3/5}/(m_1 + m_2)^{1/5}$. This means that an assumption of equal masses of binary components is not restrictive. Additionally, as first recognised by Schutz (1986), it allows the luminosity distance, d_L, to a chirping source to be inferred by a gravitational wave observation.

To illustrate this, we note that $\mathcal{M}$ is related to the strain amplitude, h, frequency, f, and frequency change, $\dot{f}$, through the relations:

$$h \propto \mathcal{M}^{5/3} f^{2/3} d_L^{-1} \,,$$
$$\dot{f} \propto \mathcal{M}^{5/3} f^{11/3} \,. \tag{2.14}$$

Because $\mathcal{M}$ enters these two relations in the same way, measurements of h, f and $\dot{f}$ can provide estimates of d_L (Schutz, 1986) as described below. The constants of proportionality implied by expression (2.14) depend on a number of factors, including the angle between the system's angular momentum axis and the line of sight to the detector, and the sky position of the binary. With a single axis detector the polarisation axis of an incoming signal is indeterminate. However, the ambiguity can be fully resolved with a network of three or more detectors with different orientations in three dimensions. Then the amplitude ratio of the two gravitational wave polarisations ($h+$ and $h\times$) encodes the inclination of the plane of the binary orbit with respect to the line of sight from the Earth. Once the orbital inclination is defined, the frequency evolution of the gravity wave signal contains a complete description of the system, and the observed amplitude therefore encodes the distance of the source. As the inspiral phase is well modelled, d_L can be determined to an accuracy determined by the signal-to-noise ratio.

By providing distance measurements to sources, in analogy to electromagnetic *standard candles*, coalescing compact binary signals are termed *standard sirens*. A direct consequence of this is that gravitational wave detection, coupled with a redshift measurement from an electromagnetic counterpart, allows astronomers to determine cosmological parameters. If the gravitational wave source is a black hole binary it will probably have no electromagnetic counterpart. Even this problem is surmountable, however, if the gravitational wave detector network has sufficient angular resolution to identify the host galaxy for the event. Then the redshift can be determined by proxy, and the data can still be used to probe cosmology.

The gravitational wave detection of coalescing binaries can be used to complement Tyle Ia supernovae, which are currently the best available cosmological probes. Dalal *et al.* (2006) has shown that such inspiral observations can be used to determine the Hubble constant and constrain the quantum state for dark energy.

The suggested association of neutron star binaries and neutron star–black hole binaries with short-hard gamma-ray bursts (discussed in Section 2.7) could provide such observations. The initial gamma-ray burst is typically followed by a longer-lasting *afterglow* emitted in the X-ray, optical, and radio wavelengths. The afterglows of short-hard gamma-ray bursts are generally fainter and shorter-lived than those produced by the long-duration bursts. Localisations obtained from X-ray observations can allow ground-based follow-up observations to obtain precise localisation of sources on the sky to enable redshift estimations. For neutron star binary and neutron star–black hole binary inspirals such independent sky localisation helps to minimise the degeneracies that otherwise may limit the determination of d_L through gravitational wave observations alone.

For a short gamma-ray burst/gravitational wave event at ~ 200 Mpc, observed with a four-detector network, Dalal *et al.* (2006) showed that an improvement in the accuracy of d_L, to around 6%, is achievable. If advanced networks of gravitational wave detectors are combined with all-sky gamma-ray burst monitors it is realistic to expect the Hubble constant to be measured to within 2–3% once advanced ground-based detectors are in operation.

An additional scenario may occur if gamma-ray bursts are beamed away from Earth, rendering the event invisible to gamma-ray surveys. An afterglow observed without the burst is known as an *orphan afterglow* (Rhoads, 1997; Nakar *et al.*, 2002). Orphan afterglows are thought to be much more common, but are difficult to detect without a gravitational wave trigger providing localisation information. Such coordinated observations could enable a much larger population of coalescence events to be used for cosmology, as long as the follow-up process is efficient and effective.

2.7 Gravitational waves and gamma-ray bursts

Cosmological gamma-ray bursts are the most luminous transient events in the Universe in terms of electromagnetic radiation per unit solid angle. These highly intense beamed emissions, lasting from 10^{-3} s to around 10^3 s, can radiate total energy equivalent to that emitted by the Sun in its entire 10 Gyr lifetime. Such high luminosities mean that gamma-ray bursts can be seen to cosmological distances, allowing them to be used as a probe of the high redshift Universe ($z > 5$). Gamma-ray bursts have generally been divided into two categories: long-soft gamma-ray bursts and short-hard gamma-ray bursts. Generally, the short-hard type have shorter durations and are composed of higher energy photons (harder) than the long variety.

Observations have confirmed that the long-soft bursts are associated with the deaths of massive stars (Hjorth, 2003). One strongly favoured scenario for their emission is described by the collapsar model (Woosley *et al.*, 1999). In this model, the inner part of a progenitor Wolf–Rayet star collapses to form a rapidly rotating black hole. High angular momentum enables the infalling matter to form an accretion disc, which in turn powers an ultra-relativistic jet that blasts through the stellar envelope. Some authors have alternatively suggested that the central engines of the long-soft category of bursts may consist of highly magnetised neutron stars, known as magnetars (Duncan and Thompson, 1992; Usov, 1992; Bucciantini *et al.*, 2008).

The leading model for short-hard gamma-ray bursts is the merger of neutron star binaries or neutron star–black hole binaries. For black hole–neutron star systems, if the component mass ratio $M_{\rm NS}/M_{\rm BH} > 0.1$, the neutron star will be tidally disrupted before it is swallowed by the black hole, producing an accretion disc of $0.01 - 0.3$ solar masses (Nakar, 2007). Smaller values of $M_{\rm NS}/M_{\rm BH}$ will result in the neutron star plunging into the black hole without leaving any residual disc (Miller, 2005). For binary neutron star systems the disc, resulting from the merged neutron stars, will be created around a central black hole.

The suggested association of such compact object coalescences with the short-hard type of burst is of great importance to the gravitational wave community. Searching for gravitational wave signals with associated electromagnetic counterparts can increase the sensitivity for detection by over a factor of 2 (Cutler and Thorne, 2002) since the gravitational wave–electromagnetic coincidence allows some relaxation of the statistical criterion for detection by gravitational waves alone. Additionally, the fact that short gamma-ray bursts have been observed at much lower redshifts than the long-soft bursts makes them a definite target for ground-based detectors. We will see later on that multi-messenger observations of two or more types of radiation allow gamma-ray bursts to be used as cosmological probes.

Neutron star–black hole coalescences, whether or not they are associated with gamma-ray bursts, can allow neutron star structure and microphysics to be probed because the break point in the coalescence waveform is set by the tidal disruption of the neutron star (Hinderer *et al.*, 2010). This depends strongly on the neutron star radius and equation of state. Thus, the observation of black hole–neutron star coalescence can probe the properties of nuclear matter.

2.8 Continuous gravitational wave sources

Continuous gravitational wave sources emit radiation over very long observational periods and are expected to have well modelled quasi-monochromatic signals. In the frequency range of ground-based interferometric detectors, rapidly rotating neutron stars are the most likely sources of periodic, persistent gravitational waves (Brady *et al.*, 1998).

The inner cores of massive stars can have high angular momentum and theory suggests that neutron stars could be born rotating at near the maximum value they can endure without breaking up, which is of order 10^3 times per second (Fryer and Woosley, 2001). This is much faster than the observed spin rates of the majority of known pulsars. One explanation for this discrepancy is that angular momentum is removed though gravitational radiation.

Rotating neutron stars could radiate gravitationally due to asymmetries. Deformations could be introduced due to a build-up of strain in the crust, through the magnetic field or from accretion. Approximations of the characteristic amplitude h_c of gravitational waves from rotating neutron stars take the form:

$$h_c \sim \left(\frac{G}{c^4}\right)\left(\frac{If^2\delta}{d}\right), \tag{2.15}$$

where I is the neutron star's moment of inertia and f the gravitational wave frequency (twice the rotational frequency). The distance to the source is given by d, and δ is a measure of the star's deviation from axis-symmetry. Equation (2.15) shows that the strength of the radiation depends on how rapidly the source rotates and the degree of the distortions. Although the amplitudes are expected to be weak – $h_c \sim 10^{-26}$ – matched filtering can be applied over long observational periods to build up the signal-to-noise ratio. This technique, to be discussed in Chapter 4, is complicated by modulations from the Earth's rotational and orbital motion, as well as pulsar spin-downs.

Proposed mechanisms for producing the distortions include the Chandrasekhar–Friedman–Schutz instability, initially discovered by Chandrasekhar (1970), and later revived by Friedman and Schutz (1978). This is an instability driven by gravitational wave back reaction. The instability creates and maintains hydrodynamic waves in a neutron star's fluid components, termed *r-modes*, which propagate in the opposite direction to that of the star's rotation, producing gravitational radiation.

Neutron star binaries occur in various forms, from the neutron star–neutron star binaries such as the Hulse–Taylor pulsar PSR1913+16, to rather more common systems in which neutron stars have white dwarf or main sequence star companions. The latter often occur as interacting binaries – low mass or high mass X-ray binaries, in which X-ray emission occurs due to mass transfer on to the neutron star. The Rossi X-Ray Timing Explorer provided evidence that Low-Mass X-ray Binary (LMXB) systems contain neutron stars with a very small range of spins around (250–320) Hz. This is despite estimates, based on the spin-up time-scales of neutron stars in LMXBs, that suggest they are capable of reaching frequencies of ~ 1 kHz. To account for the narrow observed distribution, Bildsten (1998) suggested that an accretion-induced density gradient could convert angular momentum into gravitational radiation, thus limiting the rotation rates. One intriguing mechanism to achieve quadrupolar asymmetry is through non-axisymmetric magnetically confined mountains (Payne and Melatos, 2006). As LMXBs emit continuously at periods and sky positions that are well known from X-ray timing, known systems, such as the closest accreting neutron star Scorpius X-1, have been targeted in continuous gravitational wave searches (Abbott *et al.*, 2007).

2.9 Low-frequency sources

The physics of binary black holes is independent of black hole mass. However, the frequency scales inversely with the mass. Thus, the waveforms for the coalescence of black holes are identical at high or low frequency – only the masses must increase. However, the astrophysical context is different. At the low frequencies accessible by the space-based LISA detectors, we are interested in black holes of $10^5 - 10^9$ solar masses. For such large masses, there is also an opportunity to obscure extreme mass ratio binaries, where a low-mass object behaves more like a test mass as it samples and traces the extreme curvature of a massive black hole. In this situation the gravitational wave frequency is set by the high-mass black hole, while the amplitude is set by the mass of the low-mass object.

There is very strong evidence that massive black holes exist in the cores of galaxies, with masses ranging from 10^6 to 10^9 solar masses. Galactic mergers are likely to give rise to black hole mergers, so one estimate of the rate of powerful low frequency gravity wave events can be obtained by estimating the rate of galactic mergers.

Estimates for the rate of coalescence of massive black hole binaries in the Universe vary widely from $\sim 0.1\,\mathrm{yr}^{-1}$ to $\sim 100\,\mathrm{yr}^{-1}$ (Jaffe and Backer, 2003; Enoki *et al.*, 2004; Vecchio, 2004; Islam *et al.*, 2004). Vecchio (1997) suggested that for the black hole merger rate to reach one per year, practically all galaxies out to $z = 1$ would have to contribute black holes to feed the merger process. The latter is not such a strong constraint, however, since the horizon for detecting black hole mergers could be far beyond $z = 1$. For example, if one considers mergers to $z = 3$, only a few per cent of galaxies are required to have a central black hole to achieve one event per year.

As mentioned above, potentially detectable low frequency gravitational waves can be generated by low-mass objects orbiting massive black holes. The low-mass objects could be smaller black holes or neutron stars, white dwarfs or even main sequence stars. Such *extreme mass ratio inspirals* are likely in the nucleus of our own Galaxy, and could in principle be detectable well beyond the Virgo cluster (which multiplies up the number of potential sources by several thousand) (Amaro-Seoane *et al.*, 2007).

The key astrophysical question for extreme mass ratio sources isthe following: is the gravity gradient from the central black hole sufficient to tidally disrupt the incoming object? Clearly main sequence stars will be most easily disrupted. The gravitational radiation will be strong only if the central black hole is capable of swallowing the incoming object. If it is tidally disrupted the infalling object will form an accretion disc, and slowly feed the central black hole, producing negligible gravitational wave emission. To avoid tidal disruption the Schwarzschild radius of the black hole must be large compared with the radius of the infalling object. Roughly, the central black hole must be $10^7 - 10^8$ solar masses to avoid disrupting main sequence stars, 10^4 solar masses for white dwarfs, and 10 solar masses for neutron stars.

The gravitational potential of the central black hole creates a cusp-like stellar density profile. It is difficult to estimate the space density of objects in the central cusp. It depends on star formation in the central high density region of galaxies. However, the population in the cusp will never achieve dynamic equilibrium because stars approaching too close to the central black hole will be lost into the hole. The estimated rate of accretion of stars for a galaxy like the Milky Way is $\sim 10^{-8} - 10^{-6}$ per year (Amaro-Seoane *et al.*, 2007). This means that a realistic detectable event rate (say one per year) requires observations of $\sim 10^8$ galaxies, corresponding to distances of $\sim 1\,\mathrm{Gpc}$.

Short period binary systems can create interesting amplitudes of gravitational waves in the 10^{-1}–10^{-5} Hz range. Such binaries exist in several classes. One of the most definite classes consist of the W Ursa Majoris binaries (WUMas), which are contact binary stars with orbital periods of hours. They are generally low-mass systems. About one star in 150 with mass >0.6 solar masses is a contact binary (Rucinski, 1994). Even lower-mass binaries are difficult to detect; however, it is suggested that there could be a smaller population of systems with even shorter orbital periods (Verbunt, 1997).

A second important class of short period binary stars is the cataclysmic variables, consisting of a main sequence star interacting with a dwarf. Cataclysmic variables have orbital periods ranging from 1000 seconds to one day. The shortest period systems are probably white dwarf-white dwarf systems.

2.10 Stochastic background from the era of early star formation

Because of the large number of gravitational wave sources in the universe, we can be certain that there is a stochastic background of gravitational waves due to the multiple effects of continuous wave sources, burst sources and coalescing binary systems. As discussed below, such backgrounds can be detectable as individual sources within about 30 Mpc. We shall emphasise the background from core-collapse supernovae, but it is straightforward to extend the analysis to neutron star and black hole coalescence events.

The dominant contribution to the astrophysical stochastic background is set by the product of total gravitational wave energy available per source, times their coalescence rate. Coalescing binary systems have a low occurrence rate but relatively large available energy. Supernovae are more frequent, but are expected to have much smaller gravitational wave emissions.

We consider the effects of supernovae and neutron star births at cosmological distances. In this case, we are extrapolating from a the local universe ~ 10 Mpc to the much younger universe, distances $\sim$ a few Gpc and $z \sim 3$. Core-collapse supernovae result from massive stars with short lives - $\sim$Myrs. They closely track the star formation history of the universe. To estimate the rate of core-collapse supernovae we need to estimate the star formation rate in the Universe. Studies generally suggest that the star formation rate was much higher in the past, peaking between $z = 1.5$ and $z = 3$ (Hopkins and Beacom, 2006; Porciani and Madau, 2000; Rowan-Robinson, 1999). Out to $z \sim 1$, ultraviolet observations from the Sloan Digital Sky Survey (SDSS), the Galaxy Evolution Explorer (GALEX), and Classifying Objects by Medium-Band Observations (COMBO17) have constrained the star formation rate to within 30–50% (Hopkins and Beacom, 2006). While the above results are uncertain, they imply that the core-collapse supernovae rate is ~ 10–100 per second, sufficient for supernovae to create a nearly continuous stochastic background.

If we consider a simple case we can suppose that all supernovae have a gravitational wave burst duration τ, and a mean rate of occurrence within some horizon distance (say 3 Gpc) of R bursts per second. Then the mean duty cycle D of supernova bursts is given by $D = R\tau$. If, for simplicity, we assume $\tau \sim 10^{-3}$ seconds (roughly corresponding to the core bounce stage of a core-collapse supernova), then D reaches unity when R reaches

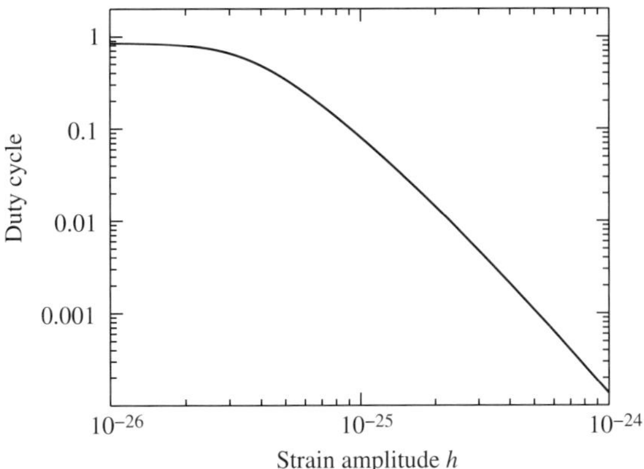

Figure 2.3 Duty cycle versus strain amplitude for supernova-generated gravitational radiation.

3×10^{10} per year, or 10^3 per second. When D reaches unity the supernova bursts create an effectively continuous stochastic background. To make an accurate determination of the supernova stochastic background, one needs to take into account both evolutionary effects and cosmological effects. If every population I star was the result of a single prior supernova from an earlier generation of population II stars, then there would need to have been $\sim 10^{21}$ supernovae to create the observed number of population I stars. This corresponds to an average rate exceeding 1000 supernovae per second. Redshifts both stretch the pulses and the mean time between bursts. The luminosity distance in non-Euclidean geometry changes the observed amplitude of each burst Blair and Ju (1996). Since the above estimates were made, various studies have predicted core-collapse predict event rates of the order 10–100 per second (Burman *et al.*, 1999; Ferrari *et al.*, 1999a; Coward *et al.*, 2004; Howell *et al.*, 2004; Buonanno *et al.*, 2005).

Figure 2.3 gives a plausible result using the star formation rate history of Hopkins and Beacom (2006). It shows the burst amplitude versus duty cycle for supernovae assumed to have a duration 5 ms and an amplitude $h_0 = 10^{-22}$ at 10 Mpc. The background is only continuous for $h \sim 2 \times 10^{-26}$. Nearer sources are stronger, but less frequent. The continuous signal is due to events occurring during early star formation. This amplitude is characteristic of sources at $z = 2-3$. As discussed in Section 2.1, modern simulations suggest that greater emissions occur post-bounce over durations of hundreds of milliseconds. Therefore, extending the simple analysis shown above, one would find that a continuous signal would be produced from less frequent sources at $z < 2$.

Like all stochastic backgrounds, the supernova background can in principle be detected by cross-correlation of signals between nearby detectors (less than half a wavelength separation: for $\tau \sim 10^{-3}$ seconds, they should be less than 100 km apart)(Flanagan, 1993). As discussed above, the signal-to-noise ratio is increased as the 1/4 power of the effective measurement time defined by the cross-correlation integration time. For 10^{10} cycles, this represents a 300-fold improvement. Thus the combined effects of all supernovae is to create a signal that can be detected at a signal-to-noise ratio comparable to that of an individual supernova at 20–30 Mpc. Thus, a detector capable of detection extragalactic supernovae can, with cross-correlation, detect a stochastic background produced at a distance 30 times greater. There are other aspects of the supernova stochastic background which are worth mentioning. Its spectrum represents the average spectrum over all supernovae, but it will be reddened according to the contribution of high redshift supernovae. The duty cycle is clearly amplitude dependent. Nearer sources will create less frequent, larger amplitude bursts. At low duty cycle the background is described as popcorn noise, while for $D > 1$ it approximates Gaussian noise. There is an important difference in this regard. The presence of a popcorn noise component (Allen, 1997; Coward and Regimbau, 2006) means that, unlike true white noise, the individual short bursts create broadband intensity correlations which might allow more powerful digging into the noise. For example, the broadband correlations might allow the background to be detected as spectral intensity correlations within a single detector. This might be combined with cross-correlation between two detectors to dig still deeper into the noise.

The energy density of the supernovae background, expressed as a fraction of closure density, is given by

$$\Omega_{\rm sn} = \Omega_{\rm m}\, f_{\rm s}\, f_{\rm sn}\, \bar{\varepsilon} \tag{2.16}$$

where $\Omega_{\rm m}$ is the mass density of baryonic matter in the Universe, $f_{\rm s}$ is the fraction of this matter that forms into stars in a Hubble time, $f_{\rm sn}$ is the fraction of stars that undergo a supernova event, and $\bar{\varepsilon}$ is the mean gravity wave conversion efficiency for supernovae. Assuming $\Omega_{\rm m} f_{\rm s} \sim 10^{-2}$ and $f_{\rm sn} \sim 10^{-1}$, it is possible that $\Omega_{\rm sn}$ could be in the range 10^{-6}–10^{-8}. However if $\bar{\varepsilon}$ is low, $\Omega_{\rm sn}$ could be $\ll 10^{-10}$. Numerical simulations generally predict $\bar{\varepsilon}$ of $\sim 10^{-9}-10^{-8}$ (Ott *et al.*, 2007). Based on these low emissions, most estimates of $\Omega_{\rm sn}$ are of order $10^{-12}-10^{-11}$ (Ferrari *et al.*, 1999a; Coward *et al.*, 2004; Howell *et al.*, 2004; Buonanno *et al.*, 2005).

If the majority of the gravitational waves are generated in relatively long duration spin-downs of neutron stars (Regimbau and de Freitas Pacheco, 2006a; Owen *et al.*, 1998; Ferrari *et al.*, 1999b) or binary inspiral (Regimbau and de Freitas Pacheco, 2006b) the spectrum will be dominated by a continuous stochastic component. If it is emitted in short supernova bursts or coalescence events, the popcorn component will dominate. If the energy density of the early star formation stochastic background is $\Omega_{g\omega} \sim 10^{-8}$, then it should be eventually detectable by pairs of advanced or third generation detectors.

2.11 Cosmological gravitational waves from the Big Bang

Various sources of gravitational waves from the early Universe have been hypothesised. These may be thought of as the gravitational wave analogue of the microwave cosmic background radiation. The cosmic microwave background originated at the epoch of last scattering, at a redshift $z \sim 10^3$ when neutral gas first formed in the Universe. Thus the microwave background probes the Universe when it was $\sim 10^5$ years old (Penzias and Wilson, 1965). A similar background of neutrinos should also exist, a relic from their epoch of last scattering, about 0.1 seconds after the Big Bang, at a redshift $z \sim 10^{10}$. Due to the weak coupling of gravitational waves with matter, their epoch of release would have been much earlier still, at around the Planck time $\sim 10^{43}$ seconds, or $z \sim 10^{30}$.

Thus primordial gravitational waves offer the tantalising possibility of probing the Universe very near to the moment of creation (Grishchuk, 1975). There are a number of mechanisms for the production of primordial gravitational waves. Vacuum fluctuations are predicted by inflationary models and could be a valuable probe of processes occurring in the Planck era. These fluctuations would have been amplified to produce a stochastic background with a flat spectrum $\Omega_{\mathrm{gw}} \sim 10^{-15}$ across a frequency bandwidth $(10^{-15}-10^{15})$ Hz (Turner, 1997). First order phase transitions could have generated gravitational waves via turbulence. Two proposed mechanisms are injection energy into the primordial plasma or bubble collisions (Kosowsky *et al.*, 1992). If the transition takes place at temperatures corresponding to the electroweak scale, $\sim 100\,\mathrm{GeV}$, Apreda *et al.* (2002) have shown that the frequency would lie within the band of space-based gravitational wave detectors. If there was no process to enhance the background amplitude, then we need consider simply a thermal background that was in equilibrium at the extremely high energies of the Planck era. The background will then have been redshifted like any other radiation. Today this radiation would be in the microwave regime and have an amplitude $h \sim 10^{35}$, which is beyond the possibility of detection.

If the Universe contained an initial inhomogeneity of amplitude h_g (Sazhin, 1988), then today it would have an amplitude at frequency f given roughly by:

$$h \sim 10^{-20} h_g / f . \tag{2.17}$$

We have little idea of the initial amplitude, except for limits set by the cosmic microwave background, which implies that inhomogeneities traced by matter had an amplitude $\sim 10^{-5}$. This would imply that the cosmological background amplitude could be 10^{-28} at 1 kHz (which is beyond terrestrial experiments) and 10^{-21} at 10^{-4} Hz (which is experimentally accessible by space-based laser interferometers.)

A constraint on the cosmological background is set by considering the energy density, and relating it to cosmological models. Thus cosmological backgrounds are often parameterised in terms of the closure density fraction Ω_g. If the spectrum contains equal energy in each decade, it has a slope of -1 on a log h–log f plot. For example, for the Universe to be closed by gravitational waves, $\Omega_g = 1$, the amplitude h_g would be 10^{-13} at 10^{-5} Hz, falling to 10^{-21} at 1 kHz. For $\Omega_g = 10^{-6}$, h_g falls by three orders of magnitude to 10^{-16} at 10^{-5} Hz and 10^{-24} at 1 kHz.

It would appear most unlikely that the Universe closed by gravitational waves, and it would be surprising if gravitational waves contributed significantly to the missing mass. The Hulse–Taylor binary pulsar, in addition to confirming the emission of gravitational waves, also allows limits to be set on the stochastic background of gravitational waves, and on the time variation of the gravitational constant. Taylor and co-workers (Taylor and Weisberg, 1989) used pulsar timing of the binary pulsar PSR 1913+16 to set limits on cosmological backgrounds. The method only works at very low frequencies, where both pulsars and atomic frequency standards have sufficient frequency precision that measurement of the arrival time variations of pulsar signals gives useful sensitivity to gravitational

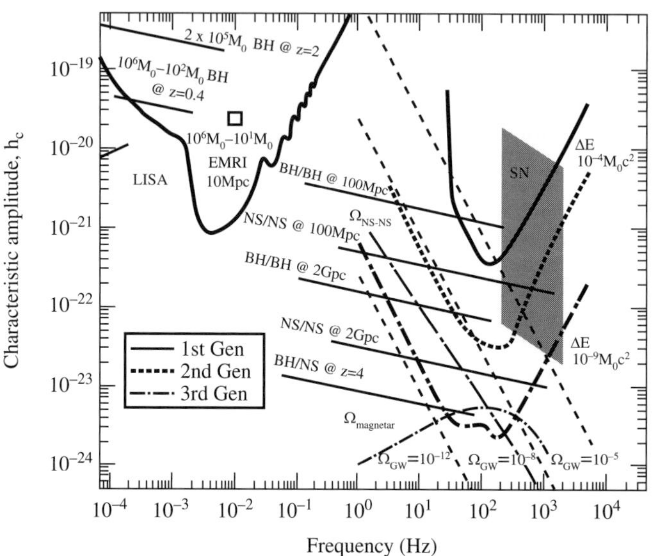

Figure 2.4 Different sources of gravitational waves are shown along with the past and future sensitivities of ground-based and space-based interferometric gravitational wave detectors. An important target for terrestrial detectors will be coalescing systems consisting of neutron stars (NSs) and black holes (BHs). Stochastic gravitational wave background signals (Ω_{GW}) are shown here assuming 10^8 s integration between two optimally orientated and co-located detectors. For LISA, sources should include the coalescence of massive black holes and extreme mass ratio inspirals (EMRIs).

waves. Taylor's timing measurements over several years set a limit to $\Omega_g = 4 \times 10^{-2}$ in the frequency range 10^{-9} to 10^{-12} Hz.

Big Bang nucleosynthesis, the period a few minutes after the Big Bang during which the first atomic nuclei were formed, also sets limits on the stochastic background of gravitational waves (Maggiore, 2000). The primordial abundances of light elements are predicted by assuming that the only contributions to the total energy density come from particles of the standard model. This is without any contribution from gravitational waves. In order to conserve this model, any contribution from gravitational waves cannot exceed $\Omega_g \sim 5.6 \times 10^{-6}$. A larger value would have produced too rapid an acceleration of the Universe during this period, resulting in abundances of light elements very different from the values observed today.

Data from the fifth LIGO science run (S5, ran between 5 November 2005 and 30 September 2007) yielded a 90% confidence upper limit of $\Omega_g \sim 3.1 \times 10^{-6}$ for the stochastic gravitational wave background (LSC Virgo Collaboration, 2009). This value improved on limits set by nucleosynthesis and set the tightest limits on Ω_g at around 100 Hz. As advanced LIGO is expected to achieve sensitivities comparable to a level $\Omega_g \sim 10^{-8}$, gravitational wave detectors may soon be able to place experimental constraints on theoretical models of stochastic backgrounds from both early Universe processes and astrophysical events.

Finally, to summarise our discussion of sources, we plot in Figure 2.4 estimates of various events along with the sensitivities of three generations of ground-based interferometric detectors and the space-based detector, LISA. We see how the sources will span over 10 decades of frequency. The comparison is approximate, as it compares various sources detected by various techniques. Supernova signals are estimated using equation (2.7) at a distance of 10 kpc – we use here a wide range of gravitational wave energies and peak frequencies to encompass the variety of plausible emission mechanisms (Ott, 2009). The figure illustrates how binary coalescence signals should sweep across the frequency bandwidths of gravitational wave detectors as they approach merger – the plot assumes black holes of 10 solar masses and neutron stars of 1.4 solar masses. Low frequency sources for LISA should include the coalescence of massive black holes and extreme mass ratio inspirals discussed in Section 2.9.

To show the signals on the same scale, which relates to detectability, stochastic sources are assumed to have been integrated up for times $\sim 10^8$ seconds, binary coalescences have been integrated over the coalescence frequency range, while the burst sources signal strengths are the only ones representing the instantaneous signal amplitude.

Acknowledgements

The authors wish to acknowledge financial support from the West Australian Government Centres of Excellence scheme, the Australian Research Council and the University of Western Australia.

References

Abadie, *et al.* 2010. *Class. Quant. Grav.*, **27**, 173001.

Abbott, B., *et al.* 2007. *Phys. Rev. D*, **76**, 082001.

Allen, B. 1997. *In Relativistic Gravitation and Gravitational Radiation, Proceedings of the Les Houches School of Physics, 26 Sept-6 Oct*, edited by J. A. Marck and J. P. Lasota, Cambridge University Press, Cambridge.

Amaro-Seoane, P., *et al.* 2007. *Class. Quant. Grav.*, **24**, R113–R170.

Apreda, R., *et al.* 2002. *Nucl. Phys. B*, **631**, 342.

Baiotti, L., *et al.* 2007. *Phys. Rev. D*, **75**, 044023.

Baiotti, L., Giacomazzo, B., and Rezzolla, L. 2008. *Phys. Rev. D*, **78**, 084033.

Baker, J., *et al.* 2006. *Astrophys. J. Lett.*, **653**, L93–L96.

Belczynski, K., *et al.* 2010a. *Astrophys. J.*, **714**, 1217–1226.

Belczynski, K., *et al.* 2010b. *Astrophys. J.*, **715**, L138–L141.

Bildsten, L. 1998. *Astrophys. J. Lett.*, **501**, L89–L93.

Blair, D. and Ju, L. 1996. *Mon. Not. R. Astron. Soc.*, **283**, 648.

Blondin, J. M., Mezzacappa, A., and DeMarino, C. 2003. *Astrophys. J.*, **584**, 971–980.

Blondin, J. M. and Mezzacappa, A. 2007. *Nature*, **445**, 58–60.

Brady, P. R., *et al.* 1998. *Phys. Rev. D*, **57**, 2101–2116.

Bucciantini, N., *et al.* 2008. *Mon. Not. R. Astron. Soc.*, **383**, L25–L29.

Bulik, T., Belczynski, K. and Prestwich, A. 2011, *Astrophys. J.*, **730**, 140.

Buonanno, A., *et al.* 2005. *Phys. Rev. D*, **72**, 084001.

Burgay, M., *et al.* 2003. *Nature*, **426**, 531.

Burman, R. R., Blair, D. G., and Woodings, S. J. 1999. Page 1092 of: T. Piran & R. Ruffini (ed), *Recent Developments in Theoretical and Experimental General Relativity, Gravitation, and Relativistic Field Theories*, World Scientific Publishers.

Burrows, A., *et al.* 2006. *Astrophys. J.*, **640**, 878.

Campanelli, M., *et al.* 2006. *Phys. Rev. Lett.*, **96**.

Centrella, J. M., *et al.* 2001. *Astrophys. J. Lett.*, **550**, L193.

Chandrasekhar, S. 1969. Ellipsoidal figures of equilibrium in *The Silliman Foundation Lectures*, New Haven: Yale University Press.

Chandrasekhar, S. 1970. *Phys. Rev. Lett.*, **24**, 611–615.

Corvino, G., *et al.* 2010. *Class. Quant. Grav.*, **27**, 114104.

Coward, D. and Regimbau, T. 2006. *New Astron. Rev.*, **50**, 461–467.

Coward, D., Burman, R., and Blair, D. 2004. *Mon. Not. R. Astron. Soc.*, **324**, 1015.

Crowther, P. A., *et al.* 2010. *Mon. Not. R. Astron. Soc.*, **403**, L41.

Cutler, C. 1998. *Phys. Rev. D*, **57**, 7089.

Cutler, C. and Thorne, K. 2002. *Proceedings of the 16th International Conference on General Relativity and Gravitation*, edited by N.T Bishop and S. N. Maharaj, World Scientific.

Dalal, N., *et al.* 2006. *Phys. Rev. D*, **74**.

Dimmelmeier, H., *et al.* 2008. *Phys. Rev. D*, **78**, 064056.

Duncan, R. C. and Thompson, C. 1992. *ApJL*, **392**, L9–L13.

Enoki, M., *et al.* 2004. *Astrophys. J.*, **615**, 19–28.

Ferrari, V., Matarrese, S., and Schneider, R. 1999a. *Mon. Not. R. Astron. Soc.*, **303**, 247.

Ferrari, V., Matarrese, S., and Schneider, R. 1999b. *Mon. Not. R. Astron. Soc.*, **303**, 258.

Flanagan, E. E. 1993. *Phys. Rev. D*, **48**, 2389.

Font, J. A., *et al.* 2000. *Phys. Rev. D.*, **61**, 044011.

Friedman, J. and Schutz, B. 1978. *Astrophys. J.*, **222**, 281–296.

Fryer, C. and Woosley, S. 2001. *Nature*, **411**, 31–34.

Fryer, C. L., Holz, D. E., and Hughes, S. A. 2002. *Astrophys. J.*, **565**, 430.

Giacomazzo, B., Rezzolla, L., and Baiotti, L. 2011. *Phys. Rev. D*, **83**, 044014.

Grishchuk, L. 1975. *Sov. Phys. JETP*, **40**, 409–415.

Hawking, S. W. 1971. *Phys. Rev. Lett.*, **26**, 1344.

Hils, D., Bender, P., and Webbink, R. 1990. *Astrophys. J.*, **360**, 75–94.

Hinderer, T., *et al.* 2010. *Phys. Rev. D*, **81**, 123016.

Hjorth, J. 2003. *Nature*, **423**, 847.

Hopkins, A. M. and Beacom, J. F. 2006. *Astrophys. J.*, **651**, 142–154.

Houser, J. L., Centrella, J. M., and Smith, S. C. 1994. *Phys. Rev. Lett.*, **72**, 9.

Howell, E., *et al.* 2004. *Mon. Not. R. Astron. Soc.*, **351**, 1237.

Islam, R. R., Taylor, J. E. and Silk, J. 2004. *Mon. Not. R. Astron. Soc.*, **354**, 629–640.

Jaffe, A. H. and Backer, D. C. 2003. *Astrophys. J.*, **583**, 616–631.

Kosowsky, A., Turner, M. S. and Watkins, R. 1992. *Phys. Rev. Lett.*, **69**, 2026.

Lai, D. and Shapiro, S. L. 1995. *Astrophys. J.*, **442**, 259–272.

LSC Virgo Collaboration. 2009. *Nature*, **1**, 1.

Maggiore, M. 2000. *Phys. Rep.*, **331**, 6.

Manca, G. M., *et al.* 2007. *Class. Quantum Grav.*, **24**, S171–S186.

Marek, A., Janka, H.-T. and Müller, E. 2009. *Astron. Astrophys.*, **496**, 475–494.

Merritt, D. M. M. 2005. *Living Reviews in Relativity*, **8**.

Miller, M., Suen, W., and Tobias, M. 2001. *Phys. Rev. D.*, **63**, 121501.

Miller, M. C. 2005. *Astrophys. J. Lett.*, **626**, L41–L44.

Mirabel, I. F. and Rodrigues, I. 2003. *Science*, **300**, 1119–1121.

Nakar, E. 2007. *Phys.Rept*, **442**, 166–236.

Nakar, E., Piran, T. and Granot, J. 2002. *Astrophys. J.*, **579**, 699–705.

Nelemans, G., Yungelson, L. and Portegies Zwart, S. 2001. *Astron. Astrophys.*, **375**, 890–898.

New, K. C. B., Centrella, J. M. and Tohline, J. E. 2000. *Phys. Rev. D*, **62**, 064019.

Oohara, K. and Nakamura, T. 1989. *Prog. Theo. Phys*, **82**, 535–554.

Ott, C. D. 2009. *Class. Quant. Grav.*, **26**, 063001.

Ott, C. D., *et al.* 2005. *Astrophys. J. Lett.*, **625**, L119.

Ott, C. D., *et al.* 2006. *Phys. Rev. Lett.*, **96**, 201102.

Ott, C. D., *et al.* 2007. *Class. Quant. Grav.*, **24**, S139–S154.

Owen, B. J., *et al.* 1998. *Phys. Rev. D*, **58**, 084020.

Panter, B., *et al.* 2008. *Mon. Not. R. Astron. Soc.*, **391**, 1117–1126.

Payne, D. J. B. and Melatos, A. 2006. *Astrophys. J.*, **641**, 471–478.

Penzias, A. A. and Wilson, R. W. 1965. *Astrophys. J.*, **142**, 419–421.

Phinney, E. S. 1991. *Astrophys. J. Lett.*, **380**, L17–L21.

Porciani, C. and Madau, P. 2000. *Astrophys. J.*, **548**, 522.

Press, W. H. and Thorne, K. S. 1972. *Annu. Rev. Astron. Astrophys.*, **10**, 338.

Prestwich, A., *et al.* 2007. *Astrophys. J.*, **669**, L21.

Pretorius, F. 2005. *Phys. Rev. Lett.*, **95**, 121101.

Pretorius, F. 2009 from *Astrophysics and Space Science Library*, Vol. 359. Edited by Colpi, M., Casella, P., Gorini, V., Moschella, U., and Possenti, A. Jointly published with Canopus Publishing Limited, Bristol, UK.

Rasio, F. A. and Shapiro, S. L. 1992. *Astrophys. J.*, **401**, 226–245.

Regimbau, T. 2007. *Phys. Rev. D*, **75**, 043002.

Regimbau, T. and de Freitas Pacheco, J. A. 2006a. *Astron. Astrophys.*

Regimbau, T. and de Freitas Pacheco, J. A. 2006b. *Astrophys. J.*, **642**, 455.

Rezzolla, L., *et al.* 2010. *Class. Quantum Grav.*, **27**, 114105.

Rezzolla, L., *et al.* 2011. *Astrophys. J. Lett.*, **732**, L6.

Rhoads, J. E. 1997. *Astrophys. J. Lett.*, **487**, L1.

Rowan-Robinson, M. 1999. *Ap&SS*, **266**, 291.

Rucinski, S. M. 1994. *Publ. Ast. Soc. Pac.*, **106**, 462–471.

Sadowski, A., *et al.* 2007. *Astrophys. J.*, **676**, 1162.

Saijo, M., *et al.* 2001. *Astrophys. J.*, **548**, 919.

Sathyaprakash, B. and Schutz, B. F. 2009. *Living Rev. Relativity*, **12**, 2.

Sazhin, M. V. 1988 in *Experimental Gravitational Phys* ed. P. F. Michelson, page 179, World Scientific Publishing Co., Singapore.

Scheidegger, S., *et al.* 2008. *Astron. Astrophys.*, **490**, 231–241.

Schutz, B. F. 1986. *Nature*, **323**, 310–311.

Shibata, M. and Karino, S. 2004. *Phys. Rev. D*, **70**, 4022.

Shibata, M. and Sekiguchi, Y. 2005. *Phys. Rev. D*, **71**, 024014.

Shibata, M., Nakamura, T. and Oohara, K. 1992. *Prog. Theo. Phys.*, **88**, 1079–1095.

Shibata, M., Baumgarte, W. and Shapiro, S. 2000. *Astrophys. J.*, **542**, 453.

Shibata, M., Karino, S. and Eriguchi, Y. 2002. *Mon. Not. R. Ast. Soc*, **334**, L27–L31.

Taylor, J. H. and Weisberg, J. M. 1989. *Astrophys. J.*, **345**, 434–450.

Thorne, K. S. 1992. *Recent Advances in General Relativity*. Burkhauser, Boston. Chap. Sources of gravitational waves and prospects for their detection.

Turner, M. S. 1997. *Phys. Rev. D*, **55**, 435–439.

Usov, V. V. 1992. *Nature*, **357**, 472–474.

Vecchio, A. 1997. *Class. Quant. Grav.*, **14**, 1431–1437.

Vecchio, A. 2004. *Phys. Rev. D*, **70**, 042001.

Verbunt, F. 1997. *Class. Quant. Grav.*, **14**, 1417.

Watts, A. L., Andersson, N. and Jones, D. I. 2005. *Astrophys. J.*, **618**, L37.

Woosley, S. E. 1993. *Astrophys. J.*, **405**, 273–277.

Woosley, S. E., MacFadyen, A. I. and Heger, A. 2011. Published in: *Supernovae and Gamma-Ray Bursts, eds. M. Livio, K. Sahu, & N. Panagia* (Cambridge: Cambridge University Press), preprint:astro-ph/9909034.

Yakunin, K. N., *et al.* 2010. *Class. Quant. Grav.*, **27**, 194005.

3

Gravitational wave detectors

D. G. Blair, L. Ju, C. Zhao, H. Miao, E. J. Howell and P. Barriga

This chapter first introduces gravitational wave detection from a very general point of view, before looking at the particular methods of detection across the spectrum from nanohertz to kilohertz. It finishes by focusing specifically on terrestrial laser interferometers.

3.1 Introduction

The discovery of radio waves by Heinrich Hertz in 1886 unleashed the communications revolution which has transformed our lives. Optimisation of radio receivers required understanding and integration of two concepts. The first was the concept of the antenna, which taps energy from a wave freely propagating in space and converts it into a signal which can be amplified and detected. The second was the receiver, which processes this energy by detection (converting it to a slowly time-varying voltage), amplification (increasing its amplitude without changing its frequency) or modulation (changing its frequency).

Designing gravitational wave receivers is analogous to designing radio receivers, except that electric charges moving freely in conductors are replaced by *test masses* floating freely in space. This concept was illustrated in Figure 1.2 in Chapter 1, showing how a ring of test particles is deformed by a passing gravitational wave. The first gravitational wave receivers were constructed by Joseph Weber in the 1960s. They took the form of large test masses in which gravitational waves could induce quadrupole vibrations. Weber went on to develop the Weber bar, in which one searched for excitations in the fundamental longitudinal vibrational mode of a cylinder. In this case, the receiver can be idealised as a pair of point masses joined by a mechanical spring. The simplest mass quadrupole resonator consists of two point masses joined by a spring. Because the stiffness of the spring is important to obtain good coupling to a gravitational wave (as we discuss later), a practical approximation consists of a cylindrical bar that in its fundamental longitudinal mode vibrates like a pair of masses connected by a stiff spring. One of the first such instruments is shown in Figure 3.1.

In such a mechanically coupled receiver, the wave acts to create an acoustic signal which must then be transduced and amplified in order to convert it into an electromagnetic signal

Figure 3.1 Joseph Weber with an aluminium bar instrumented with piezoelectric crystals to read out the vibration. It can be simplified to an equivalent system consisting of two point masses separated by a spring.

which can be analysed. This was the concept of Weber's first detectors (Weber, 1960) and of later generations of cryogenic resonant mass detectors, which were either bars or spheres, cooled to reduce thermal noise and to enable the use of low noise superconducting transducers.

A network of of five cryogenically cooled resonant bar detectors of similar design, the International Gravitational Event Collaboration (IGEC) (Astone *et al.*, 2003) ran from 1997 to 2000 and consisted of: ALLEGRO (Martin *et al.*, 2005), AURIGA (Vinante *et al.*, 2006), EXPLORER and NAUTILUS (Astone *et al.*, 2006) and NIOBE (Blair *et al.*, 1995). The last of these detector's is shown in Figure 3.2. Later instruments such as MiniGRAIL (de Waard *et al.*, 2006) and Schenberg (Aguiar *et al.*, 2006) were designed using a spherical antenna, providing a nearly omnidirectional antenna pattern to allow determination of both source direction and wave polarisation (Fafone, 2006).

Early studies also considered the Earth as a gravitational wave detector in which gravitational waves would excite quadrupole vibrations of the entire planet (Dyson, 1969). Unfortunately the seismic activity of the Earth is much too large for high sensitivity to be achievable.

Today most research is focused on detectors that consist of a distribution of at least two test masses, which float in space like buoys floating on the ocean. They are coupled together by an electromagnetic field (such as a laser beam) instead of by mechanical forces. The passing wave causes relative changes in the spacing between the test masses. These changes in spacing act on the coupling field, changing the energy in the receiver (which

Figure 3.2 The cryogenic resonant bar Niobe consisted of a vibration isolated 1.5 tonne niobium bar instrumented with a microwave superconducting transducer cooled to about 5 K.

might be a photodetector looking at interference fringes). In principle, there is no fundamental difference between acoustic detectors and 'floating mass' detectors. (We will use the word *detector* in preference to *receiver* from now on.) In almost all cases an elaborate Michelson interferometer is especially useful because of its ability to measure relative displacements in orthogonal directions, whilst suppressing the noise of the light source used for the measurement.

The strength of the electromagnetic coupling field between the floating test masses fundamentally affects the detector sensitivity because, to be measured, the gravitational wave must do work on the coupling field. This work becomes the available signal energy. The coupling field between the floating masses would seem much weaker than the mechanical coupling of acoustic detectors, but we will see later that in some circumstances laser beams can provide a coupling strength that exceeds that of solid matter.

Depending on the frequency band of interest, different factors dominate the design of detectors. However, they are all linked by a few key concepts, and by the fact that all predicted signals are near the limit of measurement sensitivity and dependent on massive levels of innovation and ingenuity.

In this chapter we will begin by simply summarising the techniques available for detection across frequency bands from the nHz range to the kHz range. Then we will outline the key concepts that provide a more unified framework for understanding these detectors, followed by an expanded discussion of these concepts. Next we will summarise specific detection techniques across the spectrum. The last section focuses on the main topic of this book – the specific challenge of building terrestrial laser interferometers.

Table 3.1. *A comparison of various detectors to the standard quantum limit*

Frequency band	Technique	Projects
Nanohertz	Pulsar timing	Nanograv (Jenet *et al.*, 2009)
		Parkes Pulsar Timing Array (Hobbs *et al.*, 2009)
		European Pulsar Timing Array (Ferdman *et al.*, 2010)
		International Pulsar Timing Array (Hobbs *et al.*, 2010)
Millihertz	Space transponder interferometers	Laser Interferometric Space Antenna (LISA) (Danzmann, 2003)
Decihertz	Fabry–Perot space laser interferometer	Deci-hertz Interferometric Gravitational Wave Observatory (Kawamura *et al.*, 2006)
Audio	Terrestrial laser interferometers	LIGO and Advanced LIGO (Sigg and the LIGO Scientific Collaboration, 2008)
		Virgo and Advanced Virgo (Acernese *et al.*, 2008)
		TAMA (Ando *et al.*, 2001)
		LCGT (Kuroda and the LCGT Collaboration, 2006)
		GEO (Willke and the LIGO Scientific Collaboration, 2007)
		AIGO (Barriga *et al.*, 2010)
		ET (Punturo *et al.*, 2010)
Kilohertz	Acoustic detectors	IGEC (Astone *et al.*, 2003).
		IGEC2 (Astone *et al.*, 2010)
		MiniGRAIL (Gottardi *et al.*, 2007)
		Schenberg (Aguiar *et al.*, 2006)

3.2 Introducing gravitational wave detectors across the spectrum

At the lowest frequencies, the nanohertz band (fractions of a cycle per year), the *pulsar timing* technique uses pairs of test masses consisting of the Earth and certain distant millisecond pulsars. While quite infinitesimal, the electromagnetic field of the pulsar beam provides the coupling field for detection. Because the pulsar is so distant, it is appropriate to consider the gravitational wave acting against the incoming radiation field. Energy is again absorbed via the Doppler interaction of the gravitational wave on the pulsar beam. Detection depends on precise timing of the detected pulses and comparison of the timing between different Earth-pulsar pairs (equivalent to an interferometer). A computer-generated image of this scheme is shown in Figure 3.3. The International Pulsar Timing Array project is a pulsar timing collaboration. It includes the Parkes Pulsar Timing Array, the European Pulsar Timing Array and the Nanograv project. Further details of these projects and others discussed in this section can be found by following the references given in Table 3.1.

Figure 3.3 Pulsar timing for detection of nanohertz gravitational waves (image courtesy of David Champion, Max-Planck-Institut für Radioastronomie).

At around one cycle per hour *transponded space laser interferometry* is proposed. This technique requires sets of kg-scale test masses linked by low-power laser beams on a multimillion kilometre baseline. The Laser Interferometer Space Antenna (LISA) is a proposed NASA/ESA project. As illustrated in Figure 3.4, it consists of three spacecraft, placed at the vertices of an equilateral triangle of size 5×10^6 km, and set in solar orbit at 1 A.U., $20°$ behind the Earth. The interferometric measuring system is created through transponding, in which each spacecraft simulates reflection by transmitting a new laser frequency locked to the incoming beam.

At around 0.1–1Hz a *space laser interferometer* with 1000 km baselines has been proposed. DECIGO, the Deci-hertz Interferometer Gravitational Wave Observatory, consists of three drag-free satellites 1000 km apart. The relative displacements between the craft would be measured using three sets of Fabry–Perot Michelson interferometers (Kawamura *et al.*, 2006).

In the *terrestrial, audio frequency band* (about 10 Hz to ~3 kHz), changes in the spacing of test masses kilometres apart are measured with very high power laser interferometers. There are a number of planned or currently operational detectors around the world, all of which were listed in Table 3.1. These include the Laser Interferometer Gravitational Wave Observatory (LIGO), which has detectors situated at Hanford, Washington and Livingston, Louisiana (shown in Figure 3.5), an Italian–French detector called Virgo at Cascina, near Pisa, and the German/British GEO 600 project based in Hannover, Germany. Case studies of these three interferometric detectors, all adopting slightly different designs, will be presented in Chapters 6, 7 and 8, respectively. Third generation detectors have also been proposed. One such detector, the European 'Einstein gravitational wave Telescope' (ET), will be discussed in Chapter 16.

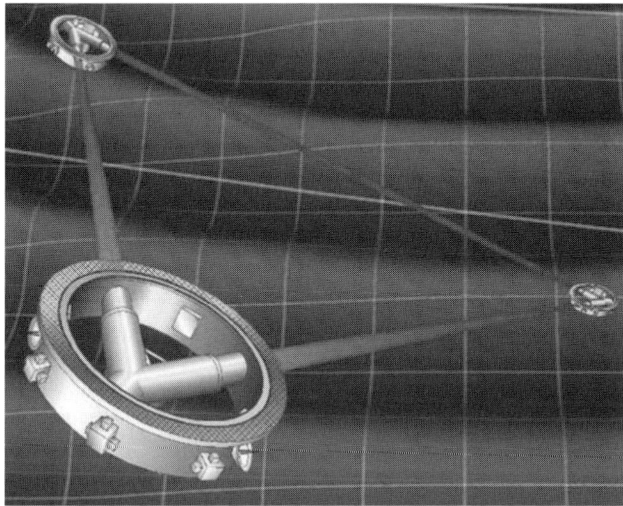

Figure 3.4 Three spacecraft carry pairs of test masses to create a multiple system of interferometers. The laser beams are sensed at the distant spacecraft. Most of the transmitted light is lost to diffraction, so return beams are created by locking another laser to the weak incoming beam – a process called transponding (courtesy of NASA and ESA).

Figure 3.5 A photograph of the LIGO detector at Livingston, Louisiana, USA (courtesy of LIGO).

It is interesting to note that acoustic detectors, such as resonant bars and spheres with single distributed metre-scale test masses, also use interferometry in the readout systems, either in the form of microwave interferometers or superconducting quantum interferometers (SQUIDS). Thus, we see an enormous variety of detectors all sharing the common principle of interferometry. Interferometry enables the detection of differential signals, suppression of common mode effects such as noise in the electromagnetic coupling beam, and high sensitivity measurement over large distances.

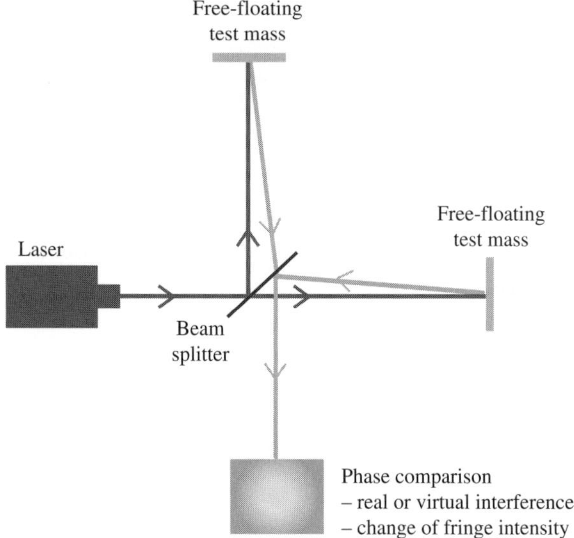

Figure 3.6 Concept of an interferometric gravitational wave detector.

In section 3.3 we will look at the key concepts in the design of the above detectors, first with a brief overview and then with more detail. To aid our discussion, Figure 3.6 provide a diagram of a general interferometer. A laser beam is the source of the electromagnetic coupling field. The relative phases (or difference in travel times) of the reflected beams are compared with the help of a beam splitter that recombines the beams. Any relative change in length of the two arms due to a passing gravitational wave is observed as an intensity variation at the beam splitter output. For pulsar timing there is no laser, and the distant test masses are the source of the coupling beams. Interference is virtual, consisting of a measurement of the relative timing difference. In some designs Fabry–Perot cavities in the arms increase the coupling field and therefore the sensitivity of the interferometer (see Figure 3.9).

Energy transfer in gravitational wave detectors

Before we embark on detailed discussions we shall discuss certain difficulties in understanding gravitational wave detectors. In the gravitational wave research community acoustic gravitational wave detectors have traditionally been understood quite differently from electromagnetically coupled detectors. Following Weber (1961), the sensitivity of acoustic detectors was estimated by considering the gravitational wave energy absorbed by the detector. However, for laser interferometer gravitational wave detectors a more convenient method of estimation was based on considering the test masses as free masses that experience the gravitational wave spatial strain h of a passing wave. If the masses are truly free, this means that their motion does no work, and by energy conservation, no energy is extracted from the wave. This free-mass approximation then leads to the idea that electromagnetically coupled gravitational wave detectors do not absorb energy from the passing wave. Indeed, when this chapter was first reviewed by experts in the community, it revealed that many people considered that laser interferometer detectors do not actually absorb energy from gravitational waves.

This point of view recalls the debate about the existence of gravitational waves that occurred from 1916 to 1957. Eddington claimed that gravitational waves travel 'at the speed of thought' and even Einstein was at one stage convinced of their non-reality (Kennefick, 2006). The proof of the reality of gravitational waves was eventually clarified by the sticky-beads thought experiment, which was publicised by Herman Bondi (Bondi, 1957). The argument, due to Richard Feynman, was presented under the pseudonym Mr Smith at a conference in Chapel Hill, North Carolina in 1957. In this thought experiment the viscous interaction of sticky beads on a rod gives rise to heat (as a gravitational wave passes), proving that gravitational waves are able to deposit energy and therefore cannot be a mathematical artifact. Equally, every practical detector must be a transducer for gravitational waves, converting wave energy into electromagnetic energy, and amplifying it to enable it to be resolved against the inevitable background of instrument noise.

The assumption that test masses are free masses is very accurate for calculating gravitational wave signals in the electromagnetically coupled detectors constructed to date, and at the lowest frequencies will always be an excellent approximation. However, new techniques that utilise very high optical power (discussed later in this chapter) make the approximation quite confusing. Most importantly, the free mass approximation gives rise to a misconception that electromagnetically coupled gravitational wave detectors do not absorb energy from the waves. Discussions of this point with our colleagues revealed that the issue had been extensively analysed by Saulson in 1997 (Saulson, 1997). This work exposed the issue but did not fully resolve the key questions on the fundamentals of detection that still remain. Saulson pointed out that the Doppler interaction on the electromagnetic coupling field due to gravitational wave strain acting between test masses gives rise to a small viscous dissipative term due to the strain-induced test mass velocity. Any test mass absorbs energy from a gravitational wave via this frequency conversion of the electromagnetic wave. However, Saulson pointed out that for a typical wave and a terrestrial detector, the power absorbed would have the extraordinarily small magnitude of $\sim 10^{-40}$ W.

As pointed out by Veitch (1991), and by Paik (1982) before him, an electromagnetically coupled detector can in general be considered as a *parametric transducer*. The word parametric refers to the fact that the signal acts on one of the frequency-determining parameters of a harmonic oscillator. The parametric aspect of the gravitational wave interaction with a laser beam ensures that the output signal is described by a product of the signal field and the coupling field, and this ensures that the signal appears as two signal sidebands at frequencies $\Omega - \omega$ and $\Omega + \omega$. These sidebands are offset from the electromagnetic frequency Ω by the gravitational wave frequency ω. Lossless parametric interactions are described by the Manley–Rowe equations (Manley and Rowe, 1956). These equations formalise energy conservation in the presence of frequency conversion. Because photon frequency is proportional to energy, the energy gain for each interaction that creates a sideband photon is either $(\Omega + \omega)/\omega$ or $(\Omega - \omega)/\omega$, or roughly Ω/ω (if $\Omega \gg \omega$). However the term for the upper sideband has opposite sign compared with the lower sideband. The interpretation of these sidebands is crucial to understanding how the detector interacts with the gravitational wave. Recently people have proposed the use of carefully tuned detectors (Braginsky, Gorodetsky and Khalili, 1997) for which the interaction is dominated by the upper sideband $\Omega + \omega$ alone. In this case it is clear that the upper sideband represents energy extraction from the wave through the work done by the test mass mirror on the radiation field. Under these circumstances the gravitational wave experiences large electromagnetic 'viscosity', and maximal energy is absorbed from the wave. The detector can then be considered as a transducer with power gain determined by the frequency ratio Ω/ω. However when the two sidebands are balanced, as in a conventional interferometer, the lower sideband $\Omega - \omega$ applies forces to the detector test masses that cancel the mechanical forces applied by the radiation field via the test mass mirror. Suddenly the energy absorbed from the wave becomes very small. The system acts as if one sideband absorbs energy from the wave while the other returns it to the wave. This interpretation is easily shown to be false however. Rather, what the second sideband does is to change the dynamical response of the test mass so that it closely resembles a free mass.

Detector designers normally consider signal-to-noise ratio and not energy coupling. For example, it is well known that increasing the optical power increases the signal-to-noise ratio at high frequency. This can be explained in terms of photon shot noise alone, but the result may often be quite consistent with increasing the signal energy through providing a stronger coupling field. Saulson explicitly calculates the energy absorption cross-section of a laser interferometer, and shows that this can be much larger than that of resonant mass detectors because it scales as the square of the detector length. He also shows the close analogy between a laser interferometer gravitational wave detector and a radio receiver.

The above discussion brings to mind another subtle issue: an apparent contradiction in the theory of resonant mass gravitational wave detectors discussed in the mid-1970s. The energy absorption cross-section of a resonant mass detector was always calculated assuming a test mass at absolute zero of temperature. However, real detectors were operated at finite temperature where the fundamental normal mode was observed to be ringing sinusoidally due to its thermal energy. A gravitational wave signal would (in principle) be detected as a sudden small change in this ringing. This change could consist of a rise or fall in amplitude, or just a phase change. This meant that the real detector might not extract energy from a wave but actually strengthen the wave, by giving up energy to the wave. This at first seemed to caste doubt on the zero temperature theory for the cross-section, until Gifford (Gifford, 1979) realised that for a large ensemble of detectors, all with random phase, the *mean* energy interaction was the same as the zero temperature result for a single detector.

Thus we see that the understanding of detectors is subtle, and sometimes a change in viewpoint helps to create a much deeper understanding. Now we shall go on to discuss the principles of detection in more detail.

3.3 Key concepts in gravitational wave detection

There are four key concepts that must be addressed in building detectors for any type of wave. The first is the fundamental issue of designing a detector with the best possible impedance matching to the wave: this determines the amount of energy that the detector receives from the wave. Radios, optical telescopes, microphones and human ears all use this principle. For example, ears use the eardrum, and intricate levers to couple sound from the low impedance of the air to the high impedance of the cochlea. Unfortunately, while electromagnetic and acoustic impedance matching can be accomplished efficiently, gravitational wave detector impedance matching is impossible because matter is almost totally transparent to gravitational waves: atomic matter represents an infinitesimal perturbation to the propagating medium.

The second factor relates to the advantage of building a detector matched to both signal frequency and wavelength. To detect the ocean swell with measurements between buoys floating on the ocean, it is best to locate the buoys half a wavelength apart (for a particular wavelength) – then they move in anti-phase and the relative motion is maximised. Television dipole antennas on rooftops are examples of such design. Unfortunately it is impossible to achieve this condition for terrestrial gravitational wave detectors because the wavelengths we are searching for in the terrestrial detection band (10 Hz to $\sim$3 kHz) correspond to detector sizes from 15 000 km to 50 km. This problem can be partially overcome if the electromagnetic wave is allowed to travel back and forth between the test masses until it has a travel time matched to half a gravitational wave period. This can be achieved by using a resonant optical cavity between the masses, but it carries with it a noise penalty because the multiple path further increases the sensitivity to test mass displacement noise.

The third factor is the problem of detector noise. Noise sources can be divided into two classes. First, technical noise, such as seismic noise or acoustic noise. These can be arbitrarily reduced with sufficient ingenuity. We should point out, however, that density fluctuations due to seismic noise at frequencies below about 1 Hz cause Newtonian gravitational force fluctuations which act on detector test masses. Such forces cannot be shielded, so set fundamental limits on the sensitivity of ground-based detectors. The second class of noise source is intrinsic noise, such as thermal noise and photon shot noise. These noises must all be minimised, but at the fundamental level we are faced with the standard *quantum limit barrier*.

Once detectors reach the standard quantum limit barrier, they become pure quantum instruments. At this point a fourth concept – measuring beyond the standard quantum limit – comes into play. Quantum measurement schemes can be devised to overcome the quantum limit barrier. A gravitational wave is a classical wave, but it interacts with a detector which, once classical noise sources have been overcome, is a pure quantum system. The interplay of quantum noise, and the classical wave offers new opportunities. We will see that some detectors can in principle achieve sensitivity limited purely by quantum noise and that measurement systems can be designed to enable detection of signals smaller than the standard quantum limit. This leads to exquisite new physics which crosses the interface between quantum measurement and the quantum–classical transition. Let us now go on and elaborate our discussion of these concepts.

Impedance matching

Impedance matching is always required for efficient energy transfer between different systems, and is routinely used in the design of electromagnetic antennas, amplifiers and transmission lines. Gravitational wave detection can be understood from this same engineering standpoint. Gravitational waves, ripples in the curvature of spacetime, propagate in a very stiff elastic medium as explained in Chapter 1. Spacetime has a characteristic impedance $\sim c^3/G$ (Blair, 1991). A gravitational wave detector is made up of test masses floating in this medium. Because of the quadrupole nature of gravitational waves, no energy is imparted to a lone isolated point mass. The wave is a wave of tidal force. It acts to change the shape of distributed systems of test masses – in the simplest case, changing the spacing between two test masses. Energy is only deposited in the system if work has to be done to change the spacing of the masses.

If two test masses are connected by a laser beam, work is done on the laser field, causing small Doppler frequency changes which optical systems can in principle be designed to detect (Saulson, 1997). If the detector were a pair of masses connected by a mechanical spring, work would be done on the spring, and this in turn could be read out by a transducer such as a piezoelectric strain gauge. The impedance matching problem arises because the impedance of any mechanical system is very low compared with free space. Resonant mass detectors such as Weber's original bars needed to maximise their acoustic impedance by using materials with high density and high velocity of sound. This ensured the maximum possible gravitational wave energy exchange with the detector. Aluminium, niobium and bronze were used for cryogenic acoustic detectors, chosen for a combination of intrinsic, practical and cost reasons. In the 1990s, synthetic sapphire, silicon, tungsten, molybdenum and silicon carbide were proposed as better materials for large acoustic gravitational wave detectors because they had high sound velocity (up to $10\,\mathrm{km\,s^{-1}}$) and high density (up to $19\,\mathrm{g\,cm^{-3}}$). However, in all cases the magnitude of the interaction is tiny because the highest acoustic impedance achievable in a detector, $\sim 10^{10}\,\mathrm{kg\,s^{-1}}$, is still 25 orders of magnitude below the impedance of spacetime $\sim 10^{35}\,\mathrm{kg\,s^{-1}}$.

The impedance mismatch ratio determines the energy coupling between the gravitational waves and the detector. The only systems to which gravitational waves could strongly couple are black holes and neutron stars, where the effective sound velocity is either c or $\sim 0.1c$, respectively. However, as we discuss below, it is possible to improve the impedance matching by using the optical spring effect in optical cavities to create media stiffer than obtainable with atomic solids.

From 1990 to 2010 better and better laser interferometer gravitational wave detectors were designed for the audio frequency band. Many improvements involved increasing the strength of the laser field between the test masses, thereby increasing the mechanical impedance for relative motion of the test masses, and reducing the impedance mismatch with the gravitational wave.

One of the most exciting developments was the realisation that radiation pressure can be used to create stiff *optical springs* in high power optical cavities. The light field in a long optical cavity creates a high intensity standing wave as it reflects back and forth between very high reflectivity mirrors. Near a resonant peak, where the circulating optical power may be hundreds of Kilowatts, fine adjustment of the cavity length (by about 1/nm) may cause the optical power to rise steeply as it becomes resonant with the incoming laser light, and then to fall back to almost zero once it is off resonance. The varying light intensity creates a varying radiation pressure force on the mirrors with magnitude $2P/c$, where P is the optical power. Because of the length dependence of optical power, the radiation pressure force is also length dependent. Near the half power point it corresponds to a linear spring. Simple calculations show that this optical spring can have a very high spring constant – typically 10 mN nm^{-1}, (or 1000 tonnes per metre). This is sufficient that the mechanical resonant frequency of a pair of test masses kilometres apart can be $\sim$100 Hz. Such mechanical stiffness is much greater than can be obtained by any atomic solid. The equivalent sound velocity (if the cavity were replaced with a solid rod) can be hundreds of kilometres per second. Combined with long length, such systems are likely to allow the best possible impedance matching to gravitational waves, although still corresponding to a very small matching ratio. The next generations of terrestrial gravitational wave detectors are likely to exploit such opto-mechanical resonances to obtain improved detector sensitivity. These ideas are discussed further in Chapter 15.

Frequency and wavelength matching

Electromagnetic dipole antennas are designed to match both the frequency and wavelength of incoming waves. This is possible because the effective wave velocity in the antenna is close to the speed of light. Thus, the dipole antenna is able to sample the entire peak to peak wavelength of a wave, and to be resonant at the wave frequency. In the case of acoustic gravitational wave detectors, acoustic resonant bars were constructed that were resonant at the expected wave frequency ($\sim$1 kHz) but with lengths about 10^{-5} of a wavelength. Thus, the gravitational wave displacement across the detector was very small compared with the 300 km wavelength of a gravitational wave at that frequency.

For detectors in space, wavelength matching can be achieved by spacing the test masses far enough apart. Note that if the test masses are far apart in comparison to the wavelength, there will be good sensitivity when the frequency is such that the test masses are an odd number of half wavelengths apart.

For terrestrial interferometric detectors, we noted earlier that frequency matching could be achieved by using optical cavities in the interferometer arms to multiply-up the effective length of the arms. Such cavities increase the optical power and, as we discussed earlier, increased power increases the coupling to the gravitational wave. However, this is at the expense of increasing sensitivity to the displacement noise of the test masses themselves.

Detector noise

We commonly think of an amplifier or a digital camera as a one-way valve. Signals or light enter the input and an amplified signal or an image appears at the output. Under everyday circumstances the one-way valve is a good approximation: whatever you do at the output does not affect the input. However, this represents a fundamentally false understanding. To understand noise in sensitive measurements it is critical to understand that *there is no such thing as a perfect one-way valve*. There is always a finite flow of energy backwards from the output to the input. As a result, every measurement system is subject to two general classes of noise: a) forward acting noise which adds to the signal, and b) back acting noise, which acts back on the system being measured.

The forward acting noise is usually called series noise or additive noise. In real detectors this may be photon shot noise arising from the random arrival of photons, or the voltage noise of an amplifier. When you connect a microphone to an amplifier, the series noise is the hiss you hear from the loudspeaker. It often arises from the Nyquist or Johnson noise of resistors, which itself is simply the low frequency limit of the Planck black body radiation law. This noise adds algebraically to the sound you hear from the loudspeaker, but can be reduced as the measurement integration time is increased (corresponding to a smaller bandwidth).

The back acting noise acts onto the system being measured. In an electronic amplifier this noise is represented by current noise: current fluctuations at the input of the amplifier. These current fluctuations act back onto the microphone, creating displacements of the microphone diaphragm. If you listen carefully to a microphone you may hear the current noise as a hiss emanating from the microphone.

Back-action noise acting on a mechanical system causes a random walk. In this case the longer the integration time, the greater the amplitude of the fluctuations, because there is more time for the noise response to accumulate. Thus a short integration time (corresponding to larger bandwidth) reduces the noise.

The presence of these two noise terms, one which increases with bandwidth, the other which decreases with bandwidth, means that there is always an optimum integration time where the noise is minimum for any signal of short duration. Equally, noise versus frequency always shows a minimum where the two noise terms cross. The same behaviour also occurs in the quantum regime, which we will consider further below.

In the case of the amplifier the product of current noise spectral density and voltage noise spectral density defines a power spectral density, measured in watts per hertz. This is a measure of the amplifier's noise energy. It can be expressed in terms of *noise temperature* T_n by dividing by the Boltzmann constant, or in terms of *noise number A*, which represents the number of noise photons at a particular frequency, per hertz of bandwidth. To determine A you simply divide the noise energy by the relevant photon energy. If the noise number is large, it means that any measurement is contaminated by a large number of noise photons.

All gravitational wave detectors depend on optimisation and minimisation of detector noise sources. For each part of the spectrum different detector technologies have been devised. Section 3.4 will discuss this in more detail. Table 3.2 summarises the dominant series and back-action noise sources for the various detection techniques. The balance between series noise sources and back-action noise sources always sets a minimum sensitivity at some optimum frequency.

For pulsar timing detection, the test masses consist of the distant millisecond pulsar and planet Earth. Their mass is so enormous and the strength of the pulsar beam is so weak that back-action noise is utterly negligible. Measurements are dominated by series noise.

At the audio frequency end of the spectrum, much of the theory of quantum measurement was motivated by the goal of reaching the quantum limit in resonant mass detectors.

Table 3.2. *A summary of the dominant series and back-action noise contributions in various type of detectors*

Detector	Series noise	Back-action noise	Quantum noise
Pulsar timing	Pulsar timing noise, plasma noise, antenna thermal noise, amplifier series noise, seismic noise	Negligible	Negligible
Acoustic detectors	Antenna thermal noise, SQUID amplifier series noise, transducer pump oscillator phase noise	SQUID amplifier current noise or transducer pump oscillator intensity noise. External force noises include Brownian, seismic and acoustic noise acting on antenna	Standard quantum limit achievable in principle but usually exceeded by classical noise sources
Laser interferometers	Photon shot noise, laser frequency noise, laser beam jitter, laser stray light noise, electronics noise	Laser intensity noise and quantum pressure fluctuations. External force noises include Brownian, seismic and acoustic noise acting on test masses	Photon shot noise dominant at high frequency. Quantum radiation pressure fluctuations dominant at low frequency in advanced detectors

However in practice, both classical series noise from amplifiers and classical back-action noise determined detector sensitivity.

For laser interferometers the high photon energy ensures that amplifier thermal noise is negligible. However the high photon energy of laser light means that photon shot noise due to the random arrival times of photons is a strong series noise limit. Much effort has enabled classical noise, such as thermal vibration of the test masses and laser noise, to be brought below the quantum noise limits across most of the spectrum. Many other technical imperfections in the laser beams and optical systems can also contribute series noise. However, the high photon energy and the high optical power required to minimise photon shot noise means that laser intensity fluctuations contribute time-varying radiation pressure forces on the test masses. These effects act on the test masses and therefore constitute back-action noise. Other external forces, such as seismic vibration, or internal forces, such as Brownian motion of the mirror suspension pendulums or the test masses themselves, often behave similarly to back-action noise, and normally increase at low frequency.

Figure 3.7 from Hobbs (2008), shows detector noise spectra across 10 decades of frequency, while Figure 3.8 shows the initial LIGO sensitivity spectrum. Both illustrate the effect of the two types of noise. At the lowest frequencies the lower cutoff is set by the duration of a single cycle compared with the maximum feasible observation time.

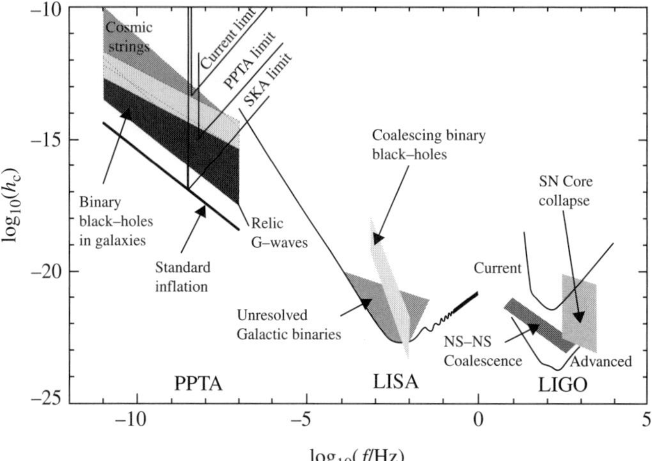

Figure 3.7 The proposed sensitivity curves and some of the possible sources for pulsar timing observations (here we show estimates for the Parkes Pulsar Timing Array (PPTA), for the proposed Laser Interferometer Space Antenna (LISA), and for initial LIGO and advanced LIGO. We assume that the sensitivity of Virgo, Advanced Virgo and other planned advanced detectors are comparable (reproduced from Hobbs, 2008).

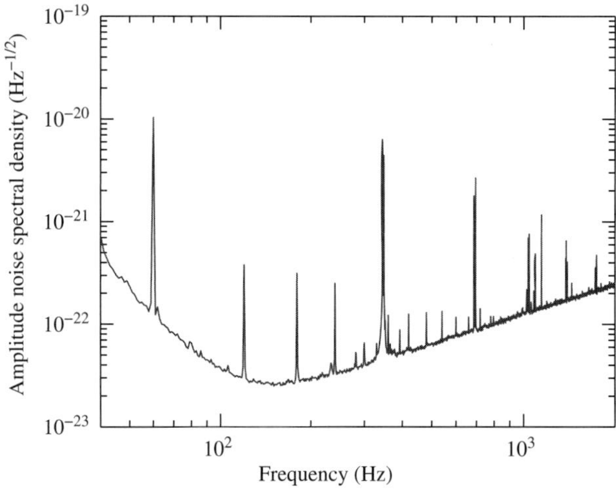

Figure 3.8 The initial LIGO strain sensitivity spectrum.

For all other detectors we see the balance of the two types of noise. For laser interferometers the noise which dominates at high frequency is photon shot noise. In initial LIGO and Virgo low frequency test mass thermal noise and seismic noise acts on the test mass, creating a minimum noise level at about 150 Hz. In the case of LISA, the sensitivity is shot noise limited at the high frequency end but not radiation pressure limited at the low 'frequency end. Other noise sources originating from the satellite play a dominant role in the low 'frequency range.

In advanced gravitational wave detectors the goal is to eliminate all technical and classical noise contributions. Then, if the measurement scheme remains conventional, the noise spectrum achieves a minimum value equal to the standard quantum limit at a single optimal frequency. In this case the back action is due to quantum radiation pressure fluctuations of the light. Sensitivity beyond this limit requires quantum measurement schemes, which we discuss further below.

Quantum measurement

When the test masses are well isolated and the other technical noises are significantly reduced, we enter the quantum regime, where the fundamental limit for the detector sensitivity arises from the *Heisenberg uncertainty principle*.

Braginsky (1968) and Giffard (1976)[1] were among early workers who investigated the quantum limit for detector sensitivity. When treating the measurement process quantum mechanically, they found that the series noise and the back-action noise, of quantum origin, impose the Heisenberg uncertainty on the sensitivity. Even though they were focusing on the case of the resonant-mass detectors, their conclusions are generally valid for any linear detectors.

Let us elaborate more on the interferometric gravitational wave detector. The corresponding series noise is the photon shot noise, which arises from quantum fluctuations of the number of photons in a given time interval. Like all random processes, this noise is proportional to the square root of the number of photons, while the signal increase is proportional to the number of photons. Thus the minimum detectable signal reduces as $1/\sqrt{N}$ and the shot noise reduces inversely as the square root of the optical power. The back-action noise is the radiation pressure noise. This arises from quantum fluctuations of the light intensity. Again, the fluctuations depend on $\sqrt{N}$, but in this case higher intensity gives a larger fluctuation. The shot noise dominates at high frequency because this corresponds to a short time interval measurement and N is smaller, while the radiation pressure noise is dominant at low frequency because this corresponds to a longer time interval measurement, $\sqrt{N}$ is larger and the radiation pressure displacement is larger. These two types of noise, if uncorrelated, set a frequency-dependent but power-independent lower bound on the sensitivity, which is the Standard Quantum Limit (SQL) (Braginsky, 1968). In terms of gravitational-wave strain, the SQL is given by

$$S_h^{\mathrm{SQL}}(\Omega) = \sqrt{\frac{2\hbar}{m\Omega^2 L^2}}, \tag{3.1}$$

where $S_h^{\mathrm{SQL}}(\Omega)$ is the frequency-dependent spectral density of gravitational wave strain at the free mass standard quantum limit. Note that S_h^{SQL} depends only on the mass m, the signal frequency Ω and the arm length L. This result assumes that the test mass is a free mass,

---

[1] Gifford treated the resonant-mass detector as a linear amplifier, of which the quantum limit on the noise in the general case was considered previously by Heffner (1962) and Haus and Mullen (1962).

which is a good approximation for all detectors constructed to date except resonant mass detectors. In the case of interferometers, the test masses are supported by soft pendulums with frequency around 1 Hz, while the detection frequency is $\sim 100\,\mathrm{Hz}$. However, future advanced detectors are planned to operate in a regime where the electromagnetic forces cause the detector sensitivity to surpass the free mass SQL.

To overcome the free mass SQL, various approaches have been discussed in the literature. The first one was proposed by Braginsky (1968) – the so-called Quantum Non-Demolition (QND) detector. Such a detector probes a conserved dynamical quantity of the test mass, instead of its linear displacement. For a free mass, the QND quantity is the momentum. Because momentum is proportional to the speed, many speed meter types of detectors have been investigated, and have shown theoretically that they may allow the free mass SQL to be surpassed (Purdue, 2002; Purdue and Chen, 2002).

A second approach uses light with modified quantum statistics, called *squeezed light*. This is light in which the quantum fluctuations of the amplitude and the phase have been altered. This is achieved by using a non-linear optical device called a squeezer. Squeezed light can be produced in such a way that either its amplitude fluctuations or its phase fluctuation are suppressed. With frequency-dependent squeezed light, we can have a small phase fluctuation at high frequencies, and small amplitude fluctuation at low frequencies. This allows both the shot noise, which dominates at high frequency, and the radiation pressure noise, which dominates at low frequency, to be reduced so that the SQL is overcome. At the time of writing, there has been significant progress in producing squeezed light and it is ready to be implemented in real detectors (McKenzie *et al.*, 2004; Vahlbruch *et al.*, 2006).

The third approach is to cancel the quantum back-action noise, which is imposed on the test mass, by filtering the output of the detector with a series of optical cavities before making the photon detection (Kimble *et al.*, 2001). This uses the fact that the amplitude fluctuations, which induce quantum back-action noise in the phase quadrature of the output, is contained in the output amplitude quadrature. If we detect a combination of the output amplitude and phase quadrature, the back action can be cancelled out at a certain frequency. By using filter cavities with proper detuning and bandwidth, the amplitude and phase quadrature can be designed to rotate in a frequency-dependent way. This allows us to detect an optimal combination of the amplitude and phase quadrature at all frequencies, and therefore completely evades the back action. This is the so-called variational readout. The only disadvantage of this approach is that it is very sensitive to optical losses and can therefore introduce uncorrelated vacuum fluctuations from the points of dissipation that destroy the quantum coherence.

The final approach is to modify the test mass dynamics with the optical spring effect. The optical spring provides a quantum approach to modifying the dynamics of the test mass. As we discussed in Section 3.3.1, an optical spring can raise the frequency of the test masses from their free pendulum frequency of a few hertz to the detection band around 100 Hz. This can significantly enhance the response of the detector to the gravitational-wave tidal force due to the mechanical resonance. This is different from classical feedback systems that can also be used to raise the mechanical frequency of a system. Classical feedback increases both the signal and the noise on an equal footing. Optical springs, however, act as

Table 3.3. *Rough comparisons of detectors to the standard quantum limit. Most detectors are quite far from the standard quantum limit, while the terrestrial laser interferometers are in the realm of being true quantum instruments*

Detector type	Frequency band (Hz)	Strain noise target $h/\sqrt{Hz}$	Quantum limit	Main noise sources
Pulsar timing	10^{-8}–10^{-9}	10^{-15}	10^{-35}	Timing noise Plasma noise
Space transponder interferometers	10^{-2}–10^{-4}	10^{-20}	10^{-22}	Readout noise Shot noise
Space optical cavity interferometers	10^{-1}–10	10^{-23}	10^{-25}	Quantum noise Control noise
Acoustic detectors	10^{3}–10^{4}	10^{-22}	10^{-25a}	Readout noise Thermal noise
Terrestrial laser interferometers	10–10^{4}	10^{-24}	10^{-24}	Quantum noise Thermal noise

[a] A quality factor of 10^{6} is assumed, and it is evaluated around the mechanical resonant frequency. Additionally, this limit requires an extremely low temperature.

quantum amplifiers, and amplify the response without adding noise (Buonanno and Chen, 2002).

The price one pays for including an optical spring is a degradation of sensitivity at frequencies below the opto-mechanical resonant frequency. This is due to the fact that the spatially separated test masses appear to be rigidly connected due to the optical rigidity or stiffness of the optical spring. Thus, the differential signal from a low frequency gravitational wave cannot be sensed properly by the light inside the arm. Fortunately, the low frequency sensitivity can be recovered. At low frequency the optical rigidity acts to make the detector optical cavities act like rigid rods. These transfer the gravitational wave strain from the far ends of the interferometer to the central mirrors where it can be measured much more easily using a low power interferometer. This is the so-called local readout scheme (Rehbein *et al.*, 2007).

All these above-mentioned approaches will be discussed in detail in Chapter 15. For now, in Table 3.3, we show the quantum limit of various detectors. It is clear that application of the above ideas is most relevant to terrestrial detectors. In the future they may be applied to space-based detectors, but they are quite irrelevant to pulsar timing detectors.

Having discussed gravitational wave detection in a single conceptual framework, we will now go on to consider the techniques used for detectors in each detector band.

3.4 Detectors from nanohertz to kilohertz

Pulsar timing detection of nanohertz gravitational waves

Detectors in this nanohertz frequency band aim to detect gravitational wave signals from supermassive black hole mergers (Hobbs *et al.*, 2010; Jenet *et al.*, 2005). They focus on timing the pulse arrival times of millisecond pulsar beams from pulsars in different parts of the sky. The test masses are the various pulsars, each paired with the Earth. Pulsars are chosen for their *stable timing residuals* after Doppler corrections have been subtracted (the residual is the change in pulse arrival time after all known pulse arrival time corrections have been applied). Fluctuations in the arrival times of pulses can be attributed to gravitational waves or to variations in the terrestrial clock. However, the effects of clock variations can be suppressed by comparing the effects for pulsars in different directions in the sky. Then the detection system becomes a virtual interferometer. For detection using pulsar timing the enormous distance to the pulsar is not important because it is thousands of gravitational wavelengths distant. It is equivalent to consider that the gravitational wave varies the Earth location relative to the incoming pulsar beam. The effective length of the detector is just half the gravitational wavelength along the incoming radio beam (all the rest of the wavelengths cancel out). Thus we have good frequency and wavelength matching. The effective length of the detector is $\sim 10^{16}$m.

In pulsar timing back-action noise is negligible because of the enormous mass of the test masses and the very low power of the electromagnetic coupling. Sensitivity is limited by various sources of series noise (not all of which are fully understood). These include receiver noise and plasma noise, which acts like a fluctuating weakly refractive medium, as well as intrinsic pulsar timing noise. These noise sources lead to a target strain amplitude $h \sim 10^{-15}$.

Plasma delay effects can be suppressed because plasma is highly dispersive. Broad bandwidth is used to capture as much signal energy as possible, combined with de-dispersion (frequency dependent timing corrections) to suppress the effects of plasma delays. Pulse arrival times for some individual millisecond pulsars can be determined to better than $\sim$100 ns, while others appear to be intrinsically noisier. By comparing timing residuals for a large number of pulsars in different directions in the sky the noise is reduced as if it were a system of interferometers, allowing noise in the timing reference clock to be separated from true signals. In this system the technical noise vastly outweighs the standard quantum limit, which is 20 orders of magnitude below the target sensitivity.

Spacecraft laser interferometers

Laser interferometers in space allow the size of detectors to be matched to gravitational waves of wavelength from thousands to millions of kilometres. Two schemes have been proposed. The LISA project proposes to place spacecraft 5 million km apart, corresponding to an optimum length for a half-minute period gravitational wave. The DECIGO project aims for frequencies in the decihertz range, but has length of $\sim 10^{-2}$ wavelengths.

Space interferometers require multiple spacecraft and must overcome the effects of buffeting by the solar wind, orbital motions, and the diffraction of their laser beams. Buffeting noise can be suppressed by using test masses suspended and protected by a housing and using micro-thrusters to make the spacecraft follow the test mass. The test mass position must be measured to enormous precision, which itself requires an optimum balance of series noise and back-action noise. Because orbital dynamics causes widely spaced objects to have relatively large differential motions, thruster forces must be applied without introducing noise.

The diffraction losses over such great distances mean that it is not practical for laser light going out from one craft to another to be reflected back directly. Instead, LISA uses transponded laser beams, in which the return beam in the interferometer is generated by another laser which is phase locked to the incoming beam. DECIGO, with its much lower baseline, proposes to use large diameter telescopes to create a system more analogous to ground-based interferometers. This enables the use of higher laser intensity but it must cope with relatively large classical radiation pressure forces which act to 'blow' the test masses and the spacecraft apart from each other.

Acoustic detectors

Resonant mass detectors were designed to measure acoustic signals induced in a large mass due to its coupling to a gravitational wave. As mentioned at the beginning of this chapter, these detectors were pioneered by Weber during the 1960s (Weber, 1960). They consisted of large vibration-isolated aluminium cylinders constructed with piezoelectric crystals glued on the surface near to the centre. These crystals are able to develop a large voltage when they undergo deformation. A low noise amplifier and lock-in amplifier allowed detection of the energy of the fundamental longitudinal resonance of the bar. A gravitational wave applies a time-varying quadrupole deformation and does mechanical work on the fundamental acoustic modes of the test mass. The absorption cross-section of the mechanical resonator to gravitational waves depends on its mass and sound velocity and is highest at the fundamental resonant frequency. The latter is linked to sound velocity and the resonator's length, since the length must be half an acoustic wavelength at the fundamental longitudinal resonance. Weber chose aluminium because of its high sound velocity, and availability in large pieces, and because it has quite low acoustic losses. Following Weber, many new resonant mass detectors using similar techniques, but with variations and improvements, were developed in the early 1970s. The absence of convincing signals and the general consensus that these detectors had insufficient sensitivity to see realistic sources led to them being abandoned. However, advanced resonant mass detectors using cryogenics and superconductivity continued to be developed.

Opfer *et al.* (1974) and Hamilton *et al.* (1974) first proposed such cryogenic detectors, and proposed cooling to millikelvin temperatures to minimise thermal noise. All sensitive resonant mass detectors have used cryogenic techniques to reduce the thermal noise and to enable the use of high sensitivity superconducting transducers.

In all resonant mass detectors the large amplitude of thermal vibration considerably exceeds the gravitational wave amplitude expected from astrophysical sources. Without methods to suppress this noise, high sensitivity would be impossible. Weber's key contribution was the realisation that in a high Q antenna – one with a low acoustic loss – the effective noise energy is reduced by a factor $\sim \tau_i/\tau_a$, where τ_i is the effective measurement integration time, and τ_a is the antenna ring-down time. The advantage of using a low acoustic loss follows directly from the fluctuation dissipation theorem (Callen and Green, 1952): the greater the dissipation, the greater the fluctuations or noise level imposed by the thermal reservoir. A high Q antenna approaches an ideal harmonic oscillator, whose motion is exactly predictable at a time in the future from the observed amplitude, frequency and phase at an earlier time: the high Q-factor gives the resonator a memory.

A high Q-factor therefore ensures that the thermal noise (which even after cryogenic cooling has a very large amplitude compared with the signal) is manifested as a highly sinusoidal, and hence predictable, waveform. In a resonant mass, a gravitational wave would appear as a small change in the amplitude and phase of this waveform. The same concept is used in the design of test masses in terrestrial laser interferometers. The test masses are designed to have very high Q-factors so that the thermal noise of acoustic modes of the test masses, which cause thermal vibration of the mirror surfaces, do not contaminate the measurement band.

The improvement of resonant mass detectors since they were first reported by Weber was enormous. Their amplitude sensitivity increased several hundred times, corresponding to about five orders of magnitude improvement in flux sensitivity. Unfortunately they are confined to operate within a rather small bandwidth of tens to hundreds of hertz around the resonant frequency. This limits the potential sources to short bursts from gravitational collapse or pulsars that may emit gravitational waves within their bandwidth. Most resonant bar detectors ceased operation once laser interferometer detectors had achieved significantly higher sensitivity. However, at the time of writing two resonant bar detectors, NAUTILUS (Astone *et al.*, 2006) and AURIGA (Vinante *et al.*, 2006), are still operating, as well as two cryogenic spherical detectors, MiniGRAIL (de Waard *et al.*, 2006) and Schenberg (Aguiar *et al.*, 2006), which are being developed to operate in the mid-kHz range.

In the next section we will take our first look at terrestrial interferometric detectors to set the scene for the rest of the book.

3.5 Introduction to terrestrial interferometers

Background

The Michelson interferometer has long been known as an extremely sensitive instrument to measure length changes (Michelson and Morley, 1887). The idea of using an interferometer as a gravitational wave detector was first suggested as early as the 1960s (Gertsenshtein and Pustovoit, 1962), and one was constructed by Robert Forward following the suggestion of his supervisor, Joseph Weber (Moss *et al.*, 1971). Further work was conducted throughout the 1970s (Weiss, 1972; Forward and Mosse, 1972; Levine and Hall, 1972; Forward, 1978) including a 30 m laser interferometer constructed by Levine and Stebbins (1972) which was used to set an upper limit on the gravitational wave emission from the Crab pulsar.

Terrestrial laser interferometers are prevented from having arbitrary length, by the sphericity of the Earth, cost and vacuum system requirements. At their optimum sensitivity their length is about 10^{-3} of the gravitational wavelength. We have seen that this difficulty is overcome by ensuring that the light undergoes multiple reflections within each interferometer arm. Figure 3.9 shows a schematic concept of a modern terrestrial interferometric detector. Pairs of mirrors are placed in the interferometer arm to create an optical resonant cavity. The optical configuration creates an effective coupled optical cavity by using a power recycling mirror at the input. This enables optical power build up. Another mirror at the output enables the storage time for the signal sidebands to be tuned to match a gravitational wave half period. Typically, however, the bandwidth is a bit smaller than half the period, but as the cavity is tuned to a frequency different from zero, the signal resonantly builds up for this frequency. This also works if the bandwidth is considerably smaller (or rather the storage time much larger) than half the gravitational wave period.

The relatively short baseline of these instruments (3 or 4 km) means that local sources of noise such as the thermal noise, of the mirrors and the test masses, and seismic noise, are larger than they would be in an instrument with a much longer baseline. Similarly the standard quantum limit scales inversely with detector length. Thus there is clearly the possibility of significant future improvements simply by building longer instruments. Such *third generation* detectors will be discussed in Chapter 16.

Sensitivity limits in terrestrial interferometers

The sensitivity of an interferometer is ultimately limited by shot noise and quantum radiation pressure noise due to photon quantum statistics. As we have seen in Section 3.3 (p. 50), the SQL for an interferometer can be obtained from the balance of these two competing quantum noise sources (Caves, 1980, 1981; Braginsky *et al.*, 1985). The first is the photon-counting error due to $\sqrt{N}$ fluctuations in the number of output photons from the interferometer. The second is the radiation pressure error. This arises from the perturbations on the end mirrors produced by fluctuating radiation pressure forces, which also scale as $\sqrt{N}$. As the input laser power P increases, the relative photon-counting error decreases as $\sqrt{N}/N$, while the radiation pressure error increases as $\sqrt{N}$. Minimising the total error with respect to P yields a minimum error of order the SQL and an optimal input power for a simple Michelson interferometer,

$$P_{\mathrm{opt}} = \frac{\lambda\, m\, c}{8\pi\, \tau^2\, \sqrt{\eta}} \tag{3.2}$$

at which the minimum error can be achieved. Here, m is the mass of an end mirror, λ is the wavelength of the light, τ is the measurement duration, c is the light velocity and η is the quantum efficiency of the photodetector.

With a reasonable set of values for interferometer parameters, $m \sim 10^2\,\mathrm{kg}$, $\tau \sim 10^{-2}\,\mathrm{s}$, $\lambda \sim 1\,\mu\mathrm{m}$, the optimum laser power P_{opt} is approximately $6 \times 10^5\,\mathrm{W}$ – a power far higher

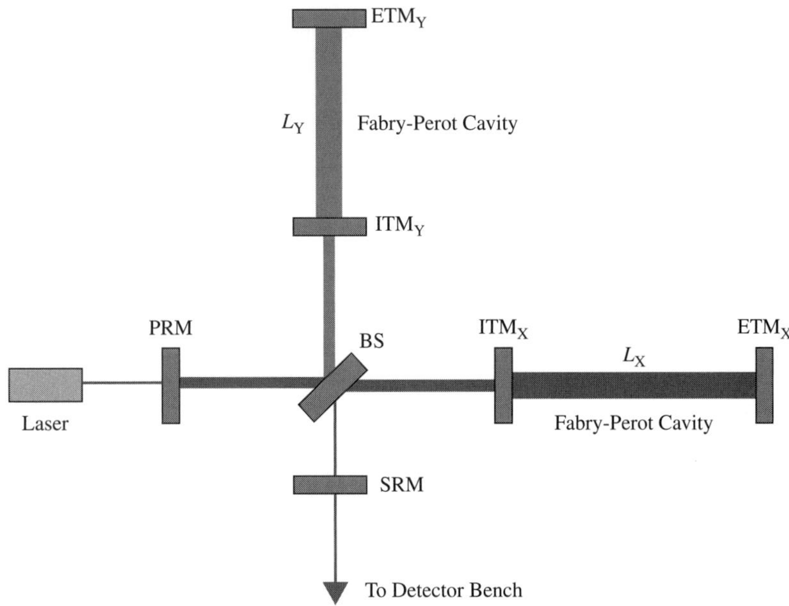

Figure 3.9 Interferometer with Fabry–Perot cavities as arms. The addition of two input test masses (ITM$_X$ and ITM$_Y$) form two Fabry–Perot cavities of lengths L_X and L_Y. The input laser light has to be resonant in the cavities in order to be able to detect gravitational waves. Power recycling is set up by the addition of power recycling (PRM) and a signal recycling mirrors (SRM). The signal recycling mirror enables resonant gain in the differential output of the interferometer, allowing detector tuning and optimisation of the signal frequency response.

than the power of present continuous-wave lasers. The low available input power means that simple interferometers used as gravitational wave detectors will be limited not by the standard quantum limit, but rather by photon-counting statistics (shot noise) which scales inversely to the square root of incident power at the beam' splitter.

As mentioned earlier, the technique of power recycling (Drever, 1983; Meers, 1988) can be used to increase the power incident at the beam splitter and improve the sensitivity of interferometer detectors. The concept is illustrated in Figure 3.9. The basic idea is that because the interferometer detector operates at a *dark fringe* output (i.e. there is complete destructive interference at the output port) almost all of the light (reduced only by losses in mirrors and beam splitters) is reflected back towards the laser, and can therefore be used again as long as it is phase coherent with the input laser beam. This technique is realised by inserting a recycling mirror between the laser and the beam splitter. The position of the recycling mirror (or the laser frequency) is then carefully adjusted so that the recycling mirror, combined with the two main cavities and the beamsplitter, form a large resonant optical cavity containing the interferometer. As we shall see later in Chapter 9, an effective laser power several 1000 times larger than the original laser may be built up inside this cavity, thus reducing the shot noise limit.

Interferometer design considerations

Laser interferometers are designed to be sensitive to the optical phase difference of two arms, and should not be sensitive to common-mode fluctuations of the input light. But in practice, because of asymmetry between the two arms, fluctuations in the input light will couple into the output signal. For various reasons some asymmetry is unavoidable. This will cause, for example, intensity fluctuations to give rise to a differential radiation pressure force that acts more strongly in one arm than another. Thus, laser intensity and frequency fluctuations must be strongly suppressed. Laser frequency stabilisation and other aspects of high power lasers are fully discussed in Chapter 13.

We discussed earlier that frequency matching could be achieved by the use of optical cavities in the interferometer arms to multiply-up the effective wavelength of the arms. Such cavities increase the optical power. The increase in power can be considered to increase the coupling of the detector to the gravitational wave, or as reducing the shot noise. Either way, the signal-to-noise ratio is improved. Two extremely important innovations in this regard have enabled practical high sensitivity detectors to be created. The first was the concept of power recycling, which we discussed above. It depends crucially on the fact that, if the detector is to be read out by a single photodetector, the optimum sensitivity occurs if the interferometer is tuned so that the optical port is a dark fringe. This necessitates very high quality optics so that the interferometer achieves a high contrast dark fringe. It also requires that interferometers have extremely low optical losses so that most of the input light is directly reflected back to the laser. If both of the above requirements are met, then a mirror placed at the input turns the entire interferometer into an optical cavity, allowing otherwise wasted power to be recycled. This concept of power recycling enables detectors using hundreds of kilowatts of optical power in their coupling field to be created using lasers of $\sim 100\,\mathrm{W}$ power (Sigg and the LIGO Scientific Collaboration, 2008).

The second innovation was the recognition that the multiple reflections in the arm cavities could be increased beyond the point where a simple analysis would tell you that they were too long, and thereby smearing out the signal by averaging over positive and negative phases of a gravitational wave. The new concepts were called duel recycling (Meers, 1988) and resonant sideband extraction (Mizuno *et al.*, 1993). It was realised that the signal sidebands from the gravitational wave could have quite different resonant properties from the carrier light. Thus, the signal could be extracted from the interferometer in an appropriate time to achieve wavelength matching, while very high power was still present in the detector.

Thermal noise

Thermal noise is one of the fundamental noise limits of gravitational wave detectors. Once the seismic noise cutoff is lowered sufficiently through the use of high performance vibration isolators, thermal noise becomes the critical source of noise in the frequency band from $\sim 10\,\mathrm{Hz}$ to several hundred Hz. Above this band laser noise dominates.

In a laser interferometer gravitational wave detector, thermal noise is mainly divided into three classes – the suspension thermal noise, the internal thermal noise of the test mass substrate, and the mirror coating thermal noise. There are also some less significant thermal noise sources such as refractive thermal noise.

Suspension thermal noise is mainly the Nyquist noise or Brownian motion of the test mass pendulum of the interferometer. The quantity that describes the loss of a harmonic resonator due to the loss angle ϕ is the quality factor $Q \sim 1/\phi$. The Q-factor of a pendulum suspension can in principle be very high because the energy storage is predominantly in the effectively lossless gravitational field. However, some elastic energy must always be stored in the flexure which supports the pendulum. The Q-factor of a pendulum is limited by the losses in this element.

Current room temperature detectors use fused silica test masses. The Q-factor of silica test masses has been observed to be $\gtrsim 10^6$ (Traeger *et al.*, 1997; Taniwaki *et al.*, 1998), although the higher reported values of order 10^8 have been observed by Ageev *et al.* (2004) for several resonant modes. Other test mass materials such as sapphire have been proposed (Ju *et al.*, 1993). The excellent thermal and mechanical properties of sapphire makes it a promising material for use in test masses, especially in cryogenic detectors (Kuroda and the LCGT Collaboration, 2010). However, present techniques cannot produce large sapphire test masses with the high homogeneity and low optical absorption of fused silica.

Silicon also has very high Q-factor $\sim 10^8$ and may be a useful material for future detectors (Nawrodt *et al.*, 2008). Unfortunately it is not transparent at the popular 1064 nm laser wavelength, so a larger wavelength $\sim 1.5\,\mu$m must be used. Silicon is a likely ideal choice for future generations of gravitational wave detectors which will require much larger test masses.

The acoustic loss in dielectric optical coatings of mirrors cause additional thermal noise. This noise is an important component of the noise budget of the next generation interferometric gravitational wave detectors. At the time of writing there is intensive research on optimising the acoustic losses of optical coatings to reduce the thermal noise. If coating losses can be reduced by a factor of 10, interferometers will be able to enter a fully quantum limited regime.

Seismic noise

At ground-based gravitational wave detector sites, the seismic motion of the ground has a typical vibration spectrum with an rms amplitude of $x_s \approx \alpha f^{-2}\,\mathrm{m\,Hz}^{-1/2}$, where, depending on the site location, the constant α can vary between 10^{-6} and 10^{-9}. Even a relatively quiet underground site has a seismic noise level with an amplitude of around $10^{-13}\,\mathrm{m\,Hz}^{-1/2}$ at 100 Hz. This is far greater than the gravitational wave-induced signals we want to measure. It is therefore essential that terrestrial gravitational wave detectors are isolated from the seismic noise background. This is achieved through high performance vibration isolation systems.

An ideal vibration isolator for a gravitational wave detector should not only suppress seismic vibration in the operational frequency passband of the detector, but also cut out the seismic noise at much lower frequencies. Such isolation ensures that the suspended test masses are effectively stationary with respect to the laser light field. If the total rms motion of the test masses were much less than an optical fringe width, the servo control requirements of the detector would be minimised. Components would then be as stable as if rigidly attached to an optical table, or placed in interplanetary space. Achieving such levels of vibration isolation would significantly simplify the operation of an interferometer.

The approaches to the problem of vibration isolation in gravitational wave detectors are varied and numerous. The main strategies come under the categories of *active isolation* (stiff system) and *passive isolation* (soft system). All detectors use combinations of those techniques with some systems more 'active' and others more 'passive'. Many innovative techniques have achieved very high performance. The two approaches will be discussed further in Chapter 11.

The worldwide network of terrestrial interferometers

The first few gravitational wave signal detections will consolidate the field of gravitational wave astronomy, revealing rich new information on the nature of different sources. This milestone will also mark the beginning of a new era, in which interferometric detectors will begin exploration of the gravitational wave sky, providing valuable data on the spatial distribution of sources, their environments and evolution. Correlating gravitational wave measurements of binary inspiral events with those obtained in the electromagnetic spectrum will prove to be a powerful probe of the Hubble law, cosmological acceleration and the equation of state of dark energy (Schutz, 1986; Finn, 1996; Dalal *et al.*, 2006).

Unfortunately, a single gravitational wave detector cannot determine the polarisation state or source direction of a transient signal. Localisation will require that the same source is observed in several separated detectors so that triangulation can determine the position of the source from the delays in signal arrival times at different detectors (Sathyaprakash and Schutz, 2009).

For good source localisation, a widely separated network of gravitational wave detectors is essential. Such a network may employ techniques such as *coincidence analysis*, in which individual events from different detectors are correlated in time, or *coherent analysis*, in which synchronised detector outputs are merged before searching for a common pattern. By effectively resolving the different times of arrival of gravitational wave events between members of a network, coherent network analysis enables a detector array to become an all-sky monitor with good angular resolution over all source directions. To maximise the time delays, and hence improve directionality, it is advantageous that a network be as geographically widely separated as possible.

Figure 3.10 illustrates how adding to a gravitational wave detector network improves the overall capacity for localisation of sources. The all-sky angular resolution map in Figure 3.10(a) shows the 90% error ellipses from the LIGO-Virgo network of detectors. Elongated ellipses are produced because these detectors are nearly co-aligned and in the same plane. Figure 3.10(b) shows the significant contraction of the error ellipses by the addition of an Australian detector. On average the increase in resolution is from around 12 square degrees to a fraction of a square degree (Blair *et al.*, 2008). The advantage in respect to source localisation is clear. Additionally, the increased ability in measuring both polarisations of

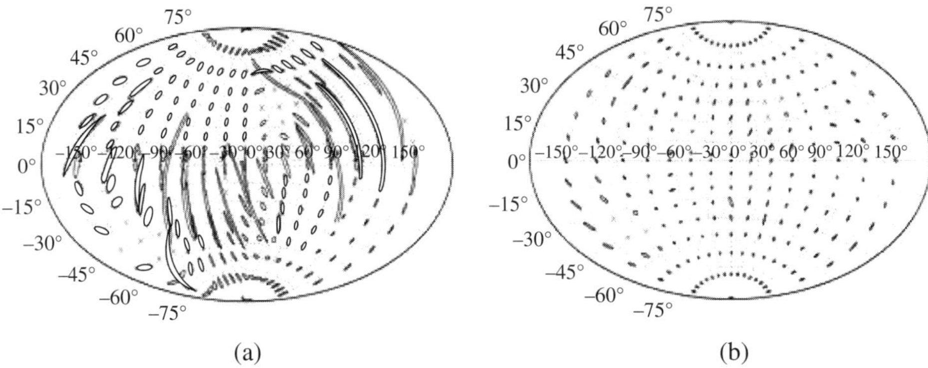

Figure 3.10 Angular resolution maps for (a) an array consisting of the LIGO and Virgo detectors, and (b) for an array with a southern hemisphere detector at Gingin, Western Australia. Each map shows the error ellipse for a point source located in a uniform grid of sky positions. The enormous advantage of adding a southern hemisphere detector is apparent.

a signal improves the determination of the luminosity distance to the source (Dalal *et al.*, 2006). The theory behind gravitational wave networks, as well as their advantages for multimessenger astronomy, will be discussed in detail in Chapter 5.

3.6 Conclusion

Like the great south land which was rumored for centuries before it was discovered, the spectrum of gravitational waves is a rumored continent, first to be detected, and then to be explored. The exploration may have begun by the time you are reading this book. Whatever happens the search will continue to motivate physicists and drive a continuing process of innovation.

In this chapter we have reviewed the common features of all types of gravitational wave detector across a spectrum from nanohertz to kilohertz. The rest of this book focuses on terrestrial gravitational wave detectors designed to hear signals that are in the audio frequency band. Chapters 6–8 will present case studies of three very different detectors: LIGO, Virgo and GEO 600. Chapters 9–13 will expand further on the technical difficulties and challenges in building and operating terrestrial laser interferometers. Chapters 14–16 will focus on advanced techniques that will be applied to the next generation of detectors.

Acknowledgements

The authors wish to acknowledge financial support from the West Australian Government Centres of Excellence scheme, the Australian Research Council and the University of Western Australia.

References

Acernese, F., *et al.* 2008. *Class. Quant. Grav.*, **25**, 184001.
Ageev, A., *et al.* 2004. *Class. Quant. Grav.*, **21**, 3887.

Aguiar, O., *et al.* 2006. *Class. Quant. Grav.*, **23**, S239–S245.

Ando, M., *et al.* 2001. *Phys. Rev. Lett.*, **86**, 3950–3954.

Astone, P., *et al.* 2003. *Phys. Rev. D*, **68**, 022001.

Astone, P., *et al.* 2006. *Class. Quant. Grav.*, **23**, S57–S62.

Astone, P., *et al.* 2010. *Phys. Rev. D*, **82**, 022003.

Barriga, P., *et al.* 2010. *Class. Quant. Grav.*, **27**, 084005.

Blair, D. G. (ed). 1991. *The Detection of Gravitational Waves*. Cambridge: Cambridge University Press.

Blair, D. G., *et al.* 1995. *Phys. Rev. Lett.*, **74**, 1908–1911.

Blair, D. G., *et al.* 2008. *Journal of Physics: Conference Series*, **122**, 012001.

Bondi, H. 1957. *Nature*, **179**, 1072–1073.

Braginsky, V. B. 1968. *JETP*, **26**, 831.

Braginsky, V. B., Mitrofanov, V. P. and Panov, V. I. 1985. *Systems with Small Dissipation*. Chicago: University of Chicago Press.

Braginsky, V. B., Gorodetshy, M. L., Khalili, F. Y., *Phys. Rev. A.*, **232**, 340, 1997.

Buonanno, A. and Chen, Y. 2002. *Phys. Rev. D*, **65**, 042001.

Callen, H. B. and Green, R. F. 1952. *Phys . Rev.*, **86**, 702.

Caves, C. M. 1980. *Phys. Rev. Lett.*, **45**, 75.

Caves, C. M. 1981. *Phys. Rev. D*, **23**, 1693.

Dalal, N., *et al.* 2006. *Phys. Rev. D*, **74** .

Danzmann, K. 2003. *Advances in Space Research*, **32**, 1233–1242. Fundamental Physics in Space.

de. Waard, A., *et al.* 2006. *Class. Quant. Grav.*, **23**, S79–S84.

Drever, R. W. P. 1983. *Gravitational Radiation*. North Holland: Amsterdam.

Dyson, F. J. 1969. *The Astrophysical Journal*, **156**, 529.

Fafone, V. 2006. *Class. Quant. Grav.*, **23**, S223–S229.

Ferdman, R. D., *et al.* 2010. *Class. Quant. Grav.*, **27**, 084014.

Finn, L. S. 1996. *Phys. Rev. D*, **53**, 2878.

Forward, R. L. 1978. *Phys. Rev. D*, **17**, 379.

Forward, R. L. and Mosse, G. E. 1972. *Bull. Am. Phys. Soc.*, **17**, 1183.

Gertsenshtein, M. E. and Pustovoit, V. I. 1962. *Soviet Physics - JETP*, **16**, 433.

Giffard, R. P. 1976. *Phys. Rev. D*, **14**, 2478–2486.

Gifford, R. 1979. Personal communication.

Gottardi, L., *et al.* 2007. *Phys. Rev. D*, **76**, 102005.

Hamilton, W. O., Pipes, P. B. and Nayar, P. S. 1974. Page 555 of: Timmerhaus, K. D., O'Sullivan, W. J., and Hammel, E. F. (eds), *Proc. Conf. Low Temperature Physics-LT13*, vol. 4. New York: Plenum Press.

Haus, H. A. and Mullen, J. A. 1962. *Phys. Rev.*, **128**, 2407–2413.

Heffner, H. 1962. *Proceedings of the IRE*, **50**, 1604 –1608.

Hobbs, G. 2008. *Class. Quant. Grav.*, **25**, 114032.

Hobbs, G., *et al.* 2009. *Publications of the Astronomical Society of Australia*, **26**, 468–475.

Hobbs, G., *et al.* 2010. *Class. Quant. Grav.*, **27**, 084013.

Jenet, F., *et al.* 2009. *ArXiv e-prints*. Astro-ph: 0909–1058.

Jenet, F. A., *et al.* 2005. *Astrophys. J.*, **625**, L123–L126.

Ju, L., Blair, D. G., and Notcutt, M. 1993. *Rev. Sci. Instrum.*, **64**, 1899.

Kawamura, S., *et al.* 2006. *Class. Quantum Grav.*, **23**, S125–S131.

Kennefick, D. 2006. *Traveling at the Speed of Thought: Einstein and the Quest for Gravitational Waves*. Princeton: Princeton University Press.

Kimble, H. J., *et al.* 2001. *Phys. Rev. D*, **65**, 022002.

Kuroda, K. and the LCGT Collaboration. 2006. Class. Quant. Grav., **23**, S215.

Kuroda, K. and the LCGT Collaboration. 2010. *Class. Quant. Grav.*, **27**, 084004.

Levine, J. and Hall, J. L. 1972. *J. Geophys. Res.*, **77**, 2595–2609.

Levine, J. and Stebbins, R. 1972. *Phys. Rev. D*, **6**, 1465–1468.

Manley, J. and Rowe, H. 1956. *Proceedings of the IRE*, **44**, 904 –913.

Martin, P., *et al.* 2005. *Class. Quant. Grav.*, **22**, S965–S973.

McKenzie, K., *et al.* 2004. *Phys. Rev. Lett.*, **93**, 161105.

Meers, B. J. 1988. *Phys. Rev. D*, **38**, 2317.

Michelson, A. A. and Morley, E. W. 1887. *American Journal of Science*, **34**, 333–345.

Mizuno, J., *et al.* 1993. *Physics Letters A*, **175**, 273–276.

Moss, G., Miller, L., and Forward, R. 1971. *Appl. Opt.*, **10**, 2495.

Nawrodt, R., *et al.* 2008. *Journal of Physics: Conference Series*, **122**, 012008.

Opfer, J. E., *et al.* 1974. Page 559 of: Timmerhaus, K. D., O'Sullivan, W. J., and Hammel, E. F. (eds.), *Proc. Conf. Low Temperature Physics-LT13*, vol. 4. New York: Plenum Press.

Paik, H. J. 1982. In: Ruffini, R. (ed.), *Proceedings of the 2nd Marcel Grossman Meeting on General Relativity*. Amsterdam, North Holland.

Punturo, M., *et al.* 2010. *Class. Quant. Grav.*, **27**, 194002.

Purdue, P. 2002. *Phys. Rev. D*, **66**, 022001.

Purdue, P. and Chen, Y. 2002. *Phys. Rev. D*, **66**, 122004.

Rehbein, H., *et al.* 2007. *Phys. Rev. D*, **76**, 062002.

Sathyaprakash, B. and Schutz, B. F. 2009. *Living Rev. Relativity*, **12**, 2.

Saulson, P. R. 1997. *Class. Quant. Grav.*, **14**, 2435.

Schutz, B. F. 1986. *Nature*, **323**, 310.

Sigg, D. and the LIGO Scientific Collaboration. 2008. *Class. Quant. Grav.*, **25**, 114041.

Taniwaki, M., *et al.* 1998. *Phys Lett. A*, **246**, 273.

Traeger, S., Willke, B. and Danzmann, K. 1997. *Phys. Lett. A*, **225**, 39.

Vahlbruch, H., *et al.* 2006. *Phys. Rev. Lett.*, **97**, 011101.

Veitch, P. J. 1991. *The Detection of Gravitational Waves*. Cambridge: Cambridge University Press. Chap. Fabry-Perot cavity gravity-wave detectors, page 186.

Vinante, A. *et al.* 2006. *Class. Quant. Grav.*, **23**, S103–S110.

Weber, J. 1960. *Phys. Rev.*, **117**, 306.

Weber, J. 1961. *General Relativity and Gravitational Waves*. Interscience, New York.

Weiss, R. 1972. *MIT Res. Lab. Electron. Q. Report.*, **105**, 54.

Willke, B. and the LIGO Scientific Collaboration 2007. *Class. Quant. Grav.*, **24**, S389.

4

Gravitational wave data analysis

B. S. Sathyaprakash and B. F. Schutz

This chapter focuses on data analysis, a central component of gravitational wave astronomy. After a short introduction to the field we discuss the techniques used to search for the three classes of gravitational wave signals. These include well predicted signals such as coalescing compact binaries, less certain signals such as those from supernovae, and the stochastic signals from gravitational wave backgrounds. We will finish by briefly discussing issues relevant to network detection.

4.1 Introduction

Observing gravitational waves requires a data analysis strategy that is in many ways different from conventional astronomical data analysis. There are several reasons why this is so. Gravitational wave antennae are essentially omni-directional, with their response better than 50% of the root mean square over 75% of the sky. Hence, data analysis systems will have to carry out all-sky searches for sources. Additionally, gravitational wave interferometers are typically broadband, covering three to four orders of magnitude in frequency. While this is obviously to our advantage, as it helps to track sources whose frequency may change rapidly, it calls for searches to be carried out over a wide range of frequencies.

In Einstein's theory, gravitational radiation has two independent states of polarisation. Measuring polarisation is of fundamental importance as there are other theories of gravity in which the number of polarisation states is more than two; in some theories dipolar and even scalar waves exist (Will, 1993). Polarisation has astrophysical implications too. For example, gravitational wave polarisation measurements can be very helpful in resolving the degeneracy that occurs in the measurement of the mass and inclination of a binary system. As is well known, electromagnetic observations of binaries cannot yield the total mass — but only the leave as combination $m \sin \iota$, where ι is the inclination of the binary's orbit to the line of sight. However, measurement of polarisation can determine the angle ι since the polarisation state depends on the binary's inclination with the line of sight. Polarisation measurements would be possible with a network of detectors, which means analysis algorithms that work with data from multiple antennae will have to be developed. This should also benefit coincidence analysis and coherent analysis and improve the efficiency of event recognition.

Unlike typical detection techniques for electromagnetic radiation from astronomical sources, most astrophysical gravitational waves are detected coherently, by following the phase of the radiation rather than just the energy. That is, the signal-to-noise ratio (SNR) is built up by coherent superposition of many wave cycles emitted by a source. The phase evolution contains more information than the amplitude does and the signal structure is a rich source of the underlying physics. Tracking a signal's phase means that searches will have to be made not only for specific sources, but over a huge region of the parameter space for each source, placing severe demands both on the theoretical understanding of the emitted waveforms as well as on the data analysis hardware.

Finally, gravitational wave detection is computationally intensive. Gravitational wave antennae acquire data continuously for many years at the rate of several megabytes per second. About 1% of this data is signal data; the rest is housekeeping data that monitors the operation of the detectors. The large parameter space mentioned above requires that the signal data be filtered many times for different searches, and this puts big demands on computing hardware and algorithms.

Data analysis for broadband detectors has been in development since the mid 1980s (Thorne, 1987; Schutz, 1991, 1989). The field has a regular series of annual Gravitational Wave Data Analysis Workshops; the published proceedings are a good place to find current thinking and challenges. Early coincidence experiments with interferometers (Nicholson *et al.*, 1996) and bars (Allen, 2000) provided the first opportunities to apply these techniques. Although the theory is now fairly well understood (Jaranowski and Królak, 2005), strategies for implementing data analysis depend on available computer resources, data volumes, astrophysical knowledge, and source modelling, and so are under constant revision.

We will begin this chapter by briefly reviewing some of the important definitions and concepts used in gravitational wave data analysis. We will then go on to discuss the matched filtering algorithm and how it is used in searches for binary coalescences and continuous wave sources. After a short discussion of suboptimal filtering methods we will finish the chapter by discussing signal detection strategies for stochastic gravitational wave backgrounds.

4.2 Source amplitudes vs sensitivity

How does one compare the gravitational wave amplitude of astronomical sources with the instrumental sensitivity and assess what sort of sources will be observable against noise? Comparisons are almost always made in the frequency domain, since stationary noise is most conveniently characterised by its power spectral density (PSD).

The simplest signal to characterise is a long-lasting periodic signal with a given fixed frequency f_0. In an observation time T, all the signal power $|\tilde{h}(f_0)|^2$, were $\tilde{h}(f_0)$ is the strain amplitude spectrum given earlier in Section 1.6, is concentrated in a single frequency bin of width $1/T$. The noise against which it competes is just the noise power spectral density in the same bin, $S_h(f_0)/T$. The power SNR is then $T|\tilde{h}(f_0)|^2/S_h(f_0)$, and the amplitude SNR is $\sqrt{T}|\tilde{h}(f_0)|/|S_h(f_0)|^{1/2}$. This improves with observation time as the square root of the time.

The reason for this is that the noise is stationary, but longer and longer observation times permit the signal to compete only with noise in smaller and smaller frequency windows.

Of course, no expected gravitational wave signal would have a single fixed frequency in the detector frame, because the detector is attached to the Earth, whose motion produces frequency modulations. But the principle of this SNR increase with time can still be maintained if one has a signal model that allows one to exclude more and more noise from competing with the signal over longer and longer periods of time. This happens with *matched filtering*, which we focus on in Section 4.3.

Short-lived signals have wider bandwidths, and long observation times are not relevant. To characterise their SNR, it is useful to define the dimensionless noise power per logarithmic bandwidth, $h_n(f) = f S_h(f)$. The signal Fourier amplitude $\tilde{h}(f) \equiv \int_{-\infty}^{\infty} dt \, h(t) e^{2\pi i f t}$ has dimensions of Hz^{-1} and so the Fourier amplitude per logarithmic frequency, which is called the *characteristic signal amplitude* $h_c = f|\tilde{h}(f)|$, is dimensionless. This quantity should be compared with $h_n(f)$ to obtain a rough estimate of the SNR of the signal at some characteristic source frequency: SNR $\sim h_c/h_n$. A more precise form, assuming an optimal filter will be given in the next section.

4.3 Matched filtering and optimal signal-to-noise ratio

Matched filtering is a data analysis technique that efficiently searches for a signal of known shape buried in noisy data (Helstrom, 1968). The technique consists in correlating the output of a detector with a waveform, variously known as a template or a filter. Given a signal $h(t)$ buried in noise $n(t)$, the task is to find an 'optimal' template $q(t)$ that would produce, on the average, the best possible SNR. In this chapter, we shall treat the problem of matched filtering as an operational exercise. However, this intuitive picture has a solid basis in the theory of hypothesis testing. The interested reader may consult any standard textbook on signal analysis, for example Helstrom (1968), for details.

Let us first fix our notation. We shall use $x(t)$ to denote the detector output, which is assumed to consist of a background noise $n(t)$ and a useful gravitational wave signal $h(t)$. The Fourier transform of a quantity $x(t)$ will be denoted $\tilde{x}(f)$ and is defined as

$$\tilde{x}(f) = \int_{-\infty}^{\infty} x(t) e^{2\pi i f t} \, dt. \tag{4.1}$$

Optimal filter

The detector output $x(t)$ is just a realisation of noise $n(t)$, i.e., $x(t) = n(t)$, when no signal is present. In the presence of a signal $h(t)$ with an *arrival time* t_a, $x(t)$ takes the form

$$x(t) = h(t - t_a) + n(t). \tag{4.2}$$

The correlation c of a template $q(t)$ with the detector output is defined as

$$c(\tau) \equiv \int_{-\infty}^{\infty} x(t) q(t + \tau) \, dt. \tag{4.3}$$

In the above equation, τ is called the *lag*; it denotes the duration by which the filter function lags behind the detector output. The purpose of the above correlation integral is to concentrate all the signal energy at one place. The following analysis reveals how this is achieved; we shall work out the *optimal* filter $q(t)$ that maximises the correlation $c(\tau)$ when a signal $h(t)$ is present in the detector output. To do this let us first write the correlation integral in the Fourier domain by substituting for $x(t)$ and $q(t)$, in the above integral, their Fourier transforms $\tilde{x}(f)$ and $\tilde{q}(f)$, i.e. $x(t) \equiv \int_{-\infty}^{\infty} \tilde{x}(f)\exp(-2\pi\mathrm{i}ft)\,\mathrm{d}f$ and $q(t) \equiv \int_{-\infty}^{\infty} \tilde{q}(t)\exp(-2\pi\mathrm{i}ft)\,\mathrm{d}f$, respectively. After some straightforward algebra, one obtains

$$c(\tau) = \int_{-\infty}^{\infty} \tilde{x}(f)\tilde{q}^*(f)\mathrm{e}^{-2\pi\mathrm{i}f\tau}\,\mathrm{d}f, \tag{4.4}$$

where $\tilde{q}^*(f)$ denotes the complex conjugate of $\tilde{q}(f)$.

Since n is a random process, c is also a random process. Moreover, correlation is a linear operation and hence c obeys the same probability distribution function as n. In particular, if n is described by a Gaussian random process with zero mean, then c is also described by a Gaussian distribution function, although its mean and variance will, in general, differ from those of n. The mean value of c, denoted by $S \equiv \bar{c}$, is clearly the correlation of the template q with the signal h, since the mean value of n is zero:

$$S \equiv \bar{c}(\tau) = \int_{-\infty}^{\infty} \tilde{h}(f)\tilde{q}^*(f)\mathrm{e}^{-2\pi\mathrm{i}f(\tau - t_a)}\,\mathrm{d}f. \tag{4.5}$$

The variance of c, denoted $N^2 \equiv \overline{(c - \bar{c})^2}$, turns out to be

$$N^2 = \overline{(c - \bar{c})^2} = \int_{-\infty}^{\infty} S_h(f)|\tilde{q}(f)|^2\,\mathrm{d}f. \tag{4.6}$$

Now the SNR ρ is defined by $\rho^2 \equiv S^2/N^2$.

The form of integrals in equations (4.5) and (4.6) leads naturally to the definition of the scalar product of waveforms. Given two functions, $a(t)$ and $b(t)$, we define their scalar product $\langle a, b \rangle$ to be (Finn, 1992; Finn and Chernoff, 1993; Chernoff and Finn, 1993)

$$\langle a, b \rangle \equiv 2\int_0^{\infty} \frac{\mathrm{d}f}{S_h(f)}\left[\tilde{a}(f)\tilde{b}^*(f) + \tilde{a}^*(f)\tilde{b}(f)\right]. \tag{4.7}$$

Noting that the Fourier transform of a real function $h(t)$ obeys $\tilde{h}(-f) = \tilde{h}^*(f)$, we can write down the SNR in terms of the above scalar product:

$$\rho^2 = \frac{\langle h\mathrm{e}^{2\pi\mathrm{i}f(\tau - t_a)}, S_h q \rangle}{\sqrt{\langle S_h q, S_h q \rangle}}. \tag{4.8}$$

From this it is clear that the template q that obtains the maximum value of ρ is simply

$$\tilde{q}(f) = \gamma\,\frac{\tilde{h}(f)\mathrm{e}^{\mathrm{i}2\pi f(\tau - t_a)}}{S_h(f)}, \tag{4.9}$$

where γ is an arbitrary constant. From the above expression for an optimal filter we note two important things. First, the SNR is maximised when the lag parameter τ is equal to the time of arrival of the signal t_a. Second, the optimal filter is not just a copy of the signal, but rather it is weighted down by the noise PSD. We will see below why this should be so.

Optimal signal-to-noise ratio

We can now work out the optimal SNR by substituting equation (4.9) for the optimal template in equation (4.8),

$$\rho_{\text{opt}} = \langle h, h \rangle^{1/2} = 2 \left[\int_0^\infty df \, \frac{\left| \tilde{h}(f) \right|^2}{S_h(f)} \right]^{1/2}. \tag{4.10}$$

We note that the optimal SNR is not just the total energy of the signal (which would be $2 \int_0^\infty df |\tilde{h}(f)|^2$), but rather the integrated signal power weighted down by the noise PSD. This is in accordance with what we would guess intuitively: the contribution to the SNR from a frequency bin where the noise PSD is high is smaller than that from a bin where the noise PSD is low. Thus, an optimal filter automatically takes into account the nature of the noise PSD.

The expression for the optimal SNR given by equation (4.10) suggests how one may compare signal strengths with the noise performance of a detector. Note that one cannot directly compare $\tilde{h}(f)$ with $S_h(f)$, as they have different physical dimensions. In gravitational wave literature one writes the optimal SNR in one of the following equivalent ways:

$$\rho_{\text{opt}} = 2 \left[\int_0^\infty \frac{df}{f} \, \frac{\left| \sqrt{f} \tilde{h}(f) \right|^2}{S_h(f)} \right]^{1/2} = 2 \left[\int_0^\infty \frac{df}{f} \, \frac{\left| f \tilde{h}(f) \right|^2}{f S_h(f)} \right]^{1/2}, \tag{4.11}$$

which facilitates the comparison of signal strengths with noise performance. One can compare the dimensionless quantities, $f|\tilde{h}(f)|$ and $\sqrt{f S_h(f)}$, or dimensionful quantities, $\sqrt{f}|\tilde{h}(f)|$ and $\sqrt{S_h(f)}$.

Signals of interest to us are characterised by several (*a priori* unknown) parameters, such as the masses of component stars in a binary, their intrinsic spins, etc., and an optimal filter must agree with both the signal shape and its parameters. A filter whose parameters are slightly mismatched with that of a signal can greatly degrade the SNR. For example, even a mismatch of one cycle in 10^4 cycles can degrade the SNR by a factor two.

When the parameters of a filter and its shape are precisely matched with that of a signal, what is the improvement brought about, as opposed to the case when no knowledge of the signal is available? Matched filtering helps in enhancing the SNR in proportion to the square root of the number of signal cycles in the detector band, as opposed to the case in which the signal shape is unknown and all that can be done is to Fourier transform the detector output and compare the signal energy in a frequency bin to noise energy in that bin. We shall see below that, in initial interferometers, matched filtering leads to an enhancement of order $30 - 100$ for compact binary inspiral signals.

4.4 Practical applications of matched filtering

Matched filtering is currently being applied to mainly two sources: detection of (1) chirping signals from compact binaries consisting of black holes and/or neutron stars and (2) continuous waves from rapidly spinning neutron stars.

Detection of coalescing binaries

In the case of chirping binaries, post-Newtonian theory (a perturbative approximation to Einstein's equations in which the relevant quantities are expanded as a power series in $1/c$, where c is the speed of light) has been used to model the dynamics of these systems to order in $(v/c)^7$, where v is the speeds of the objects in the binary (Blanchet, 2002). This is an approximation that can be effectively used to search for binaries whose component masses are equal, or nearly equal, and where the system is "far" from coalescence. When the binary is far from coalescence, the separation between the masses is large, the motion is non-relativistic and the post-Newtonian approximation works well. When the masses are equal or nearly equal spin-orbit induced precession of the orbital plane is small and again post-Newtonian approximation is good enough. For binaries whose component masses are in the range $[1, 6]M_\odot$ post-Newtonian waveforms are pretty good for use as matched filters (Buonanno *et al.*, 2009). In reality, one takes the waveform to be valid until the last stable circular orbit (LSCO), which marks the end of the inspiral phase of the coalescing system. In the case of binaries consisting of two neutron stars, or a neutron star and a black hole, tidal effects might affect the evolution significantly before reaching the LSCO. However, this is likely to occur at frequencies well above the sensitivity band of the current ground-based detectors, so that for all practical purposes post-Newtonian waveforms are a good approximation to low-mass ($M < 10\,M_\odot$) binaries.

Progress in analytical and numerical relativity has made it possible to have a set of waveforms for the merger phase of compact binaries too. A breakthrough came with Pretorius (2005) announcing the results from the first stable simulation ever, followed by further breakthroughs by two other groups (Campanelli *et al.*, 2006; Baker *et al.*, 2006) with successful simulations. The main results from numerical simulations of non-spinning black holes are rather simple. Indeed, just as the effective one-body (EOB) model (Buonanno and Damour, 1999, 2000) had predicted, and probably contrary to what many people had expected, the final merger is just a continuation of the adiabatic inspiral, leading on smoothly to merger and ring-down. In Figure 4.1 we show the results from one of the numerical simulations (right panel) and that of the EOB (left panel), both for the same initial conditions. There is also good agreement in the prediction of the total energy emitted by the system, being 5.0% ($\pm$ 0.4%) (for a review see Pretorius, 2009) and 3.1% (Buonanno and Damour, 2000), by numerical simulations and EOB, respectively, as well as the spin of the final black hole (respectively, 0.69 and 0.8) that results from the merger.

The computational cost in matched filtering the merger phase will not be high, as there will only be on the order of a few 100 cycles in this phase. But it is important to have the

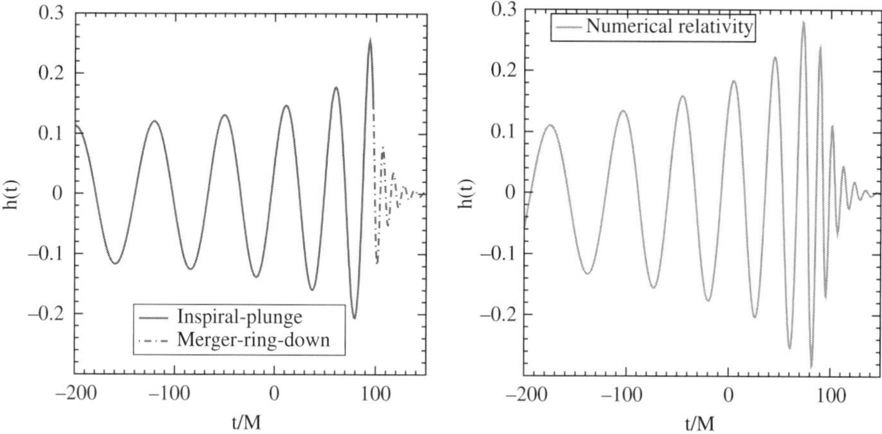

Figure 4.1 Comparison of waveforms from the analytical EOB approach (left) and numerical relativity simulations (right) for the same initial conditions. The two approaches predict very similar values for the total energy emitted in gravitational waves and the final spin of the black hole. Figure from Buonanno (2007).

correct waveforms to enhance signal visibility and, more importantly, to enable strong-field tests of general relativity.

In the general case of black hole binary inspiral, the search space is characterised by 17 different parameters. These are the two masses of the bodies, their spins, the eccentricity of the orbit, its orientation at some fiducial time, the position of the binary in the sky and its distance from the Earth, the epoch of coalescence and phase of the signal at that epoch, and the polarisation angle. However, not all these parameters are important in a matched filter search. Only those parameters that change the shape of the signal, such as the masses, orbital eccentricity and spins, or which cause a modulation in the signal due to the motion of the detector relative to the source, such as the direction to the source, are to be searched for. Others, such as the epoch of coalescence and the phase at that epoch, are simply reference points in the signal that can be determined without any significant burden on computational power.

An important question in matched filtering is the choice of spacing between templates. This is one of the most difficult problems in data analysis, especially when the dimension of the search space is large (Harry *et al.*, 2009). One might naively think of choosing templates uniformly distributed in the relevant physical parameters. For example, in the case of a search for binaries consisting of non-spinning black holes the only parameters on which the shape of the signal depends are the masses of the component black holes. One might be tempted to choose templates uniformly spaced in the component masses. The correct measure for the distance between templates, however, is not the spacing between masses but the scalar product introduced in equation (4.8). It turns out that templates uniformly spaced in component masses don't all have the same overlap with one another but they do when templates are uniformly spaced in *chirp times*.

Chirp times are post-Newtonian contributions at different orders to the duration of a signal starting from a time when the instantaneous gravitational wave frequency has a fiducial value f_L to a time when the gravitational wave frequency formally diverges and the system coalesces. For instance, the chirp times τ_0 and τ_3 at Newtonian and 1.5 PN orders, respectively, are

$$\tau_0 = \frac{5}{256\pi\,\nu\,f_L}\,(\pi M f_L)^{-5/3}\,, \qquad \tau_3 = \frac{1}{8\,\nu\,f_L}\,(\pi M f_L)^{-2/3}\,, \tag{4.12}$$

where M is the total mass and $\nu = m_1 m_2/M^2$ is the symmetric mass ratio. The above relations can be inverted to obtain M and ν in terms of the chirp times:

$$M = \frac{5}{32\pi^2\,f_L}\,\frac{\tau_3}{\tau_0}\,, \qquad \nu = \frac{1}{8\pi\,f_L\,\tau_3}\left(\frac{32\pi\,\tau_0}{5\,\tau_3}\right)^{2/3}\,. \tag{4.13}$$

There is a significant amount of literature on the computational requirements to search for compact binaries (Sathyaprakash and Dhurandhar, 1991; Dhurandhar and Sathyaprakash, 1994; Owen, 1996; Owen and Sathyaprakash, 1999). The estimates for initial detectors are not alarming and it is possible to search for these systems online. Searches for these systems by the LIGO Scientific Collaboration (LSC) (see, for example, Abbott *et al.*, 2006) employs a hexagonal lattice of templates (Cokelaer, 2007) in the two-dimensional space of chirp times. For the best LIGO detectors we need several thousand templates to search for component masses in the range $[m_{\text{low}}, m_{\text{high}}] = [1, 100]\,M_\odot$ (Owen and Sathyaprakash, 1999). Decreasing the lower end of the mass range leads to an increase in the number of templates that grows roughly as $m_{\text{low}}^{-8/3}$ and most current searches (Abbott *et al.*, 2004, 2005c) only begin at $m_{\text{low}} = 1\,M_\odot$, with the exception of a search that looked for black hole binaries of primordial origin (Abbott *et al.*, 2005d), in which the lower end of the search was $0.2\,M_\odot$.

Inclusion of spins is only important when one or both of the components is *rapidly* spinning (Apostolatos, 1995; Buonanno *et al.*, 2003). Spin effects are unimportant for neutron star binaries, for which the dimensionless spin parameter q, that is, the ratio of its spin magnitude to the square of its mass, is tiny: $q = J_{\text{NS}}/M_{\text{NS}}^2 \ll 1$. For ground-based detectors, even after including spins, the computational costs, while high, are not formidable and it should be possible to carry out the search on large computational clusters in real time (Buonanno *et al.*, 2003; Abbott *et al.*, 2008d).

Searching for continuous wave signals.

In the case of continuous waves (CWs), the signal shape is pretty trivial: a sinusoidal oscillation with small corrections to take account of the slow spin-down of the neutron star/pulsar caused by loss of angular momentum to gravitational waves and other radiation/particles. However, what leads to an enormous computational cost here is the Doppler modulation of the signal due to Earth's rotation, the motion of the Earth around the Solar System barycentre and the Moon. The number of independent patches that we have to observe so as not to lose appreciable amounts of SNR can be worked out in the following manner. The baseline of a gravitational wave detector for CW sources is essentially $L = 2 \times 1$ A.U.

$\simeq 3 \times 10^{11}$ m. For a source that emits gravitational waves at 100 Hz, the wavelength of the radiation is $\lambda = 3 \times 10^6$ m, and the angular resolution $\Delta\theta$ of the antenna at an SNR of 1 is $\Delta\theta \simeq \lambda/L = 10^{-5}$, or a solid angle of $\Delta\Omega \simeq (\Delta\theta)^2 = 10^{-10}$. In other words, the number of patches one should search for is $N_{\text{patches}} \sim 4\pi/\Delta\Omega \simeq 10^{11}$. Moreover, for an observation that lasts for about a year ($T \simeq 3 \times 10^7$ s) the frequency resolution is $\Delta f = 1/T \simeq 3 \times 10^{-8}$. Searching over a frequency band of 300 Hz, around the best sensitivity of the detector, gives the number of frequency bins to be about 10^{10}. Thus, it is necessary to search over roughly 10^{11} patches in the sky for each of the 10^{10} frequency bins. This is a formidable task and one can only perform a matched filter search over a short period (days/weeks) of the data or over a restricted region in the sky, or just perform targeted searches for known objects such as pulsars, the galactic centre, etc. (Brady *et al.*, 1998).

The severe computational burden faced in the case of CW searches has led to the development of specialised searches that look for signals from known pulsars (Abbott *et al.*, 2005b, 2007b, 2008b) using an efficient search algorithm that makes use of the known parameters (Christensen *et al.*, 2004; Dupuis and Woan, 2005) and hierarchical algorithms that add power incoherently with the minimum possible loss in signal visibility (Krishnan *et al.*, 2004; Abbott *et al.*, 2005a; Sintes and Krishnan, 2006; Abbott *et al.*, 2008a). The most ambitious project in this regard is the Einstein@Home project[1]. The goal here is to carry out coherent searches for CW signals using wasted CPUs on idle computers at homes, offices and university departments around the world. The project has been successful in attracting a large number of subscriptions and provides the largest computational infrastructure to the LSC for the specific search of CW signals and the first scientific results from such have been published by the LSC (Abbott *et al.*, 2008c).

4.5 Suboptimal filtering methods

It is not always possible to compute the shape of the signal from a source. For instance, there is no computation, numerical or analytical, that reliably gives us the highly relativistic and non-linear dynamics of gravitational collapse, the supernova that follows it and the emitted gravitational signal. The biggest problem here is the unknown physical state of the pre-supernova star and the complex physics that is involved in the collapse and explosion. Thus, matched filtering cannot be used to detect signals from supernovae.

Even when the waveform is known, the great variety in the shape of the emitted signals might render matched filtering ineffective. In binaries, in which one of the component masses is much smaller than the other, the smaller body will evolve on a highly precessing and in some cases eccentric orbit, due to strong spin–orbit coupling. Moreover, the radiation back-reaction effects, which in the case of equal mass binaries are computed in an approximate way by averaging over an orbital time-scale, should be computed much more accurately. The resulting motion of the small body in the Kerr spacetime of the larger body is extremely complicated, leading to a waveform that is rather complex and matched filtering would not be a practical approach.

---

[1] The Einstein@Home project homepage can found at http://einstein.phys.uwm. edu/

Suboptimal methods can be used in such cases and they have a two-fold advantage: they are less sensitive to the shape of the signal and are computationally significantly cheaper than matched filtering. Of course, the price is a loss in the SNR. The best suboptimal methods are sensitive to signal amplitudes a factor of two to three larger than that required by matched filtering and a factor of 10 to 30 in volume.

Most suboptimal techniques are one form of time–frequency transform or the other. They determine the presence or absence of a signal by comparing the power over a small volume in the time–frequency plane in a given segment of data to the average power in the same volume over a large segment of data. The time–frequency transform $q(\tau, f)$ of data $x(t)$ using a window $w(t)$ is defined as

$$q(\tau, f) = \int_{-\infty}^{\infty} w(t - \tau)x(t)e^{2\pi i f t} \, dt. \tag{4.14}$$

The window function $w(t - \tau)$ is centred at $t = \tau$, and one obtains a time–frequency map by moving the window from one end of a data segment to the other. The window is not unique and the effectiveness of a window depends on the signal one is looking for. Once the time–frequency map is constructed, one can look for excess power (compared to average) in different regions (Anderson *et al.*, 2000), or look for certain patterns.

The method followed depends on the signal one is looking for. For instance, when looking for unknown signals, all that can be done is to look for a departure from averaged behaviour in different regions of the map (Anderson *et al.*, 2000). However, when some knowledge of the spectral and temporal content of the signal is known, it is possible to tune the algorithm to improve efficiency. The wavelet-based *waveburst* algorithm is one such example (Klimenko and Mitselmakher, 2004) that has been applied to search for unstructured bursts in LIGO data (Abbott *et al.*, 2007a).

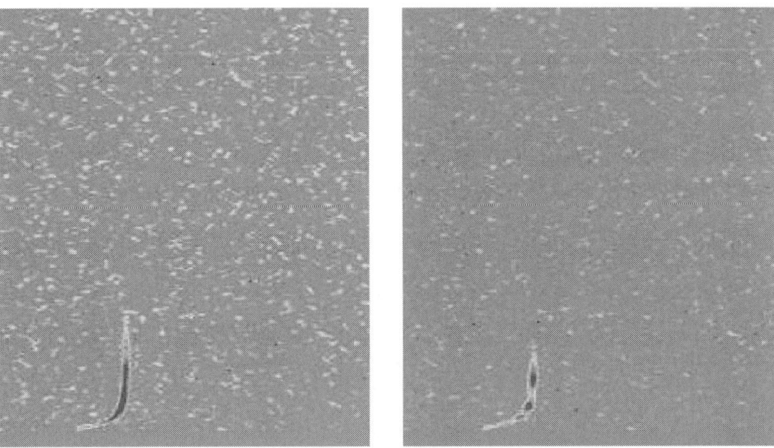

Figure 4.2 Time–frequency maps showing the track left by the inspiral of a small black hole falling into an SMBH as expected in LISA data. The left panel is for a central black hole without spin and the right panel is for a central black hole whose dimensionless spin parameter is $q = 0.9$.

One can employ strategies that improve detection efficiency over a simple search for excess power. For example, chirping signals will leave a characteristic track in the time–frequency plane, with increasing frequency and power as a function of time. A time–frequency map of a chirp signal buried in noisy data is shown in Figure 4.2. An algorithm that optimises the search for specific shapes in the time–frequency plane is discussed in Heng *et al.* (2004). These and other methods have been applied to understand how to analyse LISA data (Gair and Jones, 2007; Wen and Gair, 2005).

4.6 False alarms, detection threshold and coincident observation

If gravitational wave event rates are expected to be low, as was the case for initial interferometers, one has to set a high threshold, so that the noise-generated false alarms mimicking an event are negligible.

For a detector output sampled at 1 kHz and processed through a large number of filters, say 10^3, one has $\sim 3 \times 10^{13}$ instances of noise in a year. If the filtered noise is Gaussian, then the probability $P(x)$ of observing an amplitude in the range of x to $x + \mathrm{d}x$ is

$$P(x)\,\mathrm{d}x = \frac{1}{\sqrt{2\pi}\,\sigma} \exp\left(\frac{-x^2}{2\sigma^2}\right)\,\mathrm{d}x, \qquad (4.15)$$

where σ is the standard deviation. The above probability-distribution function implies that the probability that the noise amplitude is greater than a given threshold η is

$$P(x|x \geq \eta) = \int_{\eta}^{\infty} P(x)\,\mathrm{d}x = \frac{1}{\sqrt{2\pi}\,\sigma} \int_{\eta}^{\infty} \exp\left(\frac{-x^2}{2\sigma^2}\right)\,\mathrm{d}x. \qquad (4.16)$$

Demanding that no more than one noise-generated false alarm occurs in 100 years of observation means that $P(x|x \geq \eta) = 1/(3 \times 10^{15})$. Solving this equation for η, one finds that $\eta \simeq 7.8\sigma$ in order that false alarms are negligible in 100 years of observation. Therefore, a source is detectable only if its amplitude is significantly larger than the effective noise amplitude, i.e. $f\tilde{h}(f) \gg h_n(f)$.

The reason for accepting only such high-sigma events is that the event rate of a transient source, i.e. a source lasting for a few seconds to minutes, such as a binary inspiral, could be as low as a few per year, and so the false alarm rate should be significantly smaller than the expected rate of events. Setting higher thresholds for detection helps in detecting signals with a high confidence. However, signal enhancement techniques (cf. Section 4.3) make it possible to detect a signal of relatively low amplitude [i.e. when the signal amplitude $h(t)$ is significantly smaller than the rms value of noise $n(t)$], provided there are a large number of wave cycles and the shape of the wave is known accurately.

Real detector noise is neither Gaussian nor stationary and therefore the filtered noise cannot be expected to obey these properties either. One of the most challenging problems is how to remove or veto the false alarms generated by a non-Gaussian and/or non-stationary background. There has been some effort to address the issue of non-Gaussianity (Creighton, 1999) and non-stationarity (Mohanty, 2000); a lot of work has also been done in the context of LIGO Scientific Collaboration and Virgo in developing vetoes based on the correlation of the GW strain channel with the environmental and instrumental channels. Additionally, the availability of a network of gravitational wave detectors alleviates the problem to some extent. This is because a high amplitude gravitational wave event will be coincidentally observed in several detectors, although not necessarily with the same SNR, while false alarms are, in general, not coincident, as they are normally produced by independent sources located close to the detectors.

Coincident observations have helped to reduce the false alarm rate significantly by vetoing out spurious events. One can associate with each event in a given detector an ellipsoid in the parameter space of the signal, whose location and orientation depend on where in the parameter space and when the event was found. The SNR can be used to fix the size of the ellipsoid (Robinson *et al.*, 2008). One is associating with each event a 'sphere' of influence in the multi-dimensional space of masses, spins, arrival times, etc., and there is a stringent demand that the spheres associated with events from different detectors should overlap each other in order to claim a detection. Since random triggers from a network of detectors are less likely to be consistent with one another, this method serves as a very powerful veto.

The false alarm rate can be further reduced and possibly even nullified by subjecting coincident events to further consistency checks in a detector network consisting of four or more detectors. Each gravitational wave event is characterised by five kinematic (or extrinsic) observables: the two polarisations $(h_+, h_\times)$ and the location of the source with respect to the detector, given by the luminosity distance, D, and the angles θ and φ. Each detector in a network measures a single number, say the amplitude of the wave. In addition, in a network of N detectors, there are $N - 1$ independent time delays in the arrival times of the wave at various detector locations, giving a total of $2N - 1$ observables. Thus, the minimum number of detectors needed to reconstruct the wave and its source is $N = 3$. More than three detectors in a network will have redundant information that will be consistent with the quantities inferred from any three detectors, provided the event is a true coincident event and not a chance coincidence, and most likely a true gravitational wave event. In a detector network consisting of $N(\geq 4)$ detectors, one can perform $2N - 6$ consistency checks. Such consistency checks further reduce the number of false alarms and further work is needed in implementing such vetoes.

When the shape of a signal is known, matched filtering is the optimal strategy to pull out a signal buried in Gaussian, stationary noise. The presence of high amplitude transients in the data can render the background non-stationary and non-Gaussian, therefore matched filtering is not necessarily an optimal strategy. However, the knowledge of a signal's shape, especially when it has a broad bandwidth, can be used beyond matched filtering to construct a χ^2 veto (Allen, 2005) to distinguish between triggers caused by a true signal from those caused by high amplitude transients or other artifacts. One specific implementation of the χ^2 veto compares the expected signal spectrum with the real spectrum to quantify the confidence with which a trigger can be accepted as a true gravitational wave signal and has been the most powerful method for greatly reducing the false alarm rate. In the next section we will further quantify the χ^2 veto (Allen, 2005) by using the scalar product introduced in the context of matched filtering.

χ^2 *veto*

The main problem with real data is that it can be glitchy in the form of high amplitude transients that might look like damped sinusoids. An inspiral signal and a template employed to detect it are both broadband signals. Therefore, the matched-filter SNR for such signals has contributions from a wide range of frequencies. However, the statistic of matched filtering, namely the SNR, is an integral over frequency and it is not sensitive to contributions from different frequency regions. Imagine dividing the frequency range of integration into a finite number of bins $f_k \le f < f_{k+1}$, $k = 1, \ldots, p$, such that their union spans the entire frequency band, $f_1 = 0$ and $f_{p+1} = \infty$, and further that the contribution to the SNR from each frequency bin is the same, namely,

$$4 \int_{f_k}^{f_{k+1}} \frac{|\tilde{h}(f)|^2}{S_h(f)} \mathrm{d}f = \frac{4}{p} \int_0^\infty \frac{|\tilde{h}(f)|^2}{S_h(f)} \mathrm{d}f. \tag{4.17}$$

Now, define the contribution to the matched filtering statistic coming from the k-th bin by (Allen, 2005)

$$z_k \equiv \langle q, x \rangle_k \equiv 2 \int_{f_k}^{f_{k+1}} \left[\tilde{q}^*(f)\tilde{x}(f) + \tilde{q}(f)\tilde{x}^*(f) \right] \frac{\mathrm{d}f}{S_h(f)}, \tag{4.18}$$

where, as before, $\tilde{x}(f)$ and $\tilde{q}(f)$ are the Fourier transforms of the detector output and the template, respectively. Note that the sum $z = \sum_k z_k$ gives the full matched filtering statistic (Allen, 2005):

$$z = \langle q, x \rangle \equiv 2 \int_0^\infty \left[\tilde{q}^*(f)\tilde{x}(f) + \tilde{q}(f)\tilde{x}^*(f) \right] \frac{\mathrm{d}f}{S_h(f)}. \tag{4.19}$$

Having chosen the bins and quantities z_k as above, one can construct a statistic based on the measured SNR in each bin as compared to the expected value, namely

$$\chi^2 = p \sum_{k=1}^p \left(z_k - \frac{z}{p} \right)^2. \tag{4.20}$$

When the background noise is stationary and Gaussian, the quantity χ^2 obeys the well-known chi-square distribution with $p - 1$ degrees of freedom. Therefore, the statistical properties of the χ^2 statistic are known. Imagine two triggers with identical SNRs, but one caused by a true signal and the other caused by a glitch that has power only in a small frequency range. It is easy to see that the two triggers will have very different χ^2 values; in the first case the statistic will be far smaller than in the second case. This statistic has served as a very powerful veto in the search for signals from coalescing compact binaries and it has been instrumental in cleaning up the data (see, e.g., Abbott *et al.*, 2004, 2005c).

4.7 Detection of stochastic signals by cross-correlation

As described in Chapter 2, the Universe might be filled with stochastic gravitational waves that were either generated in the primeval Universe or by a population of background sources. For point sources, although each source in a population might not be individually detectable, they could collectively produce a confusion background via a random superposition of the waves from that population. Since the waves are random in nature, it is not possible to use the techniques already described in this chapter to detect a stochastic background. However, we might use the noisy stochastic signal in one of the detectors as a 'matched filter' for the data in another detector (Thorne, 1987; Flanagan, 1993; Allen, 1997; Bruce and Romano, 1999). In other words, it should be possible to detect a stochastic background by cross-correlating the data from a pair of detectors; the common gravitational wave background will grow in time more rapidly than the random backgrounds in the two instruments, thereby facilitating the detection of the background.

If two instruments with identical spectral noise density S_h are cross-correlated over a bandwidth Δf for a total time T, the spectral noise density of the output is reduced by a factor of $(T \Delta f)^{1/2}$. Since the noise amplitude is proportional to the square root of S_h, the amplitude of a signal that can be detected by cross-correlation improves as the fourth root of the observing time. This should be compared with the square-root improvement that matched filtering gives.

The *cross-correlation technique* works well when two detectors are situated close to one another. When separated, only those waves whose wavelength is larger than or comparable to the distance between the two detectors, or which arrive from a direction perpendicular to the separation between the detectors, can contribute coherently to the cross-correlation statistic. Since the instrumental noise builds up rapidly at lower frequencies, detectors that are farther apart are less useful in cross-correlation. However, very near-by detectors (as in the case of two LIGO detectors within the same vacuum tube in Hanford) will suffer from common background noise from the control system and the environment, making it rather difficult to ascertain if any excess noise is due to a stochastic background of gravitational waves.

As described in Chapter 2, stochastic gravitational wave backgrounds are conventionally described by Ω_{gw}, which defines the energy density in gravitational waves per log frequency interval, as a fraction of the critical density ρ_c required to close the Universe. If the source of radiation is *scale-free* (which means that there is no preferred length-scale or time-scale in the process), then it will produce a power-law spectrum, i.e. one in which $\Omega_{\mathrm{gw}}(f)$ depends on a power of f. One example is inflation, which predicts a *flat* energy spectrum, one in which Ω_{gw} is essentially independent of frequency (Allen, 1997).

The energy in the cosmological background is, of course, related to the spectral density of noise that the background would produce in a gravitational wave detector. Since we describe the gravitational wave noise in terms of the amplitude rather than energy, there are scaling factors involving the frequency between the two. An isotropic gravitational wave background incident on an interferometric detector will induce a strain spectral noise density equal to (Thorne, 1987; Allen, 1997)

$$S_{\text{gw}}(f) = \frac{3H_0^2}{10\pi^2} f^{-3} \, \Omega_{\text{gw}}(f).$$

(4.21)

Note that the explicit dependence on frequency is f^{-3}: two factors come from the relation of energy and squared-strain, and one factor from the fact that Ω_{gw} is an energy distribution per unit logarithmic frequency. Note also that there are no explicit factors of c or G needed in this formula if one wants to work in non-geometrised units.

If we scale H_0 by $h_{100} = H_0/(100 \text{ km s}^{-1} \text{ Mpc}^{-1})$, and note that $100 \text{ km s}^{-1} \text{ Mpc}^{-1} = 3.24 \times 10^{-18} \text{ s}^{-1}$, then this equation implies that the strain noise is

$$S_{\text{gw}}^{1/2} = 5.6 \times 10^{-22} \text{ Hz}^{-1/2} \, \Omega_{\text{gw}}^{1/2} \left(\frac{f}{100 \text{ Hz}} \right)^{-3/2} h_{100}.$$

(4.22)

To be observed by a single gravitational wave detector, the gravitational wave noise must be larger than the instrumental noise. This is a bolometric method of detection of the background, and it requires great confidence in the understanding of the detector, in order to believe that the observed noise is external. This is how the cosmic microwave background was originally discovered in a radio telescope by Penzias and Wilson. As described above, if there are two detectors, then one may be able to get better sensitivity by cross-correlating their output. From equation (4.22) and the earlier discussion, it is straightforward to deduce that two co-located detectors, each with spectral noise density S_h and fully uncorrelated instrumental noise, observing over a bandwidth f at frequency f for a time T, can detect a stochastic background with energy density

$$\Omega_{\text{gw}}^{1/2} h_{100} = \left(\frac{S_h^{1/2}}{3.1 \times 10^{-18} \text{ Hz}^{-1/2}} \right) \left(\frac{f}{10 \text{ Hz}} \right)^{5/4} \left(\frac{T}{3 \text{ yrs}} \right)^{-1/4}.$$

(4.23)

The two LIGO detectors (separated by about 10 ms in light-travel time) are reasonably well placed for performing such correlations, particularly when upgrades push their lower frequency limit to 20 Hz or less. Two co-located first-generation LIGO instruments operating at 100 Hz could, in a one-year correlation, reach a sensitivity of $\Omega_{\text{gw}} \sim 1.7 \times 10^{-8}$. But the separation of the actual detectors takes its toll at this frequency, so that they can actually only reach $\Omega_{\text{gw}} \sim 10^{-6}$. Advanced LIGO may improve this by two or three orders of magnitude, going well below the nucleosynthesis bound. The third-generation instrument, the Einstein Telescope (ET), which features in Chapter 16, can go even deeper. Two co-located ETs, observing at 10 Hz for three years, could reach $\Omega_{\text{gw}} \sim 10^{-12}$. At this frequency the detectors could be as far apart as 5000 km without a substantial loss in correlation sensitivity.

Correlation searches are also possible between resonant detectors or between one resonant and one interferometric detector (Astone *et al.*, 1994). This has been implemented with bar detectors (Astone *et al.*, 1999) and between LIGO and the ALLEGRO bar detector (Whelan *et al.*, 2003).

LISA does not gain by a simple correlation between any two of its independent interferometers, since they share a common arm, which contributes common noise that competes with that of the background. A gravitational wave background of $\Omega_{\mathrm{gw}} \sim 10^{-10}$ would compete with LISA's expected instrumental noise. However, using all three interferometers together can improve things for LISA at low frequencies, assuming that the LISA instrumental noise is well behaved (Hogan and Bender, 2001). This might enable LISA to go below 10^{-11}.

4.8 Network detection

Gravitational wave detectors are almost omni-directional. Both interferometers and bars have good sensitivity over a large area of the sky. In this regard, gravitational wave antennae are unlike conventional astronomical telescopes, e.g. optical, radio, or infrared bands, which observe only a very small fraction of the sky at any given time. The good news is that gravitational wave interferometers will have good sky coverage and therefore only a small number (around six) are enough to survey the sky. The bad news, however, is that gravitational wave observations will not automatically provide the location of the source in the sky. It will either be necessary to observe the same source in several non-co-located detectors and triangulate the position of the source using the information from the delay in the arrival times of the signals at different detectors, or observe for a long time and use the location-dependent Doppler modulation caused by the motion of the detector relative to the source to infer the source's position in the sky. The latter is a well-known technique in radio astronomy of synthesising a long-baseline observation to gain resolution, and only possible for sources, such as rotating neutron stars or stochastic backgrounds, that last for a long enough duration.

A network of detectors is, therefore, essential for source reconstruction. Network observation is not only powerful in identifying a source in the sky, but independent observation of the same source in several detectors adds to the detection confidence, especially since the noise background is not well understood and is plagued by non-stationarity and non-Gaussianity.

The improvement in resolution that worldwide networks of gravitational wave detectors offer will be highly important for multi-messenger astronomy in which observations will be coordinated with different kinds of radiation and information carriers: electromagnetic, neutrino, cosmic ray and gravitational wave. This exciting field will be featured in the next chapter, devoted to gravitational wave network detection.

Acknowledgements

Much of the content in this chapter is based on our review in the open-access journal *Living Reviews in Relativity*, **12** (2009) 2. URL: (cited on 1 May 2011): http://www.livingreviews.org/lrr-2009-2. B. F. Schutz gratefully acknowledges the support of DFG grant SFB/TR-7. B.S.Sathyaprakash thanks the Science and Technology Facilities Council grant ST/I000887/1.

References

Abbott, B., *et al.* 2004. *Phys. Rev. D*, **69**.
Abbott, B., *et al.* 2005a. *Phys. Rev. D*, **72**.

Abbott, B., *et al.* 2005b. *Phys. Rev. Lett.*, **94**.

Abbott, B., *et al.* 2005c. *Phys. Rev. D*, **72**.

Abbott, B., *et al.* 2005d. *Phys. Rev. D*, **72**.

Abbott, B., *et al.* 2006. *Phys. Rev. D*, **73**.

Abbott, B., *et al.* 2007a. *Class. Quantum Grav.*, **24**, 5343–5370.

Abbott, B., *et al.* 2007b. *Phys. Rev. D*, **76**.

Abbott, B., *et al.* 2008a. *Phys. Rev. D*, **77**.

Abbott, B., *et al.* 2008b. *Astrophys. J. Lett.*, **683**.

Abbott, B., *et al.* 2008c. *Phys. Rev. D*, **79**.

Abbott, B., *et al.* 2008d. *Phys. Rev. D*, **78**.

Allen, B. 1997. Pages 373–418 of: Marck, J.-A. and Lasota, J.-P. (eds), *Relativistic Gravitation and Gravitational Radiation*. Cambridge Contemporary Astrophysics. Cambridge: Cambridge University Press.

Allen, B. 2005. *Phys. Rev. D*, **71**.

Allen, Z., *et al.* 2000. *Phys. Rev. Lett.*, **85**, 5046–5050.

Anderson, W., *et al.* 2000. *Int. J. Mod. Phys. D*, **9**, 303–307.

Apostolatos, T. 1995. *Phys. Rev. D*, **52**, 605–620.

Astone, P., Lobo, A. and Schutz, B. 1994. *Class. Quantum Grav.*, **11**, 2093–2112.

Astone, P., *et al.* 1999. *Astron. Astrophys.*, **343**, 19.

Baker, J., *et al.* 2006. *Phys. Rev. Lett.*, **96**.

Blanchet, L. 2002. *Living Reviews in Relativity*, **5**, 3.

Brady, P., *et al.* 1998. *Phys. Rev. D*, **57**, 2101–2116.

Bruce, A. and Romano, J. 1999. *Phys. Rev. D*, **59**.

Buonanno, A. 2007. Pages 3–52 of: Bernardeau, F., Grojean, C., and Dalibard, J. (eds), *Particle Physics and Cosmology: The Fabric of Spacetime*. Amsterdam; Oxford: Elsevier.

Buonanno, A. and Damour, T. 1999. *Phys. Rev. D*, **59**.

Buonanno, A. and Damour, T. 2000. *Phys. Rev. D*, **62**.

Buonanno, A., Chen, Y., and Vallisneri, M. 2003. *Phys. Rev. D*, **67**. Erratum-ibid. D**74**, 029904(E) (2006).

Buonanno, A., *et al.* 2009. *Phys. Rev. D.*, **80**, 084043.

Campanelli, M., *et al.* 2006. *Phys. Rev. Lett.*, **96**.

Chernoff, D. and Finn, L. 1993. *Astrophys. J.*, **411**, L5–L8.

Christensen, N., *et al.* 2004. *Phys. Rev. D*, **70**.

Cokelaer, T. 2007. *Phys. Rev. D*, **76**.

Creighton, J. 1999. *Phys. Rev. D*, **60**.

Dhurandhar, S. and Sathyaprakash, B. 1994. *Phys. Rev. D*, **49**, 1707–1722.

Dupuis, R. and Woan, G. 2005. *Phys. Rev. D*, **72**.

Finn, L. 1992. *Phys. Rev. D*, **46**, 5236–5249.

Finn, L. and Chernoff, D. 1993. *Phys. Rev. D*, **47**, 2198–2219.

Flanagan, É. 1993. *Phys. Rev. D*, **48**, 2389–2407.

Gair, J. and Jones, G. 2007. *Class. Quantum Grav.*, **27**, 1145–1168.

Harry, I. W., Allen, B., and Sathyaprakash, B. S. 2009. *Phys. Rev. D*, **80** , 104014.

Helstrom, C. 1968. *Statistical Theory of Signal Detection*. 2nd edn. International Series of Monographs in Electronics and Instrumentation, vol. 9. Oxford; New York: Pergamon Press.

Heng, I., *et al.* 2004. *Class. Quantum Grav.*, **21**, S821–S826.

Hogan, C. and Bender, P. 2001. *Phys. Rev. D*, **64**.

Jaranowski, P. and Królak, A. 2005. *Living Rev. Relativity*, **8**.

Klimenko, S. and Mitselmakher, G. 2004. *Class. Quantum Grav.*, **21**, S1819–S1830.

Krishnan, B., *et al.* 2004. *Phys. Rev. D*, **70**.

Mohanty, S. 2000. *Phys. Rev. D*, **61**.

Nicholson, D., *et al.* 1996. *Phys. Lett. A*, **218**, 175–180.

Owen, B. 1996. *Phys. Rev. D*, **53**, 6749–6761.

Owen, B. and Sathyaprakash, B. 1999. *Phys. Rev. D*, **60**.

Pretorius, F. 2005. *Phys. Rev. Lett.*, **95**.

Pretorius, F. 2009. Binary black hole coalescence. In: Colpi, M., *et al.* (eds), *Physics of Relativistic Objects in Compact Binaries: From Birth to Coalescence.* Astrophysics and Space Science Library, vol. 359. Berlin and New York: Springer.

Robinson, C., Sathyaprakash, B. and Sengupta, A. 2008. *Phys. Rev. D*, **78**.

Sathyaprakash, B. and Dhurandhar, S. 1991. *Phys. Rev. D*, **44**, 3819–3834.

Schutz, B. (ed.) 1989. NATO ASI Series C, vol. 253. Dordrecht and Boston: Kluwer.

Schutz, B. 1991. Data Processing, Analysis and Storage for Interferometric Antennas. Pages 406–452 of: Blair, D. (ed), *The Detection of Gravitational Waves.* Cambridge and New York: Cambridge University Press.

Sintes, A. and Krishnan, B. 2006. *J. Phys.: Conf. Ser.*, **32**, 206–211.

Thorne, K. 1987. Gravitational radiation. Pages 330–458 of: Hawking, S. and Israel, W. (eds), *Three Hundred Years of Gravitation.* Cambridge and New York: Cambridge University Press.

Wen, L. and Gair, J. 2005. *Class. Quantum Grav.*, **22**, S445–S452.

Whelan, J., *et al.* 2003. *Class. Quantum Grav.*, **20**, S689.

Will, C. 1993. *Theory and Experiment in Gravitational Physics.* 2nd edn. Cambridge and New York: Cambridge University Press.

5

Network analysis and multi-messenger astronomy

L. Wen and B. F. Schutz

The data from widely spaced networks of gravitational wave detectors can be combined to act as a single detector with optimum sensitivity and directional resolution. We first provide a basic mathematical framework and characterise the detection capacity and relative performance of various networks. A systematic approach is then provided for the construction of network detection statistics, stable waveform extraction, null-stream construction, and source localisation. At the end, we discuss the angular resolution of an arbitrary detector network and issues relevant to the field of multi-messenger gravitational wave astronomy.

5.1 Introduction

As discussed in the previous chapters, several large-scale interferometric gravitational wave (GW) detectors have reached or exceeded their design sensitivity, and have been coordinating to operate as a global array. These include the LIGO detectors at Louisiana and Washington states of the USA, the Virgo detector in Italy and the GEO 600 detector in Germany. The three US-based interferometric GW detectors LIGO have completed their ground-breaking fifth science run in November 2007. An integrated full year's worth of data has been accumulated with all three interferometers in coincidence. Advanced LIGO will enable a 10-fold improvement in sensitivity, allowing detectors to monitor a volume of the Universe 1000 times larger than can be achieved by current detectors. These advanced detectors are predicted to detect tens to hundreds of events per year (see Abadie *et al.*, 2010). The detection of the first GW signal is virtually assured with Advanced LIGO.

Several new detectors have also been proposed. It has been recognised (e.g. Cavalier *et al.*, 2006; Blair *et al.*, 2008; Wen and Chen, 2010) that a new GW detector in Australia will add the longest baseline to the existing detector network, improve dramatically its ability to pinpoint gravitational wave sources and make multi-wavelength follow-up observations much more likely to succeed. The potential scientific benefits of such multi-wavelength observations are outstanding. In addition, a new 3-km underground cryogenic gravitational wave detector LCGT (Kuroda *et al.*, 1999) in Japan has just received partial funding for construction.

The existence of a large cooperating network of GW detectors greatly enhances our capability to detect gravitational waves and to extract science content from the data. For instance, the confidence level of a potential GW event will be greatly enhanced if seen by a number of detectors. Data from multiple detectors can be used to veto false detections resulting from artifacts. The sky direction of a GW source is mainly determined by measuring the arrival time delays of a GW between pairs of detectors. Also, the validity of Einstein's general theory of relativity can be tested using the network data. For example a network is required to confirm the prediction of general relativity that gravitational waves (GWs) have only two polarisation modes (Will, 2006). The association of a GW event with an electromagnetic event could also verify the speed of GWs to high precision (Sathyaprakash and Schutz, 2009).

5.2 Network analysis

The principle of analysing data from a network of gravitational wave detectors is very similar to that of single detector (Pai *et al.*, 2001; Jaranowski and Królak, 2005). The observed strain of an impinging GW by an interferometric detector is a linear combination of the two wave polarisations (Sathyaprakash and Schutz, 2009, and references therein),

$$h_I(t) = f_I^+ h_+(t) + f_I^\times h_\times(t), \tag{5.1}$$

where the subscript I is a label for the Ith detector, indicating the dependence of the observed quantities on detectors, t is the time label, f^+ and $f^\times$ are the detector's antenna beam pattern functions in response to the plus and cross polarisations $(h_+, h_\times)$ of a GW. These antenna beam pattern functions depend on source sky direction, wave polarisation angle, and detector orientation at time t.

We make the same assumption as in the previous chapter that the noise in all detectors is additive, and data can be written as the linear summation of signal and noise for each detector, at each time t,

$$d_I(t) = h_I(t - \tau_{1I}) + n_I(t) \tag{5.2}$$

where $t = [0, T]$, T is the duration of the data used, τ_{1I} is the wavefront arrival time at the Ith detector relative to the reference point. The arrival time delay can be calculated as $\tau_{1I} = \hat{\mathbf{n}} \cdot \mathbf{r}_{1I}/c$, where $\mathbf{r}_{1I}$ is location vector of the Ith detector relative to the reference detector, c is the speed of light, $\hat{\mathbf{n}}$ is the unit vector along the wave propagation direction. $d_I(f)$ is denoted as the Fourier transform of time-series data from the Ith GW detector

$$d_I(f) = \int_{-\infty}^{+\infty} d_I(t) e^{i2\pi f t} dt, \tag{5.3}$$

where $d_I(t)$ is the detector output (defined as $x(t)$ in Chapter 4). Suppose we have N_d detectors in the network, we can conveniently represent the network data and noise each as a N_d-dimension vector function containing data or noise from each of the detectors. Defining $\mathbf{h}(f)$ as a two-dimensional vector function containing the two polarisations of a GW in the frequency domain, $n_I(f)$ as the Fourier transform of the noise from the Ith detector, we have

$$\mathbf{d}(f) = \begin{pmatrix} d_1(f) \\ d_2(f) \\ \cdots \\ d_{N_d}(f) \end{pmatrix}, \mathbf{h}(f) = \begin{pmatrix} h_+(f) \\ h_\times(f) \end{pmatrix}, \mathbf{n}(f) = \begin{pmatrix} n_1(f) \\ n_2(f) \\ \cdots \\ n_{N_d}(f) \end{pmatrix}. \tag{5.4}$$

The response of each detector depends of course only on the gravitational wave amplitude as projected onto it, given in equation (5.1). This h_I is what is used to compute expected signal-to-noiser ratios (SNRs), as in equations (5.6) and (5.8) below.

The corresponding one-sided noise spectral density matrix for the network $S(f)$ is defined similar to the one-detector case (section 5.4)

$$\langle \mathbf{n}^\dagger(f)\mathbf{n}(f) \rangle = \frac{1}{2}\delta(f - f')S(f). \tag{5.5}$$

The bracket $\langle \cdots \rangle$ here represents the ensemble average. When the noise in various detectors is uncorrelated, $S(f)$ is a diagonalised matrix. This is the case when the detectors in the network are widely separated, e.g. in the case of the two LIGO detectors in Washington state and in Louisiana state.

Following the usual assumption that the noise is stationary and Gaussian-distributed, the logarithmic likelihood function can be written as,

$$\log \Lambda = (\mathbf{d}|\mathbf{h}) - \frac{1}{2}(\mathbf{h}|\mathbf{h}). \tag{5.6}$$

where we define the following inner product between two vectors as section 5.4

$$(\mathbf{a}|\mathbf{b}) = 2 \sum_I \int_{-\infty}^{+\infty} df \frac{a_I^*(f)b_I(f)}{S_I(f)}. \tag{5.7}$$

The square of the optimal signal-to-noise ratio (SNR^2), based on the maximum likelihood ratio principle for the detector network, is (Finn, 2001):

$$\rho_N^2 = (\mathbf{h}|\mathbf{h}). \tag{5.8}$$

While this optimal SNR can be obtained using an extensive search over the entire parameter space, similar to the single-detector case, we actually only need to search for the 'intrinsic' parameters such as the masses. The 'extrinsic' parameters such as sky direction, inclination angles in the case of binary source, and constant initial (or coalescing phase) term can be optimised analytically (Cutler and Flanagan, 1994; Pai *et al.*, 2001; Jaranowski and Królak, 2005) without numerically searching in the parameter space.

Network antenna patterns

When comparing properties of various network configurations, it is helpful to look at the *average* response of a network to an ensemble of hypothetical sources, distributed uniformly and randomly in space, all identical to one another except for orientations. A general term in the inner product in equation (5.8) can be written, from equation (5.1), explicitly in the form

$$|h_I(t)|^2 = [f_I^+ h_+(t) + f_I^\times h_\times(t)]^2.$$

If the polarisation of the incoming wave has a random orientation, then when we average this over the ensemble of sources we get the simpler

$$< |h_I(t)|^2 > = \left(|f_I^+|^2 + |f_I^\times|^2 \right) h^2(t), \tag{5.9}$$

where $h(t) = \sqrt{(h_+^2(t) + h_\times^2(t))/2}$ is called the full amplitude of the gravitational wave (Schutz, 2011). The function

$$P_I = |f_I^+|^2 + |f_I^\times|^2 \tag{5.10}$$

is a function of the position of the source on the sky relative to the detector, and is called the *antenna pattern* of the detector. The average network SNR can now be written, using equation (5.7), as

$$< \rho_N^2 > = 2 \sum_I P_I \int_{-\infty}^{+\infty} df \frac{|h(f)|^2}{S_I(f)}. \tag{5.11}$$

If we take detector number 1 (arbitrarily chosen) as a reference for sensitivity, and then define the *sensitivity ratios* of the detectors for the given waveform

$$R_I = \frac{\int_{-\infty}^{+\infty} df \frac{|h(f)|^2}{S_I(f)}}{\int_{-\infty}^{+\infty} df \frac{|h(f)|^2}{S_1(f)}}, \tag{5.12}$$

then it is possible to separate the response into two factors, one to do with the network and the other with the response of a single detector:

$$< \rho_N^2 > = 2 \left(\sum_I R_I P_I \right) \int_{-\infty}^{+\infty} df \frac{|h(f)|^2}{S_1(f)}. \tag{5.13}$$

In other words, the mean response of the network to the ensemble of sources depends on the response of a typical detector multiplied by the *network antenna pattern* P_N:

$$P_N = \sum_I R_I P_I. \tag{5.14}$$

There is a somewhat arbitrary scale in P_N that depends on which detector was chosen as the reference. But in comparing the performance of one network with another, this scale drops out, provided the same reference sensitivity is used for both networks. Another caveat is that the sensitivity ratio R_I depends on the waveform $h(f)$ (but not on its amplitude). But if the network's detectors have roughly the same shape noise function $S(f)$ and differ mainly in its overall level, say because one detector is bigger than another, then the sensitivity ratio will be independent of waveform. Examples of the antenna patterns of two networks are given in Figure 5.1.

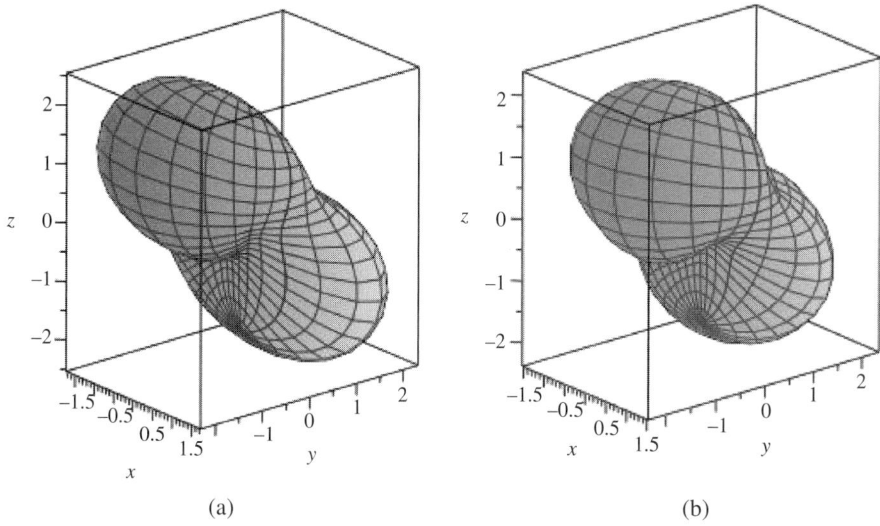

(a)   (b)

Figure 5.1 (a) Antenna pattern of the network of planned Advanced detectors: two LIGO interferometers at Hanford, one LIGO at Livingston, and Virgo in Pisa. (b) Same as (a) but with an Australian detector added to the LIGO–Virgo network in place of one of the detectors at Hanford. All detectors are assumed identical. The z-axis is aligned with the rotation axis of the Earth and the x-axis with the Greenwich meridian. The viewpoint is at longitude 40°W, latitude 20°N. The ratio of the volume of the network without Australia to the one with Australia is 1.006: the two networks cover essentially the same volume of space at a given sensitivity. The shapes are similar as well: the Australian site is nearly diametrically opposite the sites in the USA, so the individual detector antenna patterns overlap considerably.

Detection volumes and the distribution function for detected SNR

Since gravitational wave amplitudes fall off as $1/r$, setting a detection threshold on the network SNR immediately defines a volume of space, within which (on average) a source with a given intrinsic strength will be detected. If we pick a fiducial distance r_0, at which the source will have amplitude $h_0(f)$, then a source at an arbitrary distance r will have amplitude $h(f) = h_0(f)(r_0/r)$. If the threshold on ρ^2 is ρ_T^2, then the outer edge of the detection volume is at

$$r = \frac{r_0}{\rho_T} \left[2 P_N \int_{-\infty}^{+\infty} df \frac{|h_0(f)|^2}{S_1(f)} \right]^{1/2}. \tag{5.15}$$

This *detection limit* r depends on angular position on the sky only through the network antenna pattern P_N, which therefore maps onto a region of space known as the detection volume. For convenience we now assume that a source waveform $h(t)$ has been fixed, and we represent all the fixed values by a single constant κ:

$$\kappa^2 = 2 r_0^2 \int_{-\infty}^{+\infty} df \frac{|h_0(f)|^2}{S_1(f)} \tag{5.16}$$

so that the detection limit r can be written as

$$r = \kappa (P_N)^{1/2}/\rho_T. \tag{5.17}$$

Notice that surfaces of constant mean SNR ρ^2 are defined by this equation, simply by varying the threshold. As ρ^2 increases, these surfaces foliate the detection volume with surfaces of similar shape and ever decreasing size.

Since the sources are assumed uniformly distributed in space, the number of events detected is proportional to the detection volume. From this it is simple to deduce the *median SNR* of a detected population. The median SNR is the SNR that gives a detection limit r that encloses a volume just half of the full detection volume. Since volume scales as r^3, this corresponds to an SNR such that

$$\rho^2_{\text{median}} = 2^{2/3}\rho^2_T \approx 1.6\rho^2_T. \tag{5.18}$$

This is a remarkably simple result: the median SNR^2 of any detected population is always 1.6 times the detection threshold, regardless of the nature of the network. When detections are rare (as now), the expected power SNR^2 of the *first detection* is 1.6 times the threshold. We will always be working very close to the threshold!

In fact, this argument leads quickly to the *universal distribution* of SNR in any detected population, independently of the detector configuration, depending only on the network threshold ρ_T (Schutz, 2011). To show this, we compute first the detection volume explicitly. We show the dependence of the detection limit r on the polar angles (θ,ϕ) on the sky by including explicitly the angular arguments in the network antenna pattern:

$$r(\theta, \phi) = \kappa [P_N(\theta, \phi)]^{1/2}/\rho_T. \tag{5.19}$$

The furthest reach of the network is the maximum value of r over all angles,

$$r_{\max} = \max_{\theta,\phi}[r(\theta, \phi)]. \tag{5.20}$$

The volume defined by ρ_T is then

$$V = \int d\Omega \int_0^{r_{\max}} r^2 dr = \frac{1}{3}\int d\Omega\, r^3_{\max}(\theta, \phi). \tag{5.21}$$

The number of detections with SNR larger than any ρ is proportional to the detection volume with ρ_T set equal to this ρ. This scales as ρ^{-3}. This is a cumulative distribution. It is straightforward from this to show that the *universal probability density function for the distribution of detected SNR values* is

$$p(\rho) = 3\rho^3_T\rho^{-4}, \quad \rho > \rho_T \tag{5.22}$$

$$= 0, \qquad\qquad \rho < \rho_T.$$

From this universal distribution one can deduce any of the moments one wishes. For example, the mean expected amplitude SNR is $1.5\rho_T$. The mean expected power SNR^2 is $3\rho^2_T$. Both of these means are for sources stronger than the median one. That is because the universal distribution has a peak at the lowest values (at threshold) and has a long tail of strong but rare events.

Figures of merit of networks

In later sections we will make detailed comparisons of the performance of the two networks considered in Figure 5.1, especially of their angular measurement accuracies. But there are many more networks that are being talked about, with the likelihood of a large detector in Japan and the possibility of one in India. Moreover, when third-generation instruments are built, it will be important to understand how they work with other existing instruments. Therefore it is useful to have some simple numbers to characterise the relative performance of different networks. These are called *figures of merit.*

The first figure of merit in this field was introduced by Searle *et al.* (2002). This was designed to measure the detection rate of the network. Although it is not described in detail, it was computed assuming the sources were inspiralling compact binaries and all the detectors in the network were identical. This figure of merit was described in more detail in Searle *et al.* (2006), where it is computed by Monte-Carlo methods.

One of us (Schutz, 2011) has recently revisited the detection rate and defined some new figures of merit. Previous observation runs by the the LIGO Scientific Collaboration (LSC) and Virgo have shown that the duty cycle of detectors has a significant effect. Typical duty cycles of the big detectors were around 80%, which is remarkably good when one considers the complexity of the instruments, but which means that the three-detector coincidence time among Hanford, Livingston and Virgo was only about 50%. This complicates the detection rate computation when one goes to a four-detector network, because an event can be detected even when one detector is not operating. So the overall detection volume, or some such measure, is not enough to provide a realistic detection rate. One wants the detection volumes of all three-detector sub-networks, and one wants to weight these with the probability that they will be operating when the fourth detector is not. This figure of merit leads to the conclusion that, with duty cycles of 80%, if an Asian detector is added to the planned network of Advanced Detectors (LIGO and Virgo) then the detection rate roughly doubles (Schutz, 2011). This is not the case, however, with the addition of an Australian detector. There the detection rate hardly changes, although, as we show below, the angular location of sources on the sky improves dramatically.

A further figure of merit is the isotropy of a network. Antenna patterns differ greatly in their uniformity of coverage of the sky. The two shown in Figure 5.1 have very similar shapes and are rather anisotropic. If the LCGT detector in Japan is added to the standard Advanced Detector configuration of LIGO and Virgo, the pattern is much more isotropic. This might be a desirable network property if, for example, coincidences with relatively nearby astronomical events are important, or if the source population is itself anisotropic (say, neutron stars in the Milky Way). A final figure of merit was designed to measure, in an simple way, the angular position accuracy of detections, based on the formalism of Wen and Chen (2010) described later in section 5.4.

5.3 General approach for discretised data

This section gives a recipe for separating data containing signal from the so-called 'null streams' – data containing no signals – and identifying the network sensitivity at a given sky direction. A detailed explanation can be found in Wen (2008). For a given source direction, the whitened and time-delay corrected data from a network of N_d GW detectors can be written in the frequency domain, using discretised Fourier transform, as

$$\hat{\mathbf{d}}_{\mathbf{k}} = \hat{A}_k \mathbf{h}_{\mathbf{k}} + \hat{\mathbf{n}}_{\mathbf{k}}, \quad (k = 1, ..., N_k) \tag{5.23}$$

where, k is the label for frequency bins, $\hat{A}_k$ is the response matrix of the detector network at each frequency,

$$\hat{\mathbf{d}}_k = \begin{pmatrix} d_{1k}/\sigma_{1k} \\ d_{2k}/\sigma_{2k} \\ ... \\ d_{N_dk}/\sigma_{N_dk} \end{pmatrix}, \mathbf{h}_k = \begin{pmatrix} h_{+k} \\ h_{\times k} \end{pmatrix}, \hat{\mathbf{n}}_{\mathbf{k}} = \begin{pmatrix} n_{1k}/\sigma_{1k} \\ n_{2k}/\sigma_{2k} \\ ... \\ n_{N_dk}/\sigma_{N_dk} \end{pmatrix}, \tag{5.24}$$

and

$$\hat{A}_k = \begin{pmatrix} f_1^+/\sigma_{1k} & f_1^\times/\sigma_{1k} \\ f_2^+/\sigma_{2k} & f_2^\times/\sigma_{2k} \\ ... & ... \\ f_{N_d}^+/\sigma_{N_dk} & f_{N_d}^\times/\sigma_{N_dk} \end{pmatrix}, \tag{5.25}$$

where σ_{ik}^2 is the noise variance of the ith detector at the kth frequency bin. The whitened response of the detector network to gravitational waves in all frequencies can be further written in terms of vectors and matrices as,

$$\hat{\mathbf{d}} = \hat{A}\mathbf{h} + \hat{\mathbf{n}}, \tag{5.26}$$

where,

$$\hat{\mathbf{d}} = \begin{pmatrix} \hat{\mathbf{d}}_1 \\ \hat{\mathbf{d}}_2 \\ ... \\ ... \\ \hat{\mathbf{d}}_{N_k} \end{pmatrix}, \mathbf{h} = \begin{pmatrix} \mathbf{h}_1 \\ \mathbf{h}_2 \\ ... \\ ... \\ \mathbf{h}_{N_k} \end{pmatrix}, \hat{\mathbf{n}} = \begin{pmatrix} \hat{\mathbf{n}}_1 \\ \hat{\mathbf{n}}_2 \\ ... \\ ... \\ \hat{\mathbf{n}}_{N_k} \end{pmatrix}, \hat{A} = \begin{pmatrix} \hat{A}_1 & & & \\ & \hat{A}_2 & & \\ & & ... & \\ & & & ... \\ & & & & \hat{A}_{N_k} \end{pmatrix}. \tag{5.27}$$

The matrix $\hat{A}$ of dimension $N_d N_k \times 2N_k$ is therefore the response matrix of the detector network for all frequencies.

Systematic approach

The singular value decomposition of the network response matrix $\hat{A}$ can be written as,

$$\hat{A} = u s v^{\dagger}, \quad s = \begin{pmatrix} s_1 & 0 & \dots & 0 \\ 0 & s_2 & \dots & 0 \\ 0 & \dots & \dots & \dots \\ 0 & \dots & \dots & s_{2N_k} \\ 0 & 0 & \dots & 0 \\ \dots & \dots & \dots & \dots \\ 0 & 0 & \dots & 0 \end{pmatrix}, \tag{5.28}$$

where singular values are ranked with $s_1 \geq s_2 \geq \dots \geq s_{2N_k} \geq 0$. u and v are unitary matrices of dimensions $N_d N_k \times N_d N_k$ and $2N_k \times 2N_k$ respectively with $u^{\dagger}u = I, v^{\dagger}v = I$. Note that s_i^2 are eigenvalues of $\hat{A}^T \hat{A}$. Note also that the singular values of $\hat{A}$ are only ranked rearrangement of singular values from all $\hat{A}_k$.

New data streams and new signal polarisations can be constructed by linearly recombining data from different detectors using the unitary matrices resulting from the singular value decomposition of the network response matrix $\hat{A}$,

$$\hat{\mathbf{d}}' = u^{\dagger}\hat{\mathbf{d}}, \quad \mathbf{h}' = v^{\dagger}\mathbf{h}, \quad \hat{\mathbf{n}}' = u^{\dagger}\hat{\mathbf{n}}. \tag{5.29}$$

These newly constructed data streams and new signal polarisations are one-to-one related to each other via the pseudo-diagonal matrix containing singular values,

$$\begin{pmatrix} \hat{d}'_1 \\ \hat{d}'_2 \\ \dots \\ \dots \\ \dots \\ \dots \\ \hat{d}'_{N_k N_d} \end{pmatrix} = \begin{pmatrix} s_1 & 0 & \dots & 0 \\ 0 & s_2 & \dots & 0 \\ 0 & \dots & \dots & 0 \\ 0 & \dots & \dots & s_{2N_k} \\ 0 & \dots & 0 & 0 \\ \dots & \dots & \dots & \dots \\ 0 & \dots & \dots & 0 \end{pmatrix} \begin{pmatrix} h'_1 \\ h'_2 \\ \dots \\ h'_{2N_k} \end{pmatrix} + \begin{pmatrix} \hat{n}'_1 \\ \hat{n}'_2 \\ \dots \\ \dots \\ \dots \\ \dots \\ \hat{n}'_{N_k N_d} \end{pmatrix}. \tag{5.30}$$

At the presence of a GW signal, there are at most two data streams at each frequency corresponding to non-zero singular values. If the data contain pure stationary Gaussian noise only, the new data streams also follow Gaussian distribution and are statistically uncorrelated with each other. Detection algorithms can then be applied to these 'signal' data streams instead of all data.

The singular values of the response matrix $\hat{A}$ of the detector network are directly related to the statistical uncertainties in estimating waveforms from the data and can be used to derive the stable solutions to the waveforms (section 5.3.3). As shown in equation (5.30), non-zero singular values indicate the network's sensitivity at the given direction to each of the signal streams at each frequency bin. Data streams corresponding to zero singular values have null response to signals (equation (5.30)). There are at least $(N_d - 2)N_k$ 'null streams' due to the fact that there are only two polarisations for a plane GW based on Einstein's theory of general relativity. Null streams can therefore be used to test the consistency of the detected GW events and to veto against signal-like noise (Wen and Schutz, 2005; Chatterji *et al.*, 2006; Klimenko *et al.*, 2008; Sutton *et al.*, 2010). It is worth mentioning that, in addition to the null-stream-based method, another vital network veto technique using the r-statistic test for the LIGO data analysis can be found in Cadonati (2004).

Detection statistic

Detection statistics for GWs from known waveforms can be found in the literature (Cutler and Flanagan, 1994; Pai *et al.*, 2001; Jaranowski and Królak, 2005). Construction of detection statistic for GWs of *unknown waveform* using data from a network of detectors can also be found in a number of references (Flanagan and Hughes, 1998; Anderson *et al.*, 2001; Sylvestre, 2003a; Klimenko *et al.*, 2005, 2006; Mohanty *et al.*, 2006; Rakhmanov, 2006; Searle *et al.*, 2008; Wen, 2008; Klimenko *et al.*, 2008; Searle *et al.*, 2009; Sutton *et al.*, 2010). There are in general two possible approaches: (1) a direct application of maximum likelihood ratio (MLR) principle by constructing detection statistic from adding powers of all data with non-zero singular values, and (2) optimised detection statistic based on MLR together with assumptions on wave spectrum and principle of maximum SNR or similar.

Early work on optimisation of the detection statistic based on directional sensitivity of GW detectors can be found for the future space detector LISA (Nayak *et al.*, 2003) and for ground-based detectors (Klimenko *et al.*, 2005, 2006). It has been demonstrated in these papers that further optimisation from the standard MLR detection statistic can be obtained by constraining or eliminating contributions of data corresponding to weaker network sensitivity. Other work can be found in Mohanty *et al.* (2006) which is based on the SNR variability regulator, and in Wen (2008), upon which this chapter is based.

The standard detection statistic based on the MLR principle can be easily reproduced from the newly constructed data. Various optimisation strategies can then be obtained based on general assumptions of the signals, together with the singular values which represent the sensitivity of the detector network. This 'standard' solution is based on a direct application of the MLR principle for unknown waveforms. When the signal polarisations h_{+k} and $h_{\times k}$ are assumed to be independent variables for all frequencies, the components of the new signal $\mathbf{h}'$ are also independent of one another. The standard MLR detection statistic can then be constructed by adding equally all powers of 'signal' data-streams of non-zero singular values,

$$P_S^{(0)} = \sum_{i=1}^{N_p} |\hat{d}_i'|^2, \tag{5.31}$$

where N_p is the number of non-zero singular values. Note that singular values have already been ranked (equation (5.28)). In the absence of noise, $|\hat{d}_i'|^2 = s_i^2 |h'|_i^2$. In the presence of pure stationary Gaussian noise, $2|\hat{d}_i'|^2$ follows χ_2^2 distribution with variance of four. By discarding data of null response to GW signals, this detection statistic is optimal in SNRs rather than simply adding powers of all (noise-weighted) data together. However, in the presence of extremely small singular values (e.g. when two detectors are nearly aligned) or extremely weak signals, this detection statistic does not necessarily maximise the SNR.

Singular values can be used to further optimise the detection statistic by making reasonable assumptions about the signal spectrum. For instance, we can omit signal powers of small singular values with hopes that the chance is very small for the new signal component $\mathbf{h}'$ to be very strong at frequencies and polarisations where the network is not sensitive. Under this assumption, the detection statistic can be,

$$P_S^{(1)} = \sum_{i=1}^{N_p'} |\hat{\mathbf{d}}_i'|^2. \tag{5.32}$$

That is, only $N_p' \leq N_p$ number of powers of reasonable large singular values are included. For instance, if we assume that the detection statistic is a linear combination of individual powers and that the signal power $|h_i'|^2$ is the same in both frequency and new polarisations, maximisation of the SNR leads to

$$P_S^{(2)} = \sum_{i=1}^{N_p'} \alpha_i |(\hat{\mathbf{d}}')_i|^2, \tag{5.33}$$

where

$$\alpha_i = \frac{s_i^2}{\sqrt{\sum_j s_i^4}}. \tag{5.34}$$

Stable solutions to waveform extraction

The 'standard' estimator for $\mathbf{h}' = [h_1', ..., h_{N_p}']^T$ according to the maximum likelihood ratio principle of 'burst' GWs (defined as any GWs of unknown waveforms) is

$$h_i' = s_i^{-1}\hat{d}_i', \quad i = 1, ..., N_p, \tag{5.35}$$

where N_p is the number of non-zero singular values. Note the components of h' include two polarisation components for each frequency. The 'standard' solutions to wave polarisations can be extracted using the singular value decomposition components of the network response (equations (5.28) and (5.30))

$$\mathbf{h} = v\mathbf{h}'. \tag{5.36}$$

Note that the solutions of two signal polarisation components at each frequency depend on quantities within that frequency only. Faster calculations can therefore be carried out independently at each frequency. In the limit when $N_p = 2N_k$, the solution given in equation (5.36) is equivalent to that written with the Moore–Penrose inverse, $\mathbf{h} = (\hat{A}^\dagger \hat{A})^{-1}\hat{A}^\dagger \hat{\mathbf{d}}$. This is the same result as the effective one-detector strain for data from a network of GW detectors introduced by Flanagan and Hughes (1998).

The "standard" solution $\mathbf{h}$ from equation (5.36), however, can possibly lead to unstable solutions in the sense that a small error in the data can lead to large amplified error in the solution. The fisher information matrix indicates that the best possible statistical variance for the unbiasedly estimated h_k is a linear combination of $1/s_i^2$. The extracted wave polarisations h_k can contain large errors if we include data corresponding to very small singular values.

This situation occurs when the response matrix $\hat{A}$ is 'ill-conditioned' with $s_i/s_1 \ll 1$. The small singular values can result from machine truncation errors instead of zero values or from nearly degenerated solutions to the equations, e.g. in our case, when antenna beam patterns of two detectors are nearly aligned.

Regularisation is needed in order to have stable solutions to $\mathbf{h}_k$. One simple solution is to apply the 'truncated singular value decomposition' (TSVD) method by omitting data with very small singular values in equation (5.36). Thus

$$h_i^T = \sum_{j=1}^{N_T} v_{ij} h'_j, \tag{5.37}$$

where $N_T \leq N_p$ is the number of data included. The main problem is then the decision on where to start to truncate the data streams with small singular values. It depends on the accuracy requirement in waveform extraction, the type of waveforms and the type of constraints that can be put on the solutions. Truncation of terms with very small singular values can retain the least-square fit of the detector response to the data while greatly reducing the statistical errors when extracting individual signal polarisations. Recent work in this area can be found in several literatures (Klimenko *et al.*, 2005, 2006; Rakhmanov, 2006; Summerscales *et al.*, 2008; Searle *et al.*, 2008; Wen, 2008; Searle *et al.*, 2009).

Null stream construction

There are at least $(N_d - 2)N_k$ data streams with zero singular values or zero multiplication factors. These are data streams that have null response to signals. A general three-detector null stream construction can be written in the frequency domain as a linear combination of the time-delayed data from different detectors (Gürsel and Tinto, 1989; Wen and Schutz, 2005),

$$N(t) = \alpha_{23}d_1(t) + \alpha_{31}d_2(t + \tau_{12}) + \alpha_{12}d_3(t + \tau_{13}), \tag{5.38}$$

where $\alpha_{IJ} = f_I^+ f_J^\times - f_J^+ f_I^\times$ depends on the sky directions but is independent of the wave polarisation angle; it also depends on time but is conveniently assumed to be constant within the duration of the short signal. Construction of null streams for a four-detector network can be found in Tinto (1997).

The generalised null streams can be written in terms of the singular value decomposition of $\hat{A}$ (equation (5.28)) as,

$$N_i^0 = (u^\dagger \hat{\mathbf{d}})_{i+N_p}, \quad i = 1, 2, ..., N_d N_k - N_p. \tag{5.39}$$

For stationary Gaussian noise, N_i^0 follows Gaussian distribution of zero mean and unity variance and are statistically independent with each other. A detection of a true GW event therefore requires that all the null streams are consistent with 'noise-only' at the source direction. The technique of using null streams for source localisation can be found, for example, in Gürsel and Tinto (1989) and in Wen and Schutz (2005). Using null-streams as a tool for consistency check of GW events against signal-like glitches for ground-based detectors was first proposed by Wen and Schutz (2005). Further investigations on consistency check using the null streams and its relation to the null space of the network response matrix $\hat{A}$ can be found in Chatterji *et al.* (2006), Klimenko *et al.* (2008) and Sutton *et al.* (2010).

There are also 'semi-' null streams corresponding to data streams with very small but non-zero singular values,

$$N_{Wi} = (u^{\dagger}\hat{\mathbf{d}})_{i+N'_p-1}, \quad i = 1, 2, ..., N_p - N'_p, \tag{5.40}$$

where N'_p is the starting index for $1 \gg s_i > 0$. These semi-null streams exist when the equations are nearly degenerated. This happens, for example, for response of the two LIGO detectors of H1 (H2) and L1 that are designed to be nearly aligned. The semi-null stream can be also simply caused by combination of weak directional sensitivity and/or high noise level instrument of all detectors in the network. Consistency check of GW events, veto against noise, and localisation of the source can be further improved by including both the 'standard' null streams of analytically zero singular values and also these semi-null streams.

$$P_N^1 = \sum_{i=1}^{N_d N_k - N_p} |N_i^0|^2 + \sum_{i=1}^{N_p - N'_p} |N_{Wi}|^2. \tag{5.41}$$

The usage of semi-null streams depends on waveforms and therefore efficiency studies should be carried out beforehand. An early proposal of using the semi-null stream for veto against detector glitches and for source localisation can be found in Wen and Schutz (2005) for the two-detector network of H1–L1.

Localisation methods

The localisation of a GW source based on arrival time delay information requires data from at least three detectors at different sites. The principle is similar to the triangulation method used in astronomy where the arrival time delay between detectors is used to locate the source. Basically, the arrival time delay of a plane wave between two well-separated detectors give a ring in sky directions. For three well-separated detectors, the intersection of the two such rings generated from two pairs of detectors yields two degenerated source directions. In the case of the gravitational wave detector network, additional information can be obtained from amplitude information resulting from the different orientation of the antenna beam patterns for different detectors (see also discussions in section 5.4). This can be used to break the degeneracy in source localisation for a three-detector network.

For gravitational waves with known waveform, e.g. those from compact binaries of compact objects, a coherent localisation method that utilises the MLR principle which treats the angular direction as search parameters can be deduced from Pai *et al.* (2001). Such method can be time-consuming due to the large sky-grid one has to search. Less optimal but much faster methods for post-processing use the detected merger times and their delays between detectors, and the effect of arrival timing uncertainties, as well as amplitude information of GWs have been investigated (Cavalier *et al.*, 2006; Beauville *et al.*, 2006; Acernese *et al.*, 2007; Markowitz *et al.*, 2008; Fairhurst, 2009).

To localise gravitational waves of unknown waveforms, the most common methods are based on the usage of the null stream described above in this section 5.3 (Gürsel and Tinto,

1989; Tinto, 1997; Wen and Schutz, 2005; Wen, 2008; Wen *et al.*, 2008; Searle *et al.*, 2009; Wen and Chen, 2010). The idea is that at the correct source direction, the null streams should be truly null. All method suffers interference problems. A further optimisation of the null-based localisation method can be achieved by regularisation that discards data corresponding to weak network responses (Wen, 2008; Wen *et al.*, 2008; Searle *et al.*, 2009; Wen and Chen, 2010).

5.4 Angular resolution of a detector network

This section describes the intrinsic limit of the angular resolution for an arbitrary network of gravitational wave detectors. The angular resolution of an individual GW detector, arising from its antenna beam pattern, is rather poor (Thorne, 1987). However, the large baselines of the current GW-detector network facilitate much better angular resolution via triangulation.

Two approximate analytical expressions for the angular resolution can be found in literature (summarised in Sylvestre, 2003b) for a network of three gravitational wave detectors. One is an elegant approximate geometrical formula for three detectors due to Thorne (as cited in equation (8.3) of Gürsel and Tinto, 1989),

$$\Delta\Omega = \frac{2c^2 \Delta\tau_{12}\Delta\tau_{13}}{A\cos\theta}, \tag{5.42}$$

where c is the speed of light, $\Delta\tau_{12}$ and $\Delta\tau_{13}$ are the time-of-arrival accuracies between pairs of detectors, A is the area formed by the three detectors, and θ is the angle between the source direction and the normal to the plane of the three detectors. Derivations for this expression, or its underlying assumptions, however, are not available in the literature. The dependence of angular resolution on the SNR was explicated by Gürsel and Tinto (1989), by expressing the above time-of-arrival accuracy as a function of SNR and frequency (Lyne, 1989) (derived from the Fisher matrix assuming all other information of the waveform to be perfectly known). The other formula is based on the numerical result of the angular resolution of the LIGO–Virgo network around a wave incident direction normal to the plane formed by the three detector sites (Sylvestre, 2003b) — for GWs emitted from neutron star–neutron star (NS–NS) inspirals (Pai *et al.*, 2001). This particular resolution was then rescaled by the cosine of the wave incidence angle and SNR (Sylvestre, 2003b).

A more rigorous quantitative study of the angular resolution of a network of gravitational wave detectors can be found in Wen and Chen (2010). In this paper, general geometrical expressions are given for the angular resolution of an arbitrary network of interferometric GW detectors when the arrival time of a GW is unknown. The formulae show explicitly the dependence of the angular resolution on areas of quadrangles formed by projections of pairs of detectors and how they are weighted by sensitivities of individual detectors. Numerical simulations also demonstrate the capabilities of the current GW detector network. The known fact that the angular resolution is poor along the plane formed by current LIGO–Virgo detectors is confirmed. Specifically, we first follow the convention and define the solid angle uncertainties in a polar coordinate system with colatitude θ and longitude ϕ as $\Delta\Omega_s = 2\pi|\cos\theta|\sqrt{\langle\Delta\theta^2\rangle\langle\Delta\phi^2\rangle - \langle\Delta\theta\Delta\phi\rangle^2}$. For an arbitrary number of gravitational wave detectors, for short gravitational waves where the detectors' motion with respect to the Earth is negligible, is

$$\Delta\Omega_s^{(\text{short})} \propto \frac{1}{\sqrt{\sum_{J,K,L,M} \Delta_{JK}\Delta_{LM}|(\mathbf{r}_{KJ} \times \mathbf{r}_{ML}) \cdot \mathbf{n}|^2}}, \tag{5.43}$$

where indices J, K, L,M run from one to the number of detectors, $\mathbf{r}_{JK}$ is defined as the positional vector pointing from the Ith to the Jth detector. Note that $|(\mathbf{r}_{KJ} \times \mathbf{r}_{ML}) \cdot \mathbf{n}|$ is just twice the area formed by the projections of the detectors J, K, L and M onto the plane orthogonal to the wave propagation direction, Δ_{IJ} is defined in Wen *et al.* (2008) and Wen and Chen (2010). For the *worst-case* scenario where we know nothing about the signal waveform, Δ_{IJ} is proportional to the correlation of noise-weighted energy flux between detectors projected onto the null space of the signal. For the *best-case* scenario where we know exactly the signal waveform, Δ_{IJ} is proportional to the multiplication of the total noise-weighted energy flux coupled to the individual detectors I and J.

For the special case of a monochromatic GW at a frequency f, for a two-detector network,

$$\Delta\theta_s^{(2f,\text{best})} = \frac{1}{\rho_N} \frac{c}{f D_\perp} \frac{1}{\pi} \sqrt{\frac{1}{\rho_1^2\rho_2^2/\rho_N^4}}. \tag{5.44}$$

where ρ_I^2 is the optimal SNR-squared for detector I, ρ_N^2 is the optimal network SNR-squared (equation (5.8)) and $D_\perp$ is the projected baseline on the plane normal to the wave direction. For a three-detector network,

$$\Delta\Omega_s^{(3f,\text{best})} = \frac{1}{\rho_N^2} \frac{c^2}{f^2 A_\perp} \frac{1}{4\pi} \sqrt{\frac{1}{\rho_1^2\rho_2^2\rho_3^2/\rho_N^6}}, \tag{5.45}$$

where $A_\perp$ is the projected area formed by the three detector sites on a plane normal to the wave direction.

Equation (5.45) includes a feature indicating that for a given network SNR, the angular resolution is limited by the less sensitive detector. When one detector has null response to the wave (e.g. $\rho_1 \sim 0$), we have $\Delta\Omega \gg 1$, until limited by contributions from the directional derivatives of the antenna beam pattern functions that were ignored in the derivation. This is expected, as if only two detectors in the network are sensitive to the signal, only one dimension of the source direction can be resolved.

For the special case of a monochromatic gravitational wave of the frequency f in a three-detector network, at the 95% confidence level,

$$\Delta\Omega_{0.95}^{(\text{short},3f)} \approx 20 \text{ deg}^2 \left(\frac{150\,\text{Hz}}{f} \frac{10}{\rho_N}\right)^2 \frac{10^{17}\text{cm}^2}{A_N} \frac{1/27}{\rho_1^2\rho_2^2\rho_3^2/\rho_N^6} \frac{\sqrt{2}/2}{|\sin i_n|}, \tag{5.46}$$

where A_N is the triangular area formed by the three detector sites, and i_n is the angle between the wave direction and the plane formed by the three detectors. It is clear from this equation that sources along the three-detector plane cannot be localised properly (Figure 5.2). In the above equation, we adopted $A_N = 10^{17}\text{cm}^2$ for the LIGO–Virgo network. We also chose the frequency $f = 150$ Hz as, for advanced detectors, gravitational waves from the coalescing NS–NS binaries contribute most of its noise-weighted energy flux around this frequency.

The results from Wen and Chen (2010) further emphasise the fact that wide-field electromagnetic (EM) cameras with reasonable sensitivity are advantageous for follow-up observations of gravitational wave sources detected by the LIGO–Virgo network, consistent with previous knowledge (e.g. Wen *et al.*, 2007; Nuttall and Sutton, 2010). Within a typical error ellipse around 20 deg^2 for SNR $= 10$, if improved localisation (e.g. by X-ray counterpart detection) is not available, careful strategies are needed to exclude confusion noise and to identify host galaxies.

Similar results have been found for long waves for which the trajectory traced by the detectors during the signal determines the effective size of the detector network. As a simple example, we consider the situation where the detector network makes a circular motion with radius R_* and angular frequency ω_*. In the situation where the network trajectory has mapped out a size much larger than the size of the network, for a short observation of $\omega_* T \ll 1$, approximately (Wen and Chen, 2010),

$$\Delta\Omega_s^{(\text{cont},S)} = \frac{12\sqrt{15}c^2}{f^2 \rho_T^2 \pi R_*^2 (\omega_* T)^3 |\sin i_n|}, \tag{5.47}$$

where i_n is the angle between the source direction and plane of the circular motion. The term $|\sin i_n|$ accounts for the projected area formed by the detector network's trajectory, ρ_T is the optimal SNR within the observation time T. In this case, the error area decreases as a function of T^{-4}, where T^{-1} comes from the increment of SNRs with time and $(\omega_* T)^{-3}$ from the increment of the area formed by the trajectory of the entire detector network (see also Lyne, 1989; Schutz, 1991; Brady *et al.*, 1998; Prix, 2007, for arguments on T^{-4} or T^{-5} dependence). For longer observations of $\omega_* T \gg 1$, contributions from the area saturate as the maximum area is $\sim \pi R_*^2$,

$$\Delta\Omega_s^{(\text{cont},L)} = \frac{c^2}{f^2 \rho_T^2 \pi R_*^2 |\sin i_n|}. \tag{5.48}$$

That is, the error area decreases as a function of T^{-1} as a result of the increment of total SNR-squared. Multi-detectors can be treated as the one-detector case, since the area formed among multi-detectors is much less than the area formed by the detector's trajectory over the typical observation time of much longer than minutes and therefore contributes mainly to increment of SNRs.

Benefit of a larger detector network

We illustrate in Figure 5.2 the benefits of adding an Australian gravitational wave detector in improving the localisation capability of the detector network. Cumulative distributions using data from Figure 5.2 show that an Australian detector added to the LIGO–Virgo network, the medium values of error solid angles will be reduced by a factor of 6 for the *best-case* scenario and a factor of 16 for the *worst-case* scenario (Wen and Chen, 2010). The improvement in the *worst-case* scenario is more prominent than that in the *best-case*

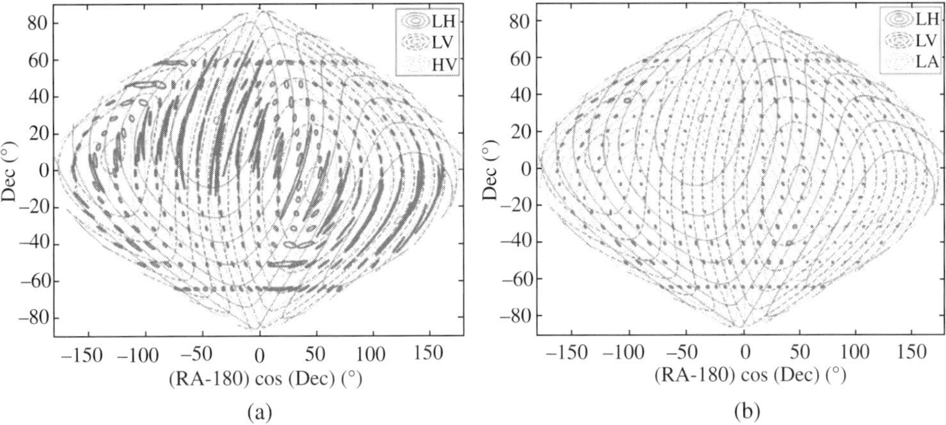

Figure 5.2 (a) All-sky map of error ellipses of angular parameters at the 95% confidence calculated from the Fisher information matrix (Wen *et al.*, 2008; Wen and Chen, 2010) for the three-detector network of the Advanced LIGO detectors (L and H) and the Advanced Virgo (V) to detect gravitational waves from bar-instability that center around 600 Hz for the *best-case* scenario where signal waveforms are known. Elongated error ellipses cluster around the plane formed by the three detector sites. These ellipses were calculated at a fixed time for a fixed optimal network SNR $\rho_N = 10$. Shown also in the background are contours of light arrival-time delays between detector pairs at a 2 ms interval for the L–H pair and 4 ms intervals for all other pairs. (b) Same as (a) but with an Australian detector added to the LIGO–Virgo network. For the Australian instrument the Advanced LIGO sensitivity is assumed.

scenario. An Australian detector would contribute by adding the longest baseline to the network, breaking the three-detector plane degeneracy for source localisation, and adding more coupled energy flux. Moreover, the dimension of the null signal space doubles when the fourth detector is added, which further improves the angular resolution for the *worst-case* scenario (Wen *et al.*, 2008; Wen and Chen, 2010). The typical error solid angle can be reduced to tens of arc-minutes for high SNR signals. This will help significantly the pointing of EM telescopes for follow-up observations and elimination of confusion sources.

5.5 Multi-messenger gravitational wave astronomy

Various astrophysical sources are expected to be detectable in both the gravitational and electromagnetic (EM) radiation. Among them, gamma ray bursts (GRBs) are the most promising. They are the brightest explosions in the Universe. Some are probably associated with very energetic supernovae, in which massive stars collapse into black holes, while others are likely the result of neutron star mergers and subsequent black hole birth. Neutron star merger events, for instance, are thought to be the sources of short-hard GRBs (with the duration of gamma-ray emission < 2 s) (e.g., Fox *et al.*, 2005; Nakar, 2007).

The benefit of joint gravitational–electromagnetic (GW–EM) observations will be enormous (e.g. Bloom *et al.*, 2009, and references therein). They will help more sensitive and confident gravitational wave detections (Kochanek and Piran, 1993). The GW–EM sources may be used as standard sirens to probe the expansion history of the Universe (Holz and Hughes, 2005). Joint GW–EM observations will help uncover the origins of violent astrophysical event such as GRBs and physics at extreme gravity.

In case of a strong EM event, follow-up searches for GW signals can be conducted in archived data in the time window of the event (e.g. Abbott *et al.*, 2005). The effort of using GW events as triggers for follow-up observations is computationally more challenging. The computational framework in search for short-duration GWs of unknown waveform can be found in Kanner *et al.* (2008). Various efforts are under way to solve the challenges for longer-duration waves of known waveforms, in particular, the NS–NS coalescing events which may last for 1000 s for advanced detectors.

All searches will ultimately face issues with angular resolution of the GW network described previously. As discussed previously, wide-field EM cameras with reasonable sensitivity and angular resolution are advantageous for follow-up observations to best catch the electromagnetic counterparts of GW sources with current LIGO–Virgo network. However, within possibly tens of degrees of angular uncertainty, if improved localisation (e.g. by observation of X-ray counterparts) is not available, careful strategies need to be designed to identify host galaxies using wide-field optical or radio telescopes (e.g. Wen *et al.*, 2007; Nuttall and Sutton, 2010).

Fast gravitational wave detection and localisation

There is a need for detecting and localising GW sources as soon as possible. As an example, the EM emissions of GRBs are complex and seem to consist of prompt emissions and delayed afterglows that probably have different emission mechanism. The optical prompt emissions are observed to rise quickly, starting tens to hundreds of seconds after the gamma-ray trigger, and are initially bright but decay steeply. More than 20 prompt optical flashes have been detected by the NASA space GRB observatory Swift and ground-based robotic optical telescopes (Klotz *et al.*, 2009). For short-hard GRBs, the prompt optical emissions are much harder to obtain, probably because of the shorter duration of their gamma-ray emission (usually < 2 s) and less energetics involved than long-duration GRBs.

Gravitational wave event triggers can be used to direct fast detections of early emissions of GRBs in X-rays or in optical light. The nearly isotropic emission of GWs are particularly useful for optical or radio transient search for the so-called 'orphan' GRB events. According to various GRB models, radiations of GRB afterglows are beamed less as they move to lower frequencies. Also, the initial beaming factor can be as large as a few hundred (Frail *et al.*, 2001; van Putten and Regimbau, 2003), meaning that there are possibly a factor of hundreds more 'orphan' GRB events that could be observed in GWs, optical or radio afterglows, but not in gamma-rays as they are beamed away from the observers.

The onset time of early X-ray afterglow or prompt optical emission for GRBs can be as soon as tens of second after the trigger of gamma-ray emission. Meanwhile, the delay between the onset time of GRBs and the end of GW emission can be as short as 0.1 s or as long as 10–100 s (Zhang and Mészáros, 2004; van Putten, 2009), since GWs are intimately connected to the formation of the central engine of GRBs. Therefore, the time window to catch the early emission of X-rays and optical light using GW triggers can be as short as tens of seconds. This poses severe constraints on how fast we need to generate reliable GW event triggers. Also, given a GW trigger, the EM telescopes need to know where to point. This requires adequate knowledge of source direction.

All these call for prompt generating of GW event triggers. The existing straightforward frequency-domain search pipelines for NS–NS coalescence are inadequate in such tasks, mainly because of latency issues. The intrinsic latency comes from the need to accumulate enough data before Fourier transforms can be performed on the data. Several efforts are underway to tackle this latency issue. The Virgo collaboration has optimised the frequency domain approach with the Multi-band Template Analysis (MBTA) program, which can achieve latency as low as around 100 s. Another low-latency inspiral detection pipeline is also under development for the LIGO Scientific Collaboration, the Low-Latency Online Inspiral Detector (LLOID). This code uses a general-purpose free-software signal-processing infrastructure called Gstreamer[1] that enables the construction of the pipeline from interchangeable analysis modules. LLOID currently achieves a latency of order tens of seconds, much less than the length of the signal. A new time-domain solution to the latency problem for detection of NS–NS coalescing GWs using a bank of infinite impulse response (IIR) filters to compute efficiently time-domain correlation can be found in Hooper *et al.* (2010). An exciting feature of this method is that the efficiency scales with power law of the starting frequency of the signal and is more advantageous for the upcoming advanced detectors that reach to much lower frequency. Furthermore, the usage of parallel filters enables embarrassingly simple parallel computing, e.g. by GPU-acceleration (Chung *et al.*, 2010).

Acknowledgements

L. Wen acknowledges support from the Australian Research Council Discovery Project and Future Fellow programs. B. F. Schutz gratefully acknowledges the support of DFG grant SFB/TR-7.

References

Abadie, J., *et al.* 2010. *Classical and Quantum Gravity*, **27**, 173001.
Abbott, B., *et al.* 2005, *Phys. Rev. D.* **72**, 04 2002.
Acernese, F., *et al.* 2007. *Classical and Quantum Gravity*, **24**, 617.
Anderson, W. G., *et al.* 2001. *Phys. Rev. D*, **63**, 042003.
Beauville, F., *et al.* 2006. *Journal of Physics Conference Series*, **32**, 212–222.
Blair, D. G., *et al.* 2008. *Journal of Physics Conference Series*, **122**, 012001.
Bloom, J. S., *et al.* 2009. *ArXiv e-prints*.
Brady, P., *et al.* 1998. *Phys. Rev. D*, **57**, 2101–2116.
Cadonati, L. 2004. *Classical and Quantum Gravity*, **21**, 1695.
Cavalier, F., *et al.* 2006. *Phys. Rev. D*, **74**, 082004.

---

[1] https://www.lsc-group.phys.uwm.edu/daswg/projects/gstlal.html

Chatterji, S., *et al.* 2006. *Phys. Rev. D*, **74**, 082005.

Chung, S. K., *et al.* 2010. *Classical and Quantum Gravity*, **27**, 135009.

Cutler, C. and Flanagan, É. E. 1994. *Phys. Rev. D*, **49**, 2658–2697.

Fairhurst, S. 2009. *New Journal of Physics*, **11**, 123006.

Finn, L. 2001. *Phys. Rev. D*, **63**, 102001.

Flanagan, É. É. and Hughes, S. A. 1998. *Phys. Rev. D*, **57**, 4535–4565.

Fox, D. B., *et al.* 2005. *Nature*, **437**, 845–850.

Frail, D. A., *et al.* 2001. *ApJ*, **562**, L55–L58.

Gürsel, Y. and Tinto, M. 1989. *Phys. Rev. D*, **40**, 3884–3938.

Holz, D. E. and Hughes, S. A. 2005. *ApJ*, **629**, 15–22.

Hooper, S., *et al.* 2010. Pages 211–214 of: *American Institute of Physics Conference Series*. American Institute of Physics Conference Series, vol. 1246.

Jaranowski, P. and Królak, A. 2005. *Living Reviews in Relativity*, **8**, 3.

Kanner, J., *et al.* 2008. *Classical and Quantum Gravity*, **25**, 184034.

Klimenko, S., *et al.* 2005. *Phys. Rev. D*, **72**, 122002.

Klimenko, S., *et al.* 2006. *Journal of Physics Conference Series*, **32**, 12–17.

Klimenko, S., *et al.* 2008. *Classical and Quantum Gravity*, **25**, 114029.

Klotz, A., *et al.* 2009. *ApJ*, **137**, 4100–4108.

Kochanek, C. S. and Piran, T. 1993. *ApJ*, **417**, L17.

Kuroda, K., *et al.* 1999. *International Journal of Modern Physics D*, **8**, 557–579.

Lyne, A. G. 1989. Page 95 of: B. F. Schutz (ed), *NATO ASIC Proc. 253: Gravitational Wave Data Analysis*.

Markowitz, J., *et al.* 2008. *Phys. Rev. D*, **78**, 122003.

Mohanty, S. D., *et al.* 2006. *Classical and Quantum Gravity*, **23**, 4799–4809.

Nakar, E. 2007. *Phys. Rep.*, **442**, 166–236.

Nayak, K. R., *et al.* 2003. *Phys. Rev. D*, **68**, 122001.

Nuttall, L. K. and Sutton, P. J. 2010. *ArXiv e-prints*.

Pai, A., Dhurandhar, S. and Bose, S. 2001. *Phys. Rev. D*, **64**, 042004.

Prix, R. 2007. *Phys. Rev. D*, **75**, 023004.

Rakhmanov, M. 2006. *Class. Quantum Grav.*, **23**, 673.

Sathyaprakash, B. S. and Schutz, B. F. 2009. *Living Reviews in Relativity*, **12**, 2.

Schutz, B. 1991. *Data Processing Analysis and Storage for Interferometric Antennas*. Cambridge: Cambridge University Press. Pages 406–452.

Schutz, B. F. 2011. *Class. Quantum Grav.* **28**, 125023.

Searle, A., Scott, S., and McClelland, D. E. 2002. *Class. Quantum Grav.*, **19**, 1465.

Searle, A. C., *et al.* 2006. *Phys. Rev. D*, **73**.

Searle, A. C., *et al.* 2008. *Class. Quantum Grav.*, **25**, 114038.

Searle, A. C., Sutton, P. J., and Tinto, M. 2009. *Class. Quantum Grav.*, **26**, 155017.

Summerscales, T. Z., *et al.* 2008. *ApJ*, **678**, 1142–1157.

Sutton, P. J., *et al.* 2010. *New Journal of Physics*, **12**, 053034.

Sylvestre, J. 2003a. *Phys. Rev. D*, **68**, 102005.

Sylvestre, J. 2003b. *ApJ*, **591**, 1152–1156.

Thorne, K. S. 1987. Pages 330–458 of: Hawking, S. W. & Israel, W. (ed), *Three Hundred Years of Gravitation*. Cambridge: Cambridge University Press.

Tinto, M. 1997. Page 251 of: I. Ciufolini & F. Fidecaro (ed), *Gravitational Waves: Sources and Detectors*. Singapore: World Scientific.

van Putten, M. H. P. M. 2009. *ArXiv e-prints*, 0905.3367.

van Putten, M. H. P. M. and Regimbau, T. 2003. *ApJ*, **593**, L15–L18.

Wen, L. 2008. *Int. J. Mod. Phys. D*, **17**, 1095–1104.

Wen, L. and Schutz, B. F. 2005. *Class. Quant. Grav.*, **22**, 1321.

Wen, L., Howell, E. and Blair, D. G. 2007. Pages 123–130 of: J. Dumarchez amd J.T.T. Van (ed.) *the XLIInd Rencontres de Moriond, "2007 Gravitational Waves and Experimental Gravity"*. The Gioi Publishers.

Wen, L., Fan, X., and Chen, Y. 2008. *Journal of Physics Conference Series*, **122**, 012038.

Wen, L. and Chen, Y. 2010. *Phys. Rev. D*, **81**, 082001.

Will, C. M. 2006. *Living Reviews in Relativity*, **9**, 3.

Zhang, B. and Mészáros, P. 2004. *Int. J. Mod. Phys. A*, **19**, 2385–2472.

Part 2

Current laser interferometer detectors
– three case studies

6

LIGO: The Laser Interferometer Gravitational-Wave Observatory

P. Fritschel on behalf of the LIGO Scientific Collaboration

This chapter features the USA-based LIGO, the Laser Interferometer Gravitational-Wave Observatory – the first of three case studies covering different worldwide interferometric gravitational wave detectors. In addition to describing the basic interferometer operation and its various components, we discuss the technological challenges that have been overcome for its successful operation.

6.1 Introduction

The prediction of gravitational waves (GWs), oscillations in the spacetime metric that propagate at the speed of light, is one of the most profound differences between Einstein's general theory of relativity and the Newtonian theory of gravity that it replaced. As discussed in Chapter 1, GWs remained a theoretical prediction for more than 50 years until the first observational evidence for their existence came with the discovery and subsequent observations of the binary pulsar PSR $1913 + 16$, by Russell Hulse and Joseph Taylor (Weisberg and Taylor, 2005). In about 300 million years, the PSR $1913 + 16$ orbit will decrease to the point where the pair coalesces into a single compact object, a process that will produce directly detectable gravitational waves. In the meantime, the direct detection of GWs will require similarly strong sources – extremely large masses moving with large accelerations in strong gravitational fields. The goal of LIGO, the Laser Interferometer Gravitational-Wave Observatory (Abramovici *et al.*, 1992), is just that: to detect and study GWs of astrophysical origin. Achieving this goal will mark the opening of a new window on the Universe, with the promise of new physics and astrophysics. In physics, GW detection could provide information about strong-field gravitation, the untested domain of strongly curved space where Newtonian gravitation is no longer even a poor approximation. In astrophysics, the sources of GWs that LIGO might detect include binary neutron stars (like PSR $1913 + 16$ but much later in their evolution); binary systems where a black hole replaces one or both of the neutron stars; a stellar core collapse which triggers a Type II supernova; rapidly rotating, non-axisymmetric neutron stars; and possibly processes in the early Universe that produce a stochastic background of GWs (Cutler and Thorne, 2002).

In the past few years the field has reached a milestone, with decades-old plans to build and operate large interferometric GW detectors now realised in several locations worldwide.

This chapter, based on a previously published article, Abbott *et al.* (2009), focuses on LIGO, which operates the most sensitive detectors yet built. We aim to describe the LIGO detectors and how they operate, explain how they have achieved their remarkable sensitivity, and review how their data can be used to learn about a variety of astrophysical phenomena.

6.2 The LIGO detectors

The oscillating quadrupolar strain pattern of a GW is well matched by a Michelson interferometer, which makes a very sensitive comparison of the lengths of its two orthogonal arms. LIGO utilises three specialised Michelson interferometers, located at two sites (see Figure 6.1): an observatory on the Hanford site in Washington houses two interferometers, the 4 km long H1 and 2 km long H2 detectors; and an observatory in Livingston, Louisiana,

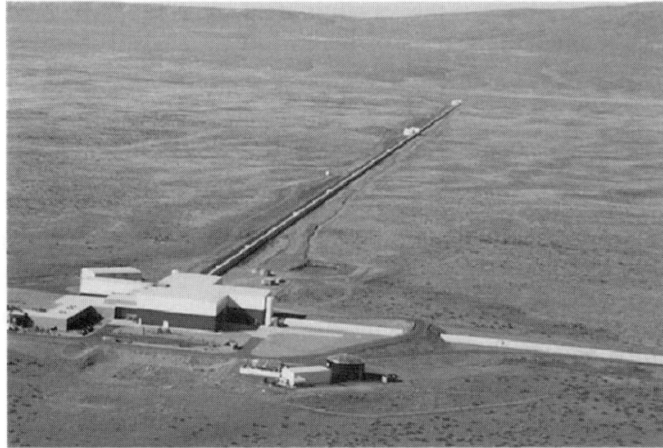

Figure 6.1 Aerial photograph of the LIGO observatories at Hanford, Washington (top) and Livingston, Louisiana (bottom). The lasers and optics are contained in the white and blue buildings. From the large corner building, evacuated beam tubes extend at right angles for 4 km in each direction (the full length of only one of the arms is seen in each photo); the tubes are covered by the arched, concrete enclosures seen here.

Table 6.1. *Parameters of the LIGO interferometers. H1 and H2 refer to the interferometers at Hanford, Washington, and L1 is the interferometer at Livingston, Louisiana*

	H1	**L1**	**H2**
Laser type and wavelength	Nd:YAG, $\lambda = 1064$ nm		
Arm cavity finesse	220		
Arm length	3995 m	3995 m	2009 m
Arm cavity storage time, τ_s	0.95 ms	0.95 ms	0.475 ms
Input power at recycling mirror	4.5 W	4.5 W	2.0 W
Power recycling gain	60	45	70
Arm cavity stored power	20 kW	15 kW	10 kW
Test mass size and mass	$\phi\, 25\,\text{cm} \times 10\,\text{cm}$, 10.7 kg		
Beam radius ITM/ETM ($1/e^2$ power)	3.6 / 4.5 cm	3.9 / 4.5 cm	3.3 / 3.5 cm
Test mass pendulum frequency	0.76 Hz		

houses the 4 km long L1 detector. Other than H2 being shorter, the three interferometers are essentially identical. Multiple detectors at separated sites are crucial for rejecting instrumental and environmental artifacts in the data, by requiring coincident detections in the analysis. Also, because the antenna pattern of an interferometer is quite wide, source localisation requires triangulation using three separated detectors.

The initial LIGO detectors were designed to be sensitive to GWs in the frequency band $40 - 7000$ Hz, and capable of detecting a GW strain amplitude as small as 10^{-21} over a 1 s integration time (Abramovici *et al.*, 1992). With funding from the National Science Foundation, the LIGO sites and detectors were designed by scientists and engineers from the California Institute of Technology, the Massachusetts Institute of Technology and the University of Florida, constructed in the late 1990s, and commissioned over the first 5 years of this decade. From November 2005 through September 2007, they operated at their design sensitivity in a continuous data-taking mode. The data from this science run, known as S5, were analysed for a variety of GW signals by a group of researchers known as the LIGO Scientific Collaboration[1] (LSC). At the most sensitive frequencies, the instrument root-mean-square (rms) strain noise reached an unprecedented level of 3×10^{-22} in a 100 Hz band. Table 6.1 lists the main design parameters of the LIGO interferometers.

6.3 Detector description

As discussed in the earlier chapters, to measure a GW strain using a Michelson interferometer, the challenge is to make the instrument sufficiently sensitive: at the targeted strain sensitivity of 10^{-21}, the resulting arm length change is only $\sim 10^{-18}$ m, a thousand times

---

[1] Homepage of the LIGO Scientific Collaboration, http://www.ligo.org

smaller than the diameter of a proton. Meeting this challenge involves the use of special interferometry techniques, state-of-the-art optics, highly stable lasers, and multiple layers of vibration isolation, all of which are described in the sections that follow. And of course a key feature of the detectors is simply their scale: the arms are made as long as practically possible to increase the signal due to a GW strain. See Table 6.1 for a list of the main design parameters of the LIGO interferometers.

Interferometer configuration

The LIGO detectors are Michelson interferometers whose mirrors also serve as gravitational test masses. A passing gravitational wave will impress a phase modulation on the light in each arm of the Michelson interferometer, with a relative phase shift of 180° between the arms. When the Michelson arm lengths are set such that the un-modulated light interferes destructively at the anti-symmetric (AS) port – the dark fringe condition – the phase-modulated sideband light will interfere constructively, with an amplitude proportional to GW strain and the input power. With dark fringe operation, the full power incident on the beam splitter is returned to the laser at the symmetric port. Only differential motion of the arms appears at the AS port; common mode signals are returned to the laser with the carrier light.

Two modifications to a basic Michelson, shown in Figure 6.2, increase the carrier power in the arms and hence the GW sensitivity. First, each arm contains a resonant Fabry–Perot optical cavity made up of a partially transmitting input mirror and a highly reflecting end mirror. The cavities cause the light to effectively bounce back and forth multiple times in the arms, increasing the carrier power and phase shift for a given strain amplitude. In the LIGO detectors the Fabry–Perot cavities multiply the signal by a factor of 100 for a 100 Hz GW. Second, a partially reflecting mirror is placed between the laser and beam splitter to implement power recycling (Meers, 1988). In this technique, an optical cavity is formed between the power recycling mirror and the Michelson symmetric port. By matching the transmission of the recycling mirror to the optical losses in the Michelson, and resonating this recycling cavity, the laser power stored in the interferometer can be significantly increased. In this configuration, known as a power recycled Fabry–Perot Michelson, the LIGO interferometers increase the power in the arms by a factor of $\approx 8,000$ with respect to a simple Michelson.

Laser and optics

The laser source is a diode-pumped, Nd:YAG master oscillator and power amplifier system, and emits 10 W in a single frequency at 1064 nm (Savage *et al.*, 1998). The laser power and frequency are actively stabilised, and passively filtered with a transmissive ring cavity (pre-mode cleaner, PMC). The laser power stabilisation is implemented by directing a sample of the beam to a photodetector, filtering its signal and feeding it back to the power amplifier; this servo stabilises the relative power fluctuations of the beam to $\sim 10^{-7}/\sqrt{\text{Hz}}$ at 100 Hz (Abbott and King, 2001). The laser frequency stabilisation is done in multiple stages that are more fully described in later sections. The first, or pre-stabilisation stage, uses the traditional technique of servo locking the laser frequency to an isolated reference

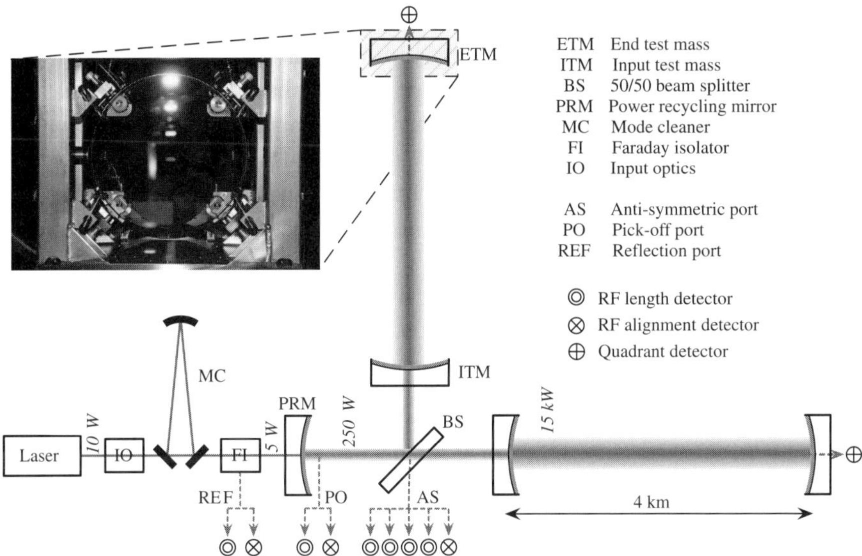

Figure 6.2 Optical and sensing configuration of the LIGO 4 km interferometers (the laser power numbers here are generic; specific power levels are given in Table 6.1). The input optics (IO) block includes laser frequency and amplitude stabilisation, and electro-optic phase modulators. The power recycling cavity is formed between the power recycling mirror (PRM) and the two input test masses (ITMs), and contains the beam splitter (BS). The inset photo shows an input test mass mirror in its pendulum suspension. The near face has a highly reflective coating for the infrared laser light, but transmits visible light. Through it one can see mirror actuators arranged in a square pattern near the mirror perimeter.

cavity using the Pound–Drever–Hall (PDH) technique (Drever *et al.*, 1983), in this case via feedback to frequency actuators on the master oscillator and to an electro-optic phase modulator. The servo bandwith is 500 kHz, and the pre-stabilisation achieves a stability level of $\sim 10^{-2}\,\mathrm{Hz}/\sqrt{\mathrm{Hz}}$ at 100 Hz. The PMC transmits the pre-stabilised beam, filtering out both any light not in the fundamental Gaussian spatial mode and laser noise at frequencies above a few MHz (Willke *et al.*, 1998). The PMC output beam is weakly phase-modulated with two radio-frequency (RF) sine waves, producing, to first-order, two pairs of sideband fields around the carrier field; these RF sideband fields are used in a heterodyne detection system described below.

After phase modulation, the beam passes into the LIGO vacuum system. All the main interferometer optical components and beam paths are enclosed in the ultra-high vacuum system ($10^{-8} - 10^{-9}$ torr) for acoustical isolation and to reduce phase fluctuations from light scattering off residual gas (Zucker and Whitcomb, 1996). The long beam tubes are particularly noteworthy components of the LIGO vacuum system. These 1.2 m diameter, 4 km long stainless steel tubes were designed to have low-outgassing so that the required vacuum could be attained by pumping only from the ends of the tubes. This was achieved by special processing of the steel to remove hydrogen, followed by an *in-situ* bakeout of the spiral-welded tubes, for approximately 20 days at 160°C.

The in-vacuum beam first passes through the mode cleaner (MC), a 12 m long, vibrationally isolated transmissive ring cavity. The MC provides a stable, diffraction-limited beam with additional filtering of laser noise above several kilohertz (Skeldon *et al.*, 1996; Yoshida *et al.*, 2000), and it serves as an intermediate reference for frequency stabilisation. The MC length and modulation frequencies are matched so that the main carrier field and the modulation sideband fields all pass through the MC. After the MC is a Faraday isolator and a reflective three-mirror telescope that expands the beam and matches it to the arm cavity mode.

The interferometer optics, including the test masses, are fused-silica substrates with multilayer dielectric coatings, manufactured to have extremely low scatter and low absorption. The test mass substrates are polished so that the surface deviation from a spherical figure, over the central 80 mm diameter, is typically 5 Å or smaller, and the surface microroughness is typically less than 2 Å (Walsh *et al.*, 1999). The mirror coatings are made using ion-beam sputtering, a technique known for producing ultralow-loss mirrors (Wei, 1989; Rempe *et al.*, 1992). The absorption level in the coatings is generally a few parts per million (ppm) or less (Ottaway *et al.*, 2006), and the total scattering loss from a mirror surface is estimated to be $60-70$ ppm.

In addition to being a source of optical loss, scattered light can be a problematic noise source, if it is allowed to reflect or scatter from a vibrating surface (such as a vacuum system wall) and recombine with the main beam (Vinet *et al.*, 1996). Since the vibrating, re-scattering surface may be moving by ~ 10 orders of magnitude more than the test masses, very small levels of scattered light can contaminate the output. To control this, various baffles are employed within the vacuum system to trap scattered light (Vinet *et al.*, 1996, 1997). Each 4 km long beam tube contains approximately 200 baffles to trap light scattered at small angles from the test masses. These baffles are stainless steel truncated cones, with serrated inner edges, distributed so as to completely hide the beam tube from the line of sight of any arm cavity mirror. Additional baffles within the vacuum chambers prevent light outside the mirror apertures from hitting the vacuum chamber walls.

Suspensions and vibration isolation

Starting with the MC, each mirror in the beam line is suspended as a pendulum by a loop of steel wire. The pendulum provides f^{-2} vibration isolation above its eigenfrequencies, allowing free movement of a test mass in the GW frequency band. Along the beam direction, a test mass pendulum isolates by a factor of nearly 2×10^4 at 100 Hz. The position and orientation of a suspended optic is controlled by electromagnetic actuators: small magnets are bonded to the optic and coils are mounted to the suspension support structure, positioned to maximise the magnetic force and minimise ground noise coupling. The actuator assemblies also contain optical sensors that measure the position of the suspended optic with respect to its support structure. These signals are used to actively damp eigenmodes of the suspension.

The bulk of the vibration isolation in the GW band is provided by four-layer mass–spring isolation stacks, to which the pendulums are mounted. These stacks provide approximately

f^{-8} isolation above ~ 10 Hz (Giaime *et al.*, 1996), giving an isolation factor of about 10^8 at 100 Hz. In addition, the L1 detector, subject to higher environmental ground motion than the Hanford detectors, employs seismic pre-isolators between the ground and the isolation stacks. These active isolators employ a collection of motion sensors, hydraulic actuators and servo controls; the pre-isolators actively suppress vibrations in the band 0.1–10 Hz, by as much as a factor of 10 in the middle of the band (Abbott *et al.*, 2004).

Sensing and controls

The two Fabry–Perot arms and power recycling cavities are essential to achieving the LIGO sensitivity goal, but they require an active feedback system to maintain the interferometer at the proper operating point (Fritschel *et al.*, 2001). The round trip length of each cavity must be held to an integer multiple of the laser wavelength so that newly introduced carrier light interferes constructively with light from previous round trips. Under these conditions the light inside the cavities builds up and they are said to be on resonance.

In addition to the three cavity lengths, the Michelson phase must be controlled to ensure that the AS port remains on the dark fringe. The four lengths are sensed with a variation of the PDH reflection scheme (Regehr *et al.*, 1995). In standard PDH, an error signal is generated through heterodyne detection of the light reflected from a cavity. The RF phase modulation sidebands are directly reflected from the cavity input mirror and serve as a local oscillator to mix with the carrier field. The carrier experiences a phase shift in reflection, turning the RF phase modulation into RF amplitude modulation, linear in amplitude for small deviations from resonance. This concept is extended to the full interferometer as follows. At the operating point, the carrier light is resonant in the arm and recycling cavities and on a Michelson dark fringe. The RF sideband fields resonate differently. One pair of RF sidebands (from phase modulation at 62.5 MHz) is not resonant and simply reflects from the recycling mirror. The other pair (25 MHz phase modulation) is resonant in the recycling cavity but not in the arm cavities.[2] The Michelson mirrors are positioned to make one arm 30 cm longer than the other so that these RF sidebands are not on a Michelson dark fringe. By design this Michelson asymmetry is chosen so that most of the resonating RF sideband power is coupled to the AS port.

In this configuration, heterodyne error signals for the four length degrees of freedom are extracted from the three output ports shown in Figure 6.2. The AS port is heterodyned at the resonating RF frequency and gives an error signal proportional to differential arm length changes, including those due to a GW. The PO port is a sample of the recycling cavity beam, and is detected at the resonating RF frequency to give error signals for the recycling cavity length and the Michelson phase (using both RF quadratures). The REF port is detected at the non-resonating RF frequency and gives a standard PDH signal proportional to deviations in the laser frequency relative to the average length of the two arms.

Feedback controls derived from these errors signals are applied to the two end mirrors to stabilise the differential arm length, to the beam splitter to control the Michelson phase, and to the recycling mirror to control the recycling cavity length. The feedback signals are

---

[2] These are approximate modulation frequencies for H1 and L1; those for H2 are about 10% higher.

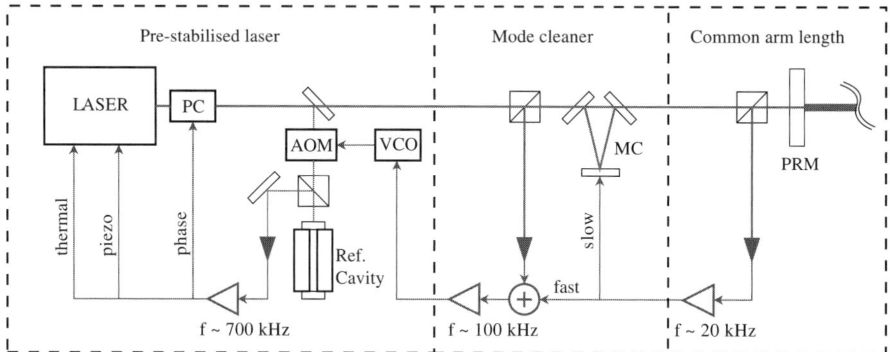

Figure 6.3 Schematic layout of the frequency stabilisation servo. The laser is locked to a fixed-length reference cavity through an acousto-optic modulator (AOM). The AOM frequency is generated by a Voltage Controlled Oscillator (VCO) driven by the MC, which is in turn driven by the common mode arm length signal from the REF port. The laser frequency is actuated by a combination of a Pockels cell (PC), piezo actuator, and thermal control.

applied directly to the mirrors through their coil–magnet actuators, with slow corrections for the differential arm length applied with longer-range actuators that move the whole isolation stack.

The common arm length signal from the REF port detection is used in the final level of laser frequency stabilisation (Adhikari, 2004) pictured schematically in Figure 6.3. The hierarchical frequency control starts with the reference cavity pre-stabilisation mentioned on page 116. The pre-stabilisation path includes an Acousto-Optic Modulator (AOM) driven by a voltage-controlled oscillator, through which fast corrections to the pre-stabilised frequency can be made. The MC servo uses this correction path to stabilise the laser frequency to the MC length, with a servo bandwidth close to 100 kHz. The most stable frequency reference in the GW band is naturally the average length of the two arm cavities, therefore the common arm length error signal provides the final level of frequency correction. This is accomplished with feedback to the MC, directly to the MC length at low frequencies and to the error point of the MC servo at high frequencies, with an overall bandwidth of 20 kHz. The MC servo then impresses the corrections onto the laser frequency. The three cascaded frequency loops – the reference cavity pre-stabilisation; the MC loop; and the common arm length loop – together provide 160 dB of frequency noise reduction at 100 Hz, and achieve a frequency stability of $5\,\mu$Hz rms in a 100 Hz bandwidth with respect to the common mode of the arm cavities.

The photodetectors are all located outside the vacuum system, mounted on optical tables. Telescopes inside the vacuum reduce the beam size by a factor of ~ 10, and the small beams exit the vacuum through high-quality windows. To reduce noise from scattered light and beam clipping, the optical tables are housed in acoustical enclosures, and the more critical tables are mounted on passive vibration isolators. Any back-scattered light along the AS port path is further mitigated with a Faraday isolator mounted in the vacuum system.

The total AS port power is typically $200-250$ mW, and is a mixture of RF sideband local oscillator power and carrier light resulting from spatially imperfect interference at the beam splitter. The light is divided equally between four length photodetectors, keeping the power on each at a detectable level of $50-60$ mW. The four length detector signals are summed and filtered, and the feedback control signal is applied differentially to the end test masses. This differential-arm servo loop has a unity-gain bandwidth of approximately 200 Hz, suppressing fluctuations in the arm lengths to a residual level of $\sim 10^{-14}$ m rms. An additional servo is implemented on these AS port detectors to cancel signals in the RF-phase orthogonal to the differential-arm channel; this servo injects RF current at each photodetector to suppress signals that would otherwise saturate the detectors. About 1% of the beam is directed to an alignment detector that controls the differential alignment of the end test masses (ETMs).

Maximal power build-up in the interferometer also depends on maintaining stringent alignment levels. Sixteen alignment degrees of freedom – pitch and yaw for each of the six interferometer mirrors and the input beam direction – are controlled by a hierarchy of feedback loops. First, alignment motion at the pendulum and isolation stack eigenfrequencies are suppressed locally at each optic using optical lever angle sensors. Second, global alignment is established with four RF quadrant photodetectors at the three output ports as shown in Figure 6.2. These RF alignment detectors measure wavefront misalignments between the carrier and sideband fields in a spatial version of PDH detection (Morrison *et al.*, 1994; Hefetz *et al.*, 1997). Together the four detectors provide five linearly independent combinations of the angular deviations from optimal global alignment (Fritschel *et al.*, 1998). These error signals feed a multiple-input multiple-output control scheme to maintain the alignment within $\sim 10^{-8}$ radians rms of the optimal point, using bandwidths between ~ 0.5 Hz and ~ 5 Hz depending on the channel. Finally, slower servos hold the beam centered on the optics. The beam positions are sensed at the arm ends using DC quadrant detectors that receive the weak beam transmitted through the ETMs, and at the corner by imaging the beam spot scattered from the beam splitter face with a CCD camera.

The length and alignment feedback controls are all implemented digitally, with a real-time sampling rate of 16384 samples per second for the length controls and 2048 samples per second for the alignment controls. The digital control system provides the flexibility required to implement the multiple-input, multiple-output feedback controls described above. The digital controls also allow complex filter shapes to be easily realised, lend the ability to make dynamic changes in filtering, and make it simple to blend sensor and control signals. As an example, optical gain changes are compensated to first order to keep the loop gains constant in time by making real-time feed-forward corrections to the digital gain based on cavity power levels.

The digital controls are also essential to implementing the interferometer *lock acquisition* algorithm. So far this section has described how the interferometer is maintained at the operating point. The other function of the control system is to acquire lock: to initially stabilise the relative optical positions to establish the resonance conditions and bring them within the linear regions of the error signals. Before lock, the suspended optics are only damped within their suspension structures; ground motion and the equivalent effect of input-light frequency fluctuations cause the four (real or apparent) lengths to fluctuate by $0.1-1$ μm rms over time-scales of $0.5-10$ s. The probability of all four degrees of freedom simultaneously falling within the ~ 1 nm linear region of the resonance points is thus extremely small and a controlled approach is required. The basic approach of the lock acquisition scheme, described in detail in Evans *et al.* (2002), is to control the degrees of freedom in sequence: first the power-recycled Michelson is controlled, then a resonance of one arm cavity is captured, and finally a resonance of the other arm cavity is captured to achieve full power build-up. A key element of this scheme is the real-time, dynamic calculation of a sensor transformation matrix to form appropriate length error signals throughout the sequence. The interferometers are kept in lock typically for many hours at a time, until lock is lost due to environmental disturbances, instrument malfunction or operator command.

Thermal effects

At full power operation, a total of $20-60$ mW of light is absorbed in the substrate and in the mirror surface of each ITM, depending on their specific absorption levels. Through the thermo-optic coefficient of fused silica, this creates a weak, though not insignificant, thermal lens in the ITM substrates (Winkler *et al.*, 1991). Thermo-elastic distortion of the test mass reflecting surface is not significant at these absorption levels. While the ITM thermal lens has little effect on the carrier mode, which is determined by the arm cavity radii of curvature, it does affect the RF sideband mode supported by the recycling cavity. This in turn affects the power build-up and mode shape of the RF sidebands in the recycling cavity, and consequently the sensitivity of the heterodyne detection signals (D'Ambrosio and Kells, 2006; Gretarsson *et al.*, 2007). Achieving maximum interferometer sensitivity thus depends critically on optimising the thermal lens and thereby the mode shape, a condition which occurs at a specific level of absorption in each ITM (approximately 50 mW). To achieve this optimum mode over the range of ITM absorption and stored power levels, each ITM thermal lens is actively controlled by directing additional heating beams, generated from CO_2 lasers, onto each ITM (Ballmer *et al.*, 2005). The power and shape of the heating beams are controlled to maximise the interferometer optical gain and sensitivity. The shape can be selected to have either a Gaussian radial profile to provide more lensing, or an annular radial profile to compensate for excess lensing.

Interferometer response and calibration

The GW channel is the digital error point of the differential-arm servo loop. In principle the GW channel could be derived from any point within this loop. The error point is chosen because the dynamic range of this signal is relatively small, since the large low-frequency fluctuations are suppressed by the feedback loop. To calibrate the error point in strain, the effect of the feedback loop is divided out, and the interferometer response to a differential arm strain is factored in (Landry and the LIGO Scientific Collaboration, 2005); this process can be done either in the frequency domain or directly in the time domain. The absolute length scale is established using the laser wavelength, by measuring the mirror drive signal required to move through an interference fringe. The calibration is tracked during operation

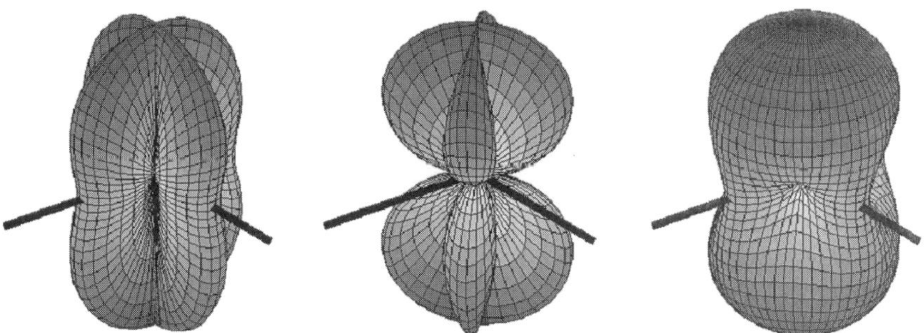

Figure 6.4 Antenna response pattern for a LIGO gravitational wave detector, in the long-wavelength approximation. The interferometer beam splitter is located at the center of each pattern, and the thick black lines indicate the orientation of the interferometer arms. The distance from a point of the plot surface to the center of the pattern is a measure of the gravitational wave sensitivity in this direction. The pattern on the left is for + polarisation, the middle pattern is for × polarisation, and the right-most one is for unpolarised waves.

with sine waves injected into the differential-arm loop. The uncertainty in the amplitude calibration is approximately ±5%. Timing of the GW channel is derived from the Global Positioning System (GPS); the absolute timing accuracy of each interferometer is better than ±10 μs.

The response of the interferometer output as a function of GW frequency is calculated in detail in Meers (1989), Fabbro and Montelatici (1995) and Rakhmanov *et al.* (2008). In the long-wavelength approximation, where the wavelength of the GW is much longer than the size of the detector, the response R of a Michelson–Fabry–Perot interferometer is approximated by a single-pole transfer function:

$$R(f) \propto \frac{1}{1 + if/f_p}, \tag{6.1}$$

where the pole frequency is related to the storage time by $f_p = 1/4\pi\tau_s$. Above the pole frequency ($f_p = 85$ Hz for the LIGO 4 km interferometers), the amplitude response drops off as $1/f$. As discussed below, the measurement noise above the pole frequency has a white (flat) spectrum, and so the strain sensitivity decreases proportionally to frequency in this region. The single-pole approximation is quite accurate, differing from the exact response by less than 1% up to ∼ 1 kHz (Rakhmanov *et al.*, 2008).

In the long-wavelength approximation, the interferometer directional response is maximal for GWs propagating orthogonally to the plane of the interferometer arms, and linearly polarised along the arms. Other angles of incidence or polarisations give a reduced response, as depicted by the antenna patterns shown in Figure 6.4. A single detector has blind spots on the sky for linearly polarised gravitational waves.

Environmental monitors

To complete a LIGO detector, the interferometers described above are supplemented with a set of sensors to monitor the local environment. Seismometers and accelerometers measure vibrations of the ground and various interferometer components; microphones monitor acoustic noise at critical locations; magnetometers monitor fields that could couple to the test masses or electronics; and radio receivers monitor RF power around the modulation frequencies. These sensors are used to detect environmental disturbances that can couple to the GW channel.

6.4 Instrument performance

Strain noise spectra

During the commissioning period, as the interferometer sensitivity was improved, several short science runs were carried out, culminating with the fifth science run (S5) at design sensitivity. The S5 run collected a full year of triple-detector coincident interferometer data during the period from November 2005 through September 2007. Since the interferometers detect GW strain amplitude, their performance is typically characterised by an amplitude spectral density of detector noise (the square root of the power spectrum), expressed in equivalent GW strain. Typical high-sensitivity strain noise spectra are shown in Figure 6.5. Over the course of S5 the strain sensitivity of each interferometer was improved, by up to 40% compared to the beginning of the run through a series of incremental improvements to the instruments.

The primary noise sources contributing to the H1 strain noise spectrum are shown in Figure 6.6. Understanding and controlling these instrumental noise components has been the major technical challenge in the development of the detectors. The noise terms can

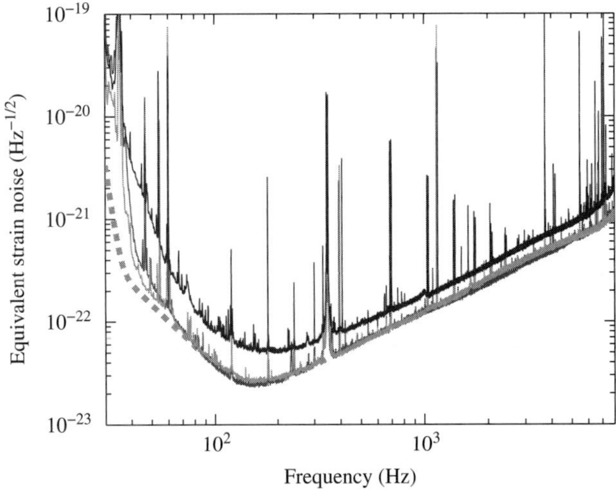

Figure 6.5 Strain sensitivities, expressed as amplitude spectral densities of detector noise converted to equivalent GW strain. The vertical axis denotes the rms strain noise in 1 Hz of bandwidth. Shown are typical high-sensitivity spectra for each of the three interferometers (darkest solid curve H2; lighter solid curves H1 and L1), along with the design goal for the 4 km detectors (partially covered dashed curve).

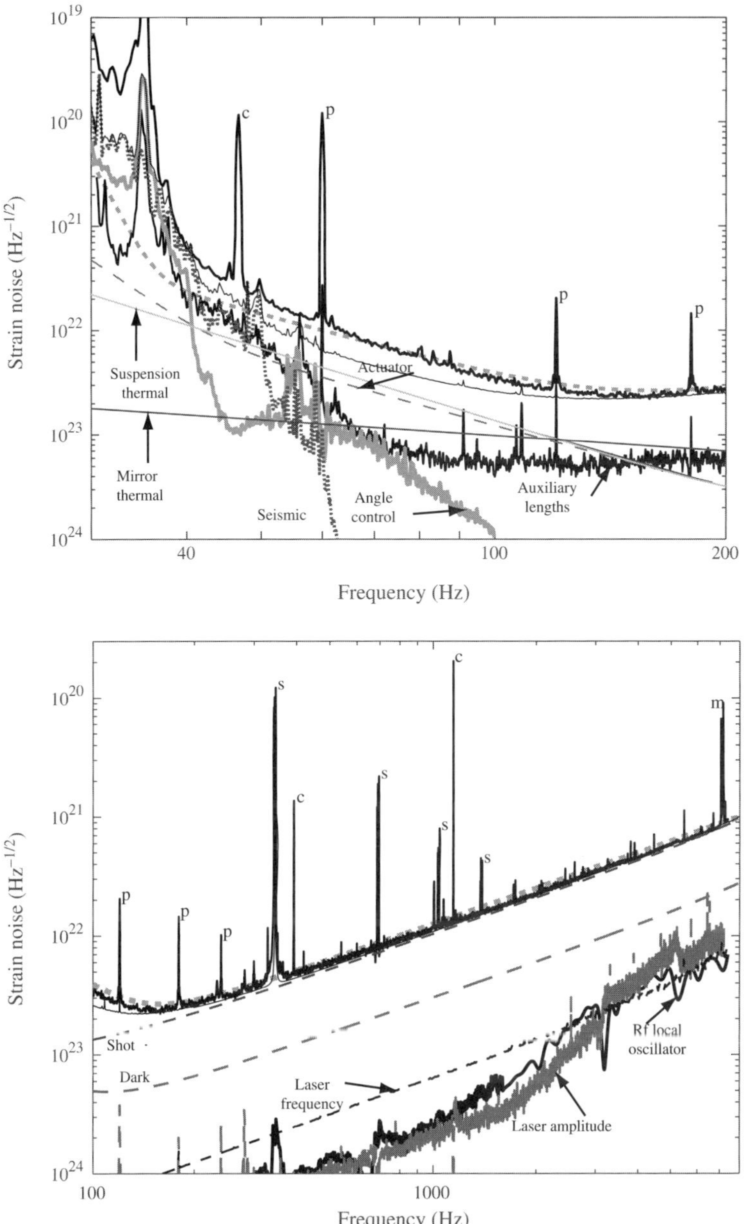

Figure 6.6 Primary known contributors to the H1 detector noise spectrum. The upper panel shows the displacement noise components, while the lower panel shows sensing noises (note the different frequency scales). In both panels, the dark upper curve is the measured strain noise (same spectrum as in Figure 6.5), the solid upper curve is the root-square-sum of all known contributors (both sensing and displacement noises) and the light dashed curve upper is the design goal. The labeled component curves are described in the text. The known noise sources explain the observed noise very well at frequencies above 150 Hz, and to within a factor of 2 in the 40 – 100 Hz band. Spectral peaks are identified as follows: c, calibration line; p, power line harmonic; s, suspension wire vibrational mode; m, mirror (test mass) vibrational mode.

be broadly divided into two classes: displacement noise and sensing noise. Displacement noises cause motions of the test masses or their mirrored surfaces. Sensing noises, on the other hand, are phenomena that limit the ability to measure those motions; they are present even in the absence of test mass motion. The strain noises shown in Figure 6.5 consists of spectral lines superimposed on a continuous broadband noise spectrum. The majority of the lines are due to power lines (60, 120, 180, ...Hz), 'violin mode' mechanical resonances (340, 680, ...Hz) and calibration lines (55, 400, and 1100 Hz). These high-Q lines are easily excluded from analysis; the broadband noise dominates the instrument sensitivity.

Sensing noise sources

Sensing noises are shown in the lower panel of Figure 6.6. By design, the dominant noise source above 100 Hz is shot noise, as determined by the Poisson statistics of photon detection. The ideal shot-noise-limited strain noise density, $\tilde{h}(f)$, for this type of interferometer is (Meers, 1988):

$$\tilde{h}(f) = \sqrt{\frac{\pi \hbar \lambda}{\eta P_{\mathrm{BS}} c}} \, \frac{\sqrt{1 + (4\pi f \tau_{\mathrm{s}})^2}}{4\pi \tau_{\mathrm{s}}}, \tag{6.2}$$

where λ is the laser wavelength, $\hbar$ is the reduced Planck constant, c is the speed of light, τ_{s} is the arm cavity storage time, f is the GW frequency, P_{BS} is the power incident on the beam splitter, and η is the photodetector quantum efficiency. For the estimated effective power of $\eta P_{\mathrm{BS}} = 0.9 \cdot 250$ W, the ideal shot-noise limit is $\tilde{h} = 1.0 \times 10^{-23}/\sqrt{\mathrm{Hz}}$ at 100 Hz. The shot-noise estimate in Figure 6.6 is based on measured photocurrents in the AS port detectors and the measured interferometer response. The resulting estimate, $\tilde{h}(100\mathrm{Hz}) = 1.3 \times 10^{-23}/\sqrt{\mathrm{Hz}}$, is higher than the ideal limit due to several inefficiencies in the heterodyne detection process: imperfect interference at the beam splitter increases the shot noise; imperfect modal overlap between the carrier and RF sideband fields decreases the signal; and the fact that the AS port power is modulated at twice the RF phase modulation frequency leads to an increase in the time-averaged shot noise (Niebauer *et al.*, 1991).

Many noise contributions are estimated using stimulus–response tests, where a sine wave or broadband noise is injected into an auxiliary channel to measure its coupling to the gravitational wave channel. This method is used for the laser frequency and amplitude noise estimates, the RF oscillator phase noise contribution, and also for the angular control and auxiliary length noise terms described below. Although laser noise is nominally common-mode, it couples to the gravitational wave channel through small, unavoidable differences in the arm cavity mirrors (Sigg, 1997; Camp *et al.*, 2000). Frequency noise is expected to couple most strongly through a difference in the resonant reflectivity of the two arms. This causes carrier light to leak out the AS port, which interferes with frequency noise on the RF sidebands to create a noise signal. The stimulus–response measurements indicate the coupling is due to a resonant reflectivity difference of about 0.5%, arising from a loss difference of tens of ppm between the arms. Laser amplitude noise can couple through an offset from the carrier dark fringe. The measured coupling is linear, indicating an effective static offset of ~ 1 pm, believed to be due to mode shape differences between the arms.

Seismic and thermal noise

Displacement noises are shown in the upper panel of Figure 6.6. At the lowest frequencies the largest such noise is seismic noise – motions of the Earth's surface driven by wind, ocean waves, human activity, and low-level earthquakes – filtered by the isolation stacks and pendulums. The seismic contribution is estimated using accelerometers to measure the vibration at the isolation stack support points, and propagating this motion to the test masses using modeled transfer functions of the stack and pendulum. The seismic wall frequency, below which seismic noise dominates, is approximately 45 Hz, a bit higher than the goal of 40 Hz, as the actual environmental vibrations around these frequencies are ~ 10 times higher than was estimated in the design.

Mechanical thermal noise is a more fundamental effect, arising from finite losses present in all mechanical systems, and is governed by the fluctuation-dissipation theorem (Saulson, 1990; Levin, 1998). It causes arm length noise through thermal excitation of the test mass pendulums (*suspension thermal noise*) (Gonzalez, 2000), and thermal acoustic waves that perturb the test mass mirror surface (*test mass thermal noise*) (Harry *et al.*, 2002). Most of the thermal energy is concentrated at the resonant frequencies, which are designed (as much as possible) to be outside the detection band. Away from the resonances, the level of thermal motion is proportional to the mechanical dissipation associated with the motion. Designing the mirror and its pendulum to have very low mechanical dissipation reduces the detection-band thermal noise. It is difficult, however, to accurately and unambiguously establish the level of broadband thermal noise *in situ*; instead, the thermal noise curves in Figure 6.6 are calculated from models of the suspension and test masses, with mechanical loss parameters taken from independent characterisations of the materials.

For the pendulum mode, the mechanical dissipation occurs near the ends of the suspension wire, where the wire flexes. Since the elastic energy in the flexing regions depends on the wire radius to the fourth power, it helps to make the wire as thin as possible to limit thermal noise. The pendulums are thus made with steel wire for its strength; with a diameter of 300 μm, the wires are loaded to 30% of their breaking stress. The thermal noise in the pendulum mode of the test masses is estimated assuming a frequency-independent mechanical loss angle in the suspension wire of 3×10^{-4} (Gillespie and Raab, 1994). This is a relatively small loss for a metal wire (Cagnoli *et al.*, 1999). Thermal noise of the test mass surface is associated with mechanical damping within the test mass. The fused-silica test mass substrate material has very low mechanical loss, of order 10^{-7} or smaller (Penn *et al.*, 2006). On the other hand, the thin-film, dielectric coatings that provide the required optical reflectivity – alternating layers of silicon dioxide and tantalum pentoxide – have relatively high mechanical loss. Even though the coatings are only a few μm thick, they are the dominant source of the relevant mechanical loss, owing to their level of dissipation and the fact that it is concentrated on the test mass face probed by the laser beam (Levin, 1998). The test mass thermal noise estimate is calculated by modeling the coatings as having a frequency-independent mechanical dissipation of 4×10^{-4} (Harry *et al.*, 2002).

Auxiliary degree of freedom noise

The auxiliary length noise term refers to noise in the Michelson and power recycling cavity servo loops which couple to the GW channel. The former couples directly to the GW channel while the latter couples in a manner similar to frequency noise. Above ~ 50 Hz the sensing noise in these loops is dominated by shot noise; since loop bandwidths of ~ 100 Hz are needed to adequately stabilise these degrees of freedom, shot noise is effectively added onto their motion. Their noise infiltration to the GW channel, however, is mitigated by appropriately filtering and scaling their digital control signals and adding them to the differential-arm control signal as a type of feed-forward noise suppression (Fritschel *et al.*, 2001). These correction paths reduce the coupling to the GW channel by $10-40$ dB.

We illustrate this more concretely with the Michelson loop. The shot-noise-limited sensitivity for the Michelson is $\sim 10^{-16}$ m/$\sqrt{\text{Hz}}$. Around 100 Hz, the Michelson servo impresses this sensing noise onto the Michelson degree of freedom (specifically, onto the beam splitter). Displacement noise in the Michelson couples to displacement noise in the GW channel by a factor of $\pi/(\sqrt{2}F) = 1/100$, where F is the arm cavity finesse. The Michelson sensing noise would thus produce $\sim 10^{-18}$ m/$\sqrt{\text{Hz}}$ of GW channel noise around 100 Hz, if uncorrected. The digital correction path subtracts the Michelson noise from the GW channel with an efficiency of 95% or more. This reduces the Michelson noise component down to $\sim 10^{-20}$ m/$\sqrt{\text{Hz}}$ in the GW channel, $5-10$ times below the GW channel noise floor.

Angular control noise arises from noise in the alignment sensors (both optical levers and wavefront sensors), propagating to the test masses through the alignment control servos. Though these feedback signals affect primarily the test mass orientation, there is always some coupling to the GW degree of freedom because the laser beam is not perfectly aligned to the center of rotation of the test mass surface (Kawamura and Zucker, 1994). Angular control noise is minimised by a combination of filtering and parameter tuning. Angle control bandwidths are 10 Hz or less and strong low-pass filtering is applied in the GW band. In addition, the angular coupling to the GW channel is minimised by tuning the center of rotation, using the four actuators on each optic, down to typical residual coupling levels of 10^{-3}–10^{-4} m rad^{-1}.

Actuation noise

The actuator noise term includes the electronics that produce the coil currents which keep the interferometer locked and aligned, starting with the digital-to-analog converters (DACs). The actuation electronics chain has extremely demanding dynamic range requirements. At low frequencies, control currents of ~ 10 mA are required to provide $\sim 5\,\mu$m of position control, and tens of mA are required to provide ~ 0.5 mrad of alignment bias. Yet the current noise through the coils must be kept below a couple of pA/$\sqrt{\text{Hz}}$ above 40 Hz. The relatively limited dynamic range of the DACs is managed with a combination of digital and analog filtering: the higher frequency components of the control signals are digitally emphasised before being sent to the DACs, and then de-emphasised following the DACs with complementary analog filters. The dominant coil current noise comes instead from the circuits that provide the alignment bias currents, followed closely by the circuits that provide the length feedback currents.

Additional noise sources

In the $50 - 100$ Hz band, the known noise sources typically do not fully explain the measured noise. Additional noise mechanisms have been identified, though not quantitatively established. Two potentially significant candidates are non-linear conversion of low-frequency actuator coil currents to broadband noise (upconversion), and electric charge build-up on the test masses. A variety of experiments have shown that the upconversion occurs in the magnets (neodymium iron boron) of the coil–magnet actuators, and produces a broadband force noise, with a f^{-2} spectral slope; this is the phenomenon known as Barkhausen noise (Cote and Meisel, 1991). The non-linearity is small but not negligible given the dynamic range involved: 0.1 mN of low-frequency (below a few hertz) actuator force upconverts of order 10^{-11} N rms of force noise in the $40 - 80$ Hz octave. This noise mechanism is significant primarily below 80 Hz, and varies in amplitude with the level of ground motion at the observatories.

Regarding electric charge, mechanical contact of a test mass with its nearby limit-stops, as happens during a large earthquake, can build up charge between the two objects. Such charge distributions are not stationary; they tend to redistribute on the surface to reduce local charge density. This produces a fluctuating force on the test mass, with an expected f^{-1} spectral slope. Although the level at which this mechanism occurs in the interferometers is not well known, evidence for its potential significance comes from a fortuitous event with L1. Following a vacuum vent and pump-out cycle partway through the S5 science run, the strain noise in the $50 - 100$ Hz band went down by about 20%; this was attributed to charge reduction on one of the test masses.

In addition to these broadband noises, there are a variety of periodic or quasi-periodic processes that produce lines or narrow features in the spectrum. The largest of these spectral peaks are identified in Figure 6.6. The groups of lines around 350 Hz, 700 Hz, etc. are vibrational modes of the wires that suspend the test masses, thermally excited with kT of energy in each mode. The power line harmonics, at 60 Hz, 120 Hz, 180 Hz, etc. infiltrate the interferometer in a variety of ways. The 60 Hz line, for example, is primarily due to the power line's magnetic field coupling directly to the test mass magnets. As all these lines are narrow and fairly stable in frequency, they occupy only a small fraction of the instrument spectral bandwidth.

6.5 Future directions

From its inception, LIGO was envisioned not as a single experiment, but as an on-going observatory. The facilities and infrastructure construction were specified, as much as possible, to accommodate detectors with much higher sensitivity. There are a set of relatively minor improvements to the first generation instruments (Adhikari *et al.*, 2006) that yield a factor of 2 increase in strain sensitivity and a corresponding factor of 8 increase in the probed volume of the Universe. The two most significant enhancements are higher laser power and a new, more efficient readout technique for the GW channel. Higher power is delivered by a new master oscillator–power amplifier system, emitting 35 W of single-frequency 1064 nm light (Frede *et al.*, 2007), 3.5 times more power than the initial LIGO lasers. For the readout, a small mode-cleaner cavity is inserted in the AS beam path, between the Faraday isolator and the length photodetectors. This cavity filters out RF sidebands and the higher-order mode content of the AS port light, reducing the shot-noise power. Instead of RF heterodyning, signal detection is done by slightly offsetting the differential arm length from the dark fringe, and using the resulting carrier field as the local oscillator in a DC homodyne detection scheme. These improvements (known collectively as Enhanced LIGO) were implemented and commissioned on H1 and L1.

Significantly greater sensitivity improvements are possible with more extensive upgrades. Advanced LIGO's significantly improved technology will achieve a factor of at least 10 in sensitivity over the initial LIGO interferometers and will lower the seismic wall frequency down to 10 Hz (Fritschel, 2003; team, 2006). Advanced LIGO has been funded by the National Science Foundation, begining in April 2008. Installation of the Advanced LIGO interferometers started in early 2011.

The Advanced LIGO interferometers are configured like initial LIGO – a power-recycled Fabry–Perot–Michelson – with the addition of a *signal recycling* mirror at the anti-symmetric output. Signal recycling gives the ability to tune the interferometer frequency response, so that the point of maximum response can be shifted away from zero frequency (Meers, 1988). The laser wavelength stays at 1064 nm, but an additional high-power stage brings the laser power up to 200 W (Willke *et al.*, 2008). The test masses will be significantly larger – 40 kg – in order to reduce radiation pressure noise and to allow larger beam sizes. Larger beams and better dielectric mirror coatings combine to reduce the test mass thermal noise by a factor of 5 compared to initial LIGO (Harry *et al.*, 2007).

The test mass suspensions become significantly more intricate to provide much better performance. They incorporate four cascaded stages of passive isolation, instead of just one, including vertical isolation comparable to the horizontal isolation at all stages except one (Robertson *et al.*, 2002). The test mass is suspended at the final stage with fused silica fibers rather than steel wires; these fibers have extremely low mechanical loss and will reduce suspension thermal noise nearly a 100-fold (Heptonstall *et al.*, 2006). The current passive seismic isolation stacks that support the suspensions are replaced with two-stage active isolation platforms (Abbott *et al.*, 2002). These stages are designed to actively reduce the ground vibration by a factor of ~ 1000 in the 1–10 Hz band, and provide passive isolation at higher frequencies. The combination of the isolation platforms and the suspensions will reduce seismic noise to negligible levels above approximately 10 Hz. The successful operation of Advanced LIGO is expected to transform the field from GW detection to GW astrophysics. We illustrate the potential using compact binary coalescences (CBCs). Detection rate estimates for CBCs can be made using a combination of extrapolations from observed binary pulsars, stellar birth rate estimates, and population synthesis models. There are large uncertainties inherent in all of these methods, however, leading to rate estimates that are uncertain by several orders of magnitude. We therefore quote a range of rates, spanning plausible pessimistic and optimistic estimates, as well as a likely rate. The rate estimates for Advanced LIGO for CBCs involving NSs of typically $1.4 M_\odot$ and/or BHs of up to $10 M_\odot$ are: $0.4 - 400 \, \mathrm{yr}^{-1}$, with a likely rate of $40 \, \mathrm{yr}^{-1}$ for NS–NS binaries; $0.2 - 300 \, \mathrm{yr}^{-1}$, with a likely rate of $10 \, \mathrm{yr}^{-1}$ for NS–BH binaries; $2 - 4000 \, \mathrm{yr}^{-1}$, with a likely rate of $30 \, \mathrm{yr}^{-1}$ for BH–BH binaries.

Acknowledgements

The content in this chapter has been adapted from Abbott *et al.* (2009) with the permission of IOP Publishing Ltd. The authors gratefully acknowledge the support of the United States National Science Foundation for the construction and operation of the LIGO Laboratory and the Particle Physics and Astronomy Research Council of the United Kingdom. The authors also gratefully acknowledge the support of the research by these agencies and by the Australian Research Council, the Natural Sciences and Engineering Research Council of Canada, the Council of Scientific and Industrial Research of India, the Department of Science and Technology of India, the Spanish Ministerio de Educacion y Ciencia, The National Aeronautics and Space Administration, the John Simon Guggenheim Foundation, the Alexander von Humboldt Foundation, the Leverhulme Trust, the David and Lucile Packard Foundation, the Research Corporation, and the Alfred P. Sloan Foundation.

References

Abbott, B. P., *et al.* 2009. *Reports on Progress in Physics*, **72**, 076901.

Abbott, R., *et al.* 2002. *Classical and Quantum Gravity*, **19**, 1591–1597.

Abbott, R., *et al.* 2004. *Class. Quant. Grav.*, **21**, S915–S921.

Abbott, R. S. and King, P. J. 2001. *Review of Scientific Instruments*, **72**, 1346–1349.

Abramovici, A., *et al.* 1992. *Science*, **256**, 325–333.

Adhikari, R. 2004. *Sensitivity and Noise Analysis of 4 km Laser Interferometric Gravitational Wave Antennae*. Ph.D. thesis, MIT.

Adhikari, R., Fritschel, P., and Waldman, S. 2006. Tech. rept. LIGO-T060156-01. LIGO Project.

Ballmer, S., *et al.* 2005. Tech. rept. LIGO-T050064-00. LIGO Project.

Cagnoli, G., *et al.* 1999. *Physics Letters A*, **255**, 230–235.

Camp, J. B., *et al.* 2000. *Journal of the Optical Society of America A*, **17**, 120–128.

Cote, P. J. and Meisel, L. V. 1991. *Phys. Rev. Lett.*, **67**, 1334. (c) 1991: The American Physical Society.

Cutler, C. and Thorne, K. S. 2002. *ArXiv General Relativity and Quantum Cosmology e-prints*.

D'Ambrosio, E. and Kells, W. 2006. *Phys. Rev.*, **D73**, 122002.

Drever, R. W. P., *et al.* 1983. *Applied Physics B: Lasers and Optics*, **31**, 97–105.

Evans, M., *et al.* 2002. *Opt. Lett.*, **27**, 598–600.

Fabbro, R. D. and Montelatici, V. 1995. *Appl. Opt.*, **34**, 4380–4396.

Frede, M., *et al.* 2007. *Opt. Express*, **15**, 459–465.

Fritschel, P. 2003. Pages 282–291 of: Cruise, M. and Saulson, P. (eds), *Society of Photo-Optical Instrumentation Engineers (SPIE) Conference Series*. Presented at the Society of Photo-Optical Instrumentation Engineers (SPIE) Conference, vol. 4856.

Fritschel, P., *et al.* 1998. *Appl. Opt.*, **37**, 6734–6747.

Fritschel, P., *et al.* 2001. *Appl. Opt.*, **40**, 4988–4998.

Giaime, J., *et al.* 1996. *Review of Scientific Instruments*, **67**, 208–214.

Gillespie, A. and Raab, F. J. 1994. *Phys. Lett.*, **A190**, 213–220.

Gonzalez, G. 2000. *Classical and Quantum Gravity*, **17**, 4409–4435.

Gretarsson, A. M., *et al.* 2007. *J. Opt. Soc. Am. B*, **24**, 2821–2828.

Harry, G. M., *et al.* 2002. *Classical and Quantum Gravity*, **19**, 897–917.

Harry, G. M., *et al.* 2007. *Classical and Quantum Gravity*, **24**, 405–415.

Hefetz, Y., Mavalvala, N. and Sigg, D. 1997. *Journal of the Optical Society of America B Optical Physics*, **14**, 1597–1605.

Heptonstall, A., *et al.* 2006. *Physics Letters A*, **354**, 353–359.

Kawamura, S. and Zucker, M. E. 1994. *Appl. Opt.*, **33**, 3912–3918.

Landry, M. and the LIGO Scientific Collaboration. 2005. *Classical and Quantum Gravity*, **22**, S985–94.

Levin, Y. 1998. *Phys. Rev.*, **D57**, 659–663.

Meers, B. J. 1988. *Phys. Rev.*, **D38**, 2317–2326.

Meers, B. J. 1989. *Phys. Lett.*, **A142**, 465–470.

Morrison, E., *et al.* 1994. *Applied Optics*, **33**, 5041–5049.

Niebauer, T. M., *et al.* 1991. *Phys. Rev. A*, **43**, 5022–5029.

Ottaway, D., *et al.* 2006. *Opt. Lett.*, **31**, 450–452.

Penn, S. D., *et al.* 2006. *Physics Letters A*, **352**, 3–6.

Rakhmanov, M., Romano, J. D., and Whelan, J. T. 2008. *Classical and Quantum Gravity*, **25**, 184017 (13pp).

Regehr, M. W., Raab, F. J., and Whitcomb, S. E. 1995. *Opt. Lett.*, **20**, 1507–1509.

Rempe, G., *et al.* 1992. *Opt. Lett.*, **17**, 363–365.

Robertson, N. A., *et al.* 2002. *Classical and Quantum Gravity*, **19**, 4043–4058.

Saulson, P. R. 1990. *Phys. Rev.*, **D42**, 2437–2445.

Savage, R., King, P., and Seel, S. 1998. *LASER PHYSICS*, **8**, 679–685.

Sigg, D. 1997. Tech. rept. LIGO-T970084-00. LIGO Project.

Skeldon, K. D., *et al.* 1996. *Review of Scientific Instruments*, **67**, 2443–2448.

team, A. L. 2006. Tech. rept. LIGO-M060056-08. LIGO Project.

Vinet, J.-Y., Brisson, V., and Braccini, S. 1996. *Phys. Rev.*, **D54**, 1276–1286.

Vinet, J.-Y., *et al.* 1997. *Phys. Rev.*, **D56**, 6085–6095.

Walsh, C. J., *et al.* 1999. *Appl. Opt.*, **38**, 2870–2879.

Wei, D. T. 1989. *Appl. Opt.*, **28**, 2813–2816.

Weisberg, J. M. and Taylor, J. H. 2005. Pages 25–32 of: Rasio, F. A. and Stairs, I. H. (eds), *Binary Radio Pulsars*. Astronomical Society of the Pacific Conference Series, vol. 328.

Willke, B., *et al.* 1998. *Opt. Lett.*, **23**, 1704–1706.

Willke, B., *et al.* 2008. *Classical and Quantum Gravity*, **25**, 114040.

Winkler, W., *et al.* 1991. *Phys. Rev. A*, **44**, 7022–7036.

Yoshida, S., *et al.* 2000. *Gravitational Wave Detection II*. Universal Academy Press, Tokyo. Chap. Recent Development in the LIGO Input Optics.

Zucker, M. E. and Whitcomb, S. E. 1996. Page 1434 of: Jantzen, R. T., Mac Keiser, G., and Ruffini, R. (eds), *Proceedings of the Seventh Marcel Grossman Meeting on Recent Developments in Theoretical and Experimental General Relativity, Gravitation, and Relativistic Field Theories*.

7

The Virgo detector

S. Braccini for the Virgo Collaboration

This chapter describes the Virgo interferometric gravitational wave detector. We first discuss the overall detector design before describing the individual subsystems in detail. We finish by outlining the commissioning and upgrades required to achieve second generation sensitivity.

7.1 Introduction

The gravitational wave detector Virgo[1], funded by CNRS (France) and INFN (Italy), is a recycled Michelson interferometer with arms replaced by 3 km long Fabry–Perot cavities. Virgo is located at the European Gravitational Observatory (EGO), close to Cascina (Pisa, Italy), and is the only ground-based antenna which, from its conception, aimed at reducing the detection band lower threshold down to 10 Hz. This goal was achieved by isolating the interferometer mirrors from ground seismic noise which, in the tens of hertz range, is many orders of magnitude larger than the small variation of the arm length induced by the gravitational wave passage.

The construction of Virgo started in the second half of 1990s and, after a commissioning phase of several years, the interferometer performed two scientific runs (VSR1, lasting four months in 2007, and VSR2, from July 2009 to January 2010), in coincidence with the antennae of the LIGO project.

Once the antenna is isolated from ground seismic noise, thermal displacements induced by pendulum dissipations in the mirror suspensions (*pendulum thermal noise*) limit the sensitivity up to a few tens of hertz. At higher frequency, the thermal noise induced by dissipation inside the mirror (*mirror thermal noise*) is dominant, up to a few hundreds of hertz. Above this frequency, the antenna sensitivity is mainly suppressed by the shot noise in the photo detection. Many other spurious mechanisms, usually named 'technical noise', can limit the interferometer performance.

During VSR2, Virgo+ has operated with a high duty cycle (above 80% operation at optimum resolution, defined as 'science mode'), very close to its design sensitivity. With this sensitivity, the antenna is able to detect, with a signal-to-noise ratio (SNR) larger than

---

[1] http://www.virgo.infn.it/

8, a signal coming from a neutron star–neutron star coalescing binary, up to 9–10 Mpc (*NS–NS horizon*). VSR2 was concluded at the beginning of 2010 in order to assemble cavity mirrors with higher reflectivity, attached to monolithic fused silica suspensions with low dissipation. The expected higher finesse and thermal noise reduction will allow Virgo+ to operate up to the end of the run (mid-2011) with a higher sensitivity in the low-frequency range, corresponding to an NS–NS horizon of a few tens of megaparsec.

Other European research groups (from Holland, Poland, Hungary and the UK) had joined the collaboration by 2010. Their contribution, already relevant for Virgo+, involved conceptualising and constructing the next-generation antenna (Advanced Virgo). The goal is to have an antenna in data-acquisition mode by 2015, in coincidence with the three Advanced LIGO detectors. The sensitivity of the advanced network, with NS–NS horizon around 300 Mpc, will certainly be sufficient to allow several detections in a few months, thus initiating the era of gravitational wave astronomy.

After a brief description of the Virgo overall design (Section 7.2), the subsystems are discussed (Section 7.3). In Section 7.4 the techniques involved in the control and commissioning of the interferometer will be outlined. The main upgrades implemented in the Virgo+ interferometer are described in Section 7.5, while the improvements necessary to meet the requirements of AdV are discussed in Section 7.6.

7.2 Virgo overall design

The optical design of Virgo is sketched in Figure 7.1. A 20 W laser beam at 1064 nm is produced by a Nd:YVO$_4$ high power laser, injection-locked to a solid state 1 W Nd:YAG

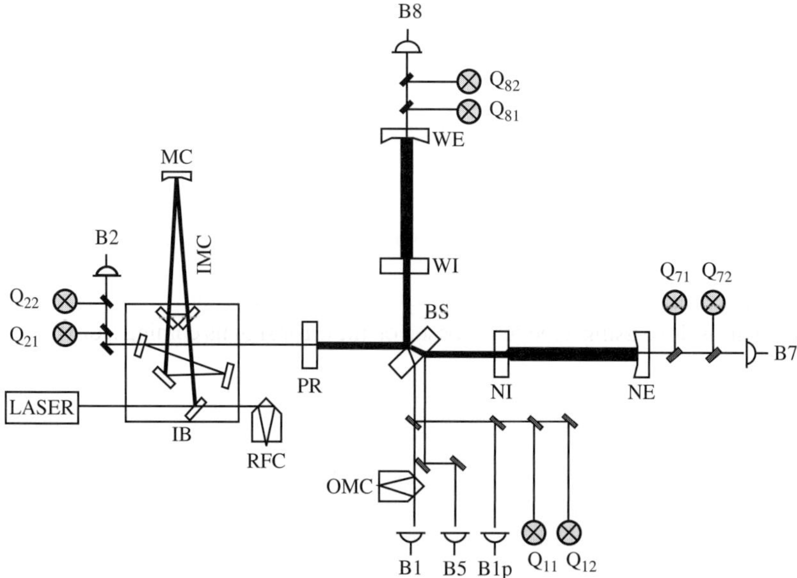

Figure 7.1 Virgo optical layout. The main symbols are defined in the text.

master laser. The beam, after passing through the optical table for its alignment, enters the vacuum system, reaching the input optical bench (IB), suspended by a reduced-size seismic-isolation system (see below). The beam is then spatially filtered by a 144 m long triangular cavity called the *input mode cleaner* (IMC) with the input and end mirrors assembled on the IB, and the intermediate *mode cleaner* (MC) mirror suspended at 144 m. This additional cavity selects the optical fundamental mode (the Gaussian TEM_{00} mode), suppressing the high-order modes. The IMC is also used as a reference to pre-stabilise the laser beam frequency. The IMC length is stabilised in the low frequency range (below a few tens of hertz, where seismic noise and other spurious mechanisms induce fluctuations of the cavity length) by using as a reference a rigid 30 cm long *reference cavity* (RFC), placed under the IB. After the IMC, the beam passes through the *power recycling mirror* (PR), is separated by the *beam splitter* (BS) and enters the two long Fabry–Perot cavities (north and west cavities). The nominal finesse of the arm cavities is 50, with an input mirror power transmittance of 11.7% and the end mirror having almost full reflectivity (power transmittance around 3×10^{-4}). The Gaussian beam radius (i.e. the distance from the optical axis at which the field amplitude and intensity drops to $1/e$ and $1/e^2$) is 6 cm at the end curved mirror, while the minimum radius (beam waist) is on the flat input mirror (2 cm). The PR is a semi-transparent mirror located between the laser source and the beam splitter and has a reflectivity towards the interferometer of around 95%. As discussed in Section 3.5, this mirror forms an additional Fabry–Perot cavity between the whole interferometer and itself. In this way the power impinging onto the BS is amplified in Virgo by a factor around 50.

The output interference signal is reconstructed by the photodiode B1, made by a set of high-quantum efficiency InGaAs photodiodes. This photodiode is assembled on a bench, outside the vacuum system. The output beam, before reaching B1, passes through a mono-lithic 2.5 cm long cavity called the *output mode cleaner* (OMC), located on a bench suspended in vacuum. The OMC is designed to filter high-order optical modes, which originate from misalignments and optical defects. The other photodiodes (labelled 'B' in Figure 7.1) are used as feedback error signals, to fix the longitudinal lengths of the interferometer (cavity resonance condition and destructive interference on B1) with a very high level of accuracy (around 10^{-12}–10^{-10} m). The longitudinal control no 7.4.1 is done using the RF-modulation of the beam light by means of the Pound–Drever technique, illustrated in section 3.5.6. In order to optimise the contrast in the interference pattern, the interferometer mirrors must be aligned with each another and with respect to the beam with a nanoradian accuracy. This is done using a feedback based on the error signals coming from the quadrant photodiodes (labelled by Q in Figure 7.1).

All six main Virgo mirrors (power recycling, beam splitter, north input, north end, west input, west end, with obvious notations in Figure 7.1) are suspended from a superattenuator (see Figure 7.2). The other optical components, whose displacement does not induce an apparent arm variation (since they are located before the beam injection to the interferometer, or after the recombination of the beams) have less stringent isolation requirements. For this reason the IB, MC and DB are suspended from shorter superattenuators, having only two mechanical filters. As shown in the following, the mirror and its reference mass are

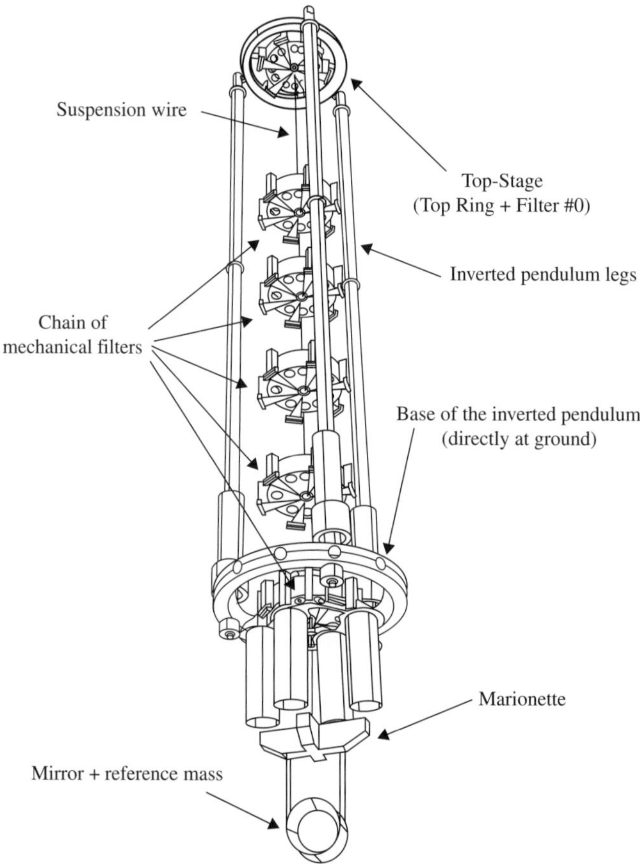

Figure 7.2 The main components of the Virgo superattenuator – a multi-stage vibration isolator for suspension of the test masses (adapted from Dattilo, 2003).

suspended in parallel from a marionette, attached to the last filter of the chain by a steel wire. Longitudinal and angular forces for the interferometer locking and alignment can be applied both to the marionette and to the mirror by coil–magnet actuators, with coils assembled on the last filter of the chain and on the reference mass. In order to reduce the variations of the refraction index, due to the gas density fluctuations, the entire interferometer must be kept in vacuum (at a pressure less than 10^{-8} mbar to meet Virgo and Virgo+ requirements, with a good safety margin). For this reason, each of the long Fabry–Perot cavities is contained in a 3 km long vacuum pipe, with a diameter of 1.2 m.

7.3 The Virgo subsystems

Injection system

The laser system is accommodated in air, on an optical bench resting on the ground. The master laser is a 1 W Nd:YAG non-planar ring oscillator (from Innolight Company). A 20

Whigh power laser from Laser Zentrum Hannover (slave laser) is locked to the light coming from the master laser. The feedback allows transferring the large frequency stability performance of the master laser to the high power slave laser. The optics located on the external injection bench, resting on the ground, and accommodated in an acoustically isolated environment, steers the beam to be injected to the IMC (Bondu *et al.*, 2002). Thanks to its high finesse (around 1000), the IMC cavity gives a strong suppression of the higher modes, different from the fundamental Gaussian mode, strongly reducing beam geometry fluctuations, in particular beam pointing jitter. The IMC also filters the laser frequency and the amplitude fluctuations, above its cut-off frequency at the cavity pole (around 500 Hz). In a 144 m long cavity, the free-spectral range (that is the frequency distance between two consecutive resonances in the transmission spectrum) is around 1 MHz. As a consequence, a long cavity make it possible to transmit to the interferometer any multiple of 1 MHz, enabling a large flexibility in choosing the RF-modulation frequencies to be transmitted to the interferometer for the longitudinal and angular control.

The locking of the laser frequency on the IMC length is made by using the Pound–Drever technique and the piezoelectric actuators of the master laser. The 30 cm long reference cavity, assembled on the IB, is made from an ultra-low thermal expansion coefficient ceramic, and supports three mirrors. As already mentioned, this high-finesse cavity (1000) is used to stabilise the IMC length in the low frequency range. A pre-stabilisation of the laser frequency, around 10^{-2} Hz Hz$^{-1/2}$ in the Virgo detection band, is achieved in this way. This pre-stabilisation is a mandatory step to achieve both the locking of the interferometer and the final frequency stabilisation; this is discussed in Section 7.4.

In Virgo only 50% of the available light of the laser (corresponding to about 8–9 W) in the IMC is transmitted to the interferometer. This is due to optical losses taking place on the various optical components. A Faraday isolator (Virgo Collaboration, 2008a), located on the IB, turned out to be necessary to prevent spurious light scattered off the interferometer from re-entering the IMC. At the output of the injection system, a small portion of the light is picked up before entering the interferometer. Its power fluctuations are monitored by a photodiode and a correction sent to the current of the slave laser. This feedback allows one to have a beam at the input of the interferometer with the required power stability, i.e. a relative intensity noise ($\delta P/P$) of about 10^{-8} Hz$^{-1/2}$. A local control system allows one to fix the position of the IB and of the MC with respect to the ground, with accuracy below 1 μm (or 1 μrad, in angle). This allows a good pre-alignment of the cavity necessary for laser locking on the IMC. After the locking, the automatic alignment system, based on quadrant photodiodes, allows low-noise angular control of the mirrors.

Seismic isolation system

The passive seismic isolation of the superattenuator is essentially provided by a five-stage pendulum. In an N-stage pendulum the horizontal displacement of the suspension point, at a frequency f much higher than its normal modes ($f_1, f_2, ..., f_n$), is transmitted to the last stage attenuated by a factor proportional to f^{2N}. In particular, the ratio between the linear spectral density of the last mass displacement (the optical component) and the linear spectral density of the suspension point displacement (where the excitation is applied), decreases as A/f^{2N} where $A = f_1^2 \cdot f_2^2 \dots \cdot f_N^2$. The 8 m long pendulum chain has all its normal modes displaced below 2 Hz. This allows it to achieve a very large attenuation of the ground horizontal seismic noise (by more than ten orders of magnitude), starting from a few hertz. Vertical seismic vibrations are not attenuated by the pendulum chain. This poses a potential problem as the vertical vibrations of the mirror are partially transferred to the laser beams, horizontal direction due to unavoidable mechanical couplings (estimated to be well below 1%). Additionally, the 3 km distant plumb lines are not parallel due to the curvature of the Earth, with a misalignment of about 3×10^{-4} rad. At least one of the two suspended cavity mirrors must be inclined with respect to the local plumb line, in order to keep them parallel. This implies a misalignment of the mirror with respect to the local vertical, and a consequent transmission of 3×10^{-4} of the mirror vertical motion to the beam direction. In order to filter the vertical ground vibrations, each mass of the multi-stage pendulum is replaced by a cylindrical mechanical filter with a set of concentric cantilever blade springs with low stiffness (Beccaria *et al.*, 1997). The blades work in parallel with a magnetic anti-spring system assembled on each filter and designed to reduce the main vertical oscillation of the blades from about 1.5 Hz down to 0.5 Hz. In particular, the blades support, through a 1 m long steel wire, the next mechanical filter, essentially forming a chain of low frequency oscillators, also in the vertical direction. Thanks to the anti-springs the chain modes are all below 2 Hz and the mirror is thus well isolated from the vertical seismic noise starting from a few hertz.

The maximum acceptable transfer function amplitude of the ground seismic vibrations is given by the ratio between the antenna sensitivity curve (expressed as a function of the arm displacement induced by the GW passage, in units of $\mathrm{m\,Hz}^{-1/2}$) and the input seismic noise linear spectral density measured on the ground. At the Virgo site, the linear spectral density of the ground seismic displacement was measured to be roughly isotropic and is approximated, within a fraction of a hertz and a few tens of hertz, by the function $10^{-7}/f^2$ $\mathrm{m\,Hz}^{-1/2}$ (Virgo Collaboration, 2004b). The requirements for seismic attenuation in Virgo, namely the maximum fraction of vertical or horizontal seismic noise that can be transmitted to the mirror along the beam direction, ranges between 10^{-9} and 3×10^{-9} in the low frequency range (10–100 Hz). The measurements of superattenuator transmission are discussed in Virgo Collaboration (2010b). They are performed by exciting the superattenuator top stage at particular frequencies, and detecting the residual mirror motion of the interferometer. The measurements (in almost all cases upper limits, since no peak was distinguished from the antenna noise floor) were found to be compliant with the requirements of Virgo and Advanced Virgo, and even the with third generation Einstein Telescope, starting from about 3 Hz.

Optical payload

The superattenuator optical payload (Bernardini *et al.*, 1995) is suspended by a steel wire from the last filter of the chain, and it is composed by the marionette, the mirror and its reference mass. The reference mass and the mirror are suspended in parallel from the marionette by two pairs of thin metal wires, with a pendulum-mode frequency of around 600 mHz. The marionette has four arms with a permanent magnet screwed to the end of each one. Each are facing four coils attached to the end of four 1 m long cylinders bolted to the bottom part of the superattenuator last filter body. This coil–magnet system steers the payload in three degrees of freedom (the translation along the beam and the rotations around the other two orthogonal axes). A fine mirror position control along these degrees of freedom is obtained by using four coil–magnet pairs, with the first element screwed on the reference mass and the second one glued on the reverse side of the mirror. Pitch and yaw motions of each suspended mirror are damped by a digital feedback (local control), with respect to the ground. The marionette and the mirror angular swings are monitored by two optical levers. A combination of the two signals is used in feedback to optimise the mirror damping. The feedback torques are applied from the marionette coils. The accomplished angular stability (of order microradians in both degrees of freedom) allows the pre-alignment of the antenna, and thus the achievement of a stable geometrical superposition between the beams interfering at the output photodiode, necessary to start the interferometer locking procedures (see Section 7.4).

In order to reduce dissipation in the mirror suspension, and thus the pendulum thermal noise, a specific kind of steel (C85) has been selected to suspend the mirror in order to optimise the ratio between the loss angle (i.e. the internal dissipation of the material) and the breaking stress (Cagnoli *et al.*, 1999). We note that a high breaking stress allows the construction of thin wires. This reduces the contribution of the dissipative part of the pendulum restoring force, due to wires, with respect to the non-dissipative part (and thus noiseless), due to gravity. The excellent performance of C85 steel is masked by the larger dissipation coming from clamps and wire–mirror contacts in the cradle suspension. An effective clamp connects the mirror suspension wire to the marionette, and the use of fused silica spacers between the wires and the mirror lateral surface has produced a pendulum quality factor of about 8×10^5 in the laboratory (Cagnoli *et al.*, 2000). The amplitude of the vertical modes of the last stage suspension wires, excited by thermal noise, allow direct measurement of the loss factor of the metal, confirming the expected value of about 1.9×10^{-4}. Above a few tens of hertz, up to a few hundreds of hertz, dissipation processes, taking place inside the substrate of the test masses (mirror thermal noise), especially in the end mirrors, play a crucial role in the limitation of the antenna sensitivity.

Inertial damping system

Below the detection band, the seismic excitation is amplified by the normal modes of the filter chain, located between 200 mHz and 2 Hz, making the mirror swing by several microns along the beam. This displacement must be actively controlled at the level of the optical payload to keep the interferometer in the longitudinal working position and within the required high level of accuracy (10^{-12}–$10^{-10}m$). Nevertheless, too large a compensation force applied close to the mirror is not acceptable even in the ultra-low frequency range. Any electro mechanical actuation system has a finite dynamics and thus induces a white noise force (affecting the entire detection band) proportional to the maximum required compensation amplitude. With the present dynamics of the digital-to-analog converter board (DAC) used for the payload control (a noise floor of $300\,\text{nV}\,\text{Hz}^{-1/2}$ over a range of $\pm 10\,\text{V}$), the maximum adjustable horizontal displacement by using the payload actuators (without affecting the detector sensitivity) is about a couple of microns. A preliminary reduction of the mirror swing induced by the chain resonances is thus necessary. This is done at the top stage of the chain. Here, a mechanical filter (filter zero), suspending the entire filter chain, is attached by means of three short metallic wires to a three-leg elastic structure (inverted pendulum). The elasticity of the inverted pendulum is obtained by a flexural joint made out of steel, through which each single leg is anchored on the bottom ring based on the ground (see Figure 7.2). By tuning the main oscillation frequencies of the inverted pendulum along the two horizontal directions at about 30–40 mHz, a good pre-attenuation is achieved in the range where the chain resonances are located. An inertial control system (inertial damping) acting on the suspension top stage is used to further decrease the mirror swing (Losurdo *et al.*, 2001). In particular, three high-sensitivity accelerometers, developed for the purpose, are used to monitor the acceleration of the suspension point in the horizontal plane (two translations and one rotation about the vertical axis). Three coil–magnet actuators are used in feedback, to keep the position of the top stage locked to an inertial frame, producing an effective damping of the superattenuator mechanical resonances. A similar approach is adopted to damp the vertical resonances. Two accelerometers and two coil–magnet actuators are used in feedback on the filter zero blades to which the chain suspension point is attached.

Mirrors

Large mirrors are necessary to reduce the pendulum thermal noise floor (the linear spectral density of mirror displacement scales with the inverse of the square root of the mass), and to accept a large size spot, due to light widening over long distances. Virgo selected a diameter of 350 mm and a mass of 20 kg for the cavity mirrors. The mirror optical losses, at the Virgo wavelength of 1.064 nm, should not exceed a power fraction of order 10^{-4}. This specification is necessary to guarantee that the total optical losses in the cavities (with the light bouncing back and forth many times in the arms) do not suppress the interferometer sensitivity. The losses can be induced by absorption (both in mirror substrate and coating), scattering, and large-scale wave-front distortion. The absorption, which takes place mainly in the coating, induces mirror heating and consequently a distortion of the optical surface and wave front (thermal lensing) index temperature dependence and radius of curvature change. The Virgo mirror substrates are made out of fused silica, exhibiting large uniformity in refractive index, low birefringence, and low losses. Tantalum pentaoxide (Ta_2O_5) was used for the coating process, because of its good optical properties in the mid-infrared region, in particular its low optical losses (less than 1 ppm). A Ta_2O_5 coating is also a good choice to minimise low mechanical dissipations. Mechanical losses are presently dominated by the coating, which determines the quality factor of the mirror resonances, and thus the mirror thermal noise level. Another precise constraint concerns the amount of scattered photons. This must be less than a few ppm in order to minimise the noise due to the scattered light (discussed in Section 7.3.8). In order to reduce the scattering losses, the Virgo mirrors have been machined with a deviation from the design curvature not exceeding a few nm (flatness specifications). The micro-roughness (i.e. the small scale defects, on distances ranging between microns and about 1 mm) is responsible of the large angle scattering, and it has been reduced down to the acceptable RMS value of 0.05 nm. These stringent requirements have been achieved by an *ad-hoc* mirror polishing.

Detection system

The optical system is designed to detect the beams coming from the interferometer output ports. As already shown (Figure 7.1), the photodiode B1 (assembled on an external bench) is placed along the interference beam, and is used to measure the gravitational wave signal. The beam reflected from the second face of the BS impinges on the photodiode B5, assembled on the same bench of B1. The two beams transmitted through the 3 km long cavities are detected by the two photodiodes B7 and B8 on the external optical benches of the terminal buildings. The beam reflected by the interferometer is monitored by the photodiode B2, assembled on the injection bench. Additional high quality optics are necessary to adapt the beam size to the photodetectors, and to handle the several beams impinging on the detection bench. The main photodiode (B1) must detect variations of light power of 10^{-10} W at the modulation frequency, corresponding to a differential deformation of the arm length, smaller than 10^{-19} m. Additionally, the residual swing of the mirror at very low frequencies (of the order of 10^{-12} m, once the interferometer is locked) causes large fluctuations of the light intensity. This puts severe constraints on the dynamic range and residual noise of the photodiodes and their readout system.

The OMC (see Figure 7.1) is a 2.5 cm long triangular optical cavity, made from silica and with a finesse of 50 (Beauville *et al.*, 2006). The large cavity bandwidth (around 75 MHz) allows the transmission of both the carrier and the sidebands in the same Airy peak. As shown in Figure 7.1, the interference pattern before the OMC is detected by the photodiode B1p. The cavity length is controlled by varying the OMC temperature, using a Peltier cell (a small solid-state heat pump made of a thin slab able to absorb or emit heat, depending on the current applied to its extremities). The OMC is kept on resonance by modulating the cavity length at 28 kHz with a piezoelectric device and detecting the error signal synchronously. In this way, the length of the cavity is controlled with an accuracy by about 10^{-10} m. Since a variation of the OMC length in the detection band would simulate the passage of a gravitational wave, the bench in vacuum is suspended from a short superattenuator, in order to filter seismic vibrations. A Faraday isolator is also assembled on the suspended bench, in order to avoid the light being back-scattered from the photodiodes and recombining with the main beam.

The position of the suspended bench is controlled in feedback in all six degrees of freedom by using a digital camera as a sensor and coil–magnet actuators. The required accuracy in the detection band of order $10^{-7}\mathrm{rad}/\mathrm{Hz}^{1/2}$ in angles and $10^{-8}\mathrm{m}/\mathrm{Hz}^{1/2}$ in the longitudinal degrees of freedom has been achieved (Beauville *et al.*, 2003).

Data acquisition and global control

About 2000 channels are acquired by the data acquisition system. More than half are acquired at 10 kHz (fast channels), with the remaining being slow monitoring channels, typically acquired with a sampling rate of a few hertz. The total acquisition rate is about 20 MBs^{-1} (8 MBs^{-1} after data compression). A GPS-based central timing system was developed to synchronise the Virgo interferometer controls and readouts with the LIGO antennas, in order to allow a coincidence analysis (Marka *et al.*, 2002). Data are digitised and acquired synchronously, due to front-end readouts. It is formatted and propagated on an Ethernet network and stored on disk. A common data format for gravitational wave detectors has been developed in the framework of the Virgo–LIGO collaboration in order to facilitate the joint analysis.

Several data are used in feedback for the control of the interferometer. In particular, data coming from photodiodes for the interferometer longitudinal control are acquired at 10 kHz, while signals coming from the quadrants used in the automatic alignment are read at 500 Hz. A hardware and software system, named Global Control (Arnaud *et al.*, 2005), has been designed to read photodiode signals (via optical link), compute the digital signal for the position corrections, and send them to the control electronics of mirror suspensions and input bench, again by optical links. The system matches the synchronisation constraint and provides an easy implementation of the algorithms needed for locking and alignment loops, as discussed in the next Section.

Vacuum system

As already mentioned, the laser beam propagates along each arm in a 3 km long cylindrical vacuum pipe of diameter 1.2 m. Such a large diameter allows up to three parallel optical cavities to be accommodated. The tubes are made of a standard stainless steel (304L), and have a thickness of 4 mm, sufficient to sustain the external atmospheric pressure with a reasonable safety margin. The pipes are located in 'external tunnels', laid on the ground to protect them from natural phenomena, and to reduce noise generated from human activities.

The refractive index of the gas, upon which the phase of the light depends, varies as the square root of the number of particles encountered by the beam along its path. As a consequence, the noise scales as the square root of the residual pressure. The result is a white noise in the entire detection band, whose linear spectral density in h, for a residual gas pressure of 10^{-9} mbar (and considering hydrogen as main component), is flat at around 10^{-25}Hz$^{-1/2}$ over the entire detection band. Additionally, the vacuum must be hydrocarbon-free to prevent mirror pollution (a partial pressure of hydrocarbons lower than 10^{-13} mbar is required). A low outgassing rate of the materials is necessary to fulfil the specifications with a sophisticated, but relatively small, pumping system. A preliminary stainless steel treatment, consisting of heating at 400 °C in air, was developed to reduce the hydrogen outgassing rate by more than two orders of magnitude. A bake-out of the whole 3 km of pipes under vacuum, at 150 °C, is required in order to eliminate the H_2O molecular layers trapped on the inner surface of the pipe. This procedure is very expensive (with 600 kW of electric power involved), thus it is only performed when the sensitivity of the interferometer is so good as to be limited only by the residual gas fluctuations. At the time of writing (2011), the level of vacuum in the Virgo pipes is 10^{-7} mbar (dominated by water), and the pipe bake-out is still not necessary.

Stainless steel absorbing baffles, with a trap conical shape, are inserted inside the pipe to suppress the light scattered off the mirrors by multiple reflections/diffusions (Vinet *et al.*, 1997). Indeed, this spurious light can impinge on another interferometer mirror and undergoes another scattering into the main beam. This light follows a different optical path, and thus joins the main beam with a random phase. Seismic and acoustic vibrations of the vacuum pipes induce a modulation of this phase, thus reintroducing seismic noise in the apparatus. The absorbing baffles are arranged inside the tube in such a way as to intercept all the photons scattered by the mirror, before they can interact with the cylindrical vacuum pipe. Montecarlo simulations (Vinet *et al.*, 1997) and analytical computations (Vinet *et al.*, 1996) have shown that the spurious light is strongly suppressed, and that the noise induced by the residual diffused light inside the pipes is negligible.

7.4 Interferometer commissioning

Interferometer locking

After the pre-commissioning of the subsystems, the interferometer is pre-aligned with the two beams that interfere on the main output photodiode (B1p). In this configuration, the interference pattern changes continuously, since the mirrors are swinging by a few microns per second along the beam axis, crossing several wavelengths and thus changing the interference conditions. During the mirror swing, the cavity crosses the resonances only at short-time intervals, producing a series of flashes on the cavity end photodiodes at 3 km (B7 and B8). Thus the interferometer must be 'locked' in the longitudinal working position, using the suspended coil–magnet actuators steering the mirrors. In particular, four longitudinal degrees of freedom must be controlled with a very high degree of accuracy (with a RMS positioning ranging from 10^{-12} m to 10^{-10} m, depending on the degree of freedom): the two Fabry–Perot cavities and the recycling cavity must be locked on the optical resonance, while the interferometer must be fixed on the dark fringe (which means tuning properly the difference of the distances between the beam splitter and the two cavity input mirrors). The radio frequency sidebands must resonate in the recycling cavity in order to produce the output signal, while only the carrier resonates in the long arm cavities, as discussed in the previous sections. The longitudinal control operates below a few tens of hertz, where the mirror swing and the interferometer noise are large. At higher frequency, the interferometer and the suspended mirrors are very stable and silent. This means that, in this frequency range, natural fluctuations do not induce a significant deviation from the interferometer longitudinal working point. The effect induced by the gravitational wave (and by any other spurious differential arm displacement) is thus not compensated at high frequency, and the signal can be detected as a variation of the interference light power at the output photodiode. Vice versa, in the low frequency range, inside the locking control band, the differential displacement of the arms, and thus the gravitational wave signal and the interferometer noise, can be measured by the feedback force used to keep the mirrors in the working position.

The longitudinal lengths are controlled by compensating the position error measured by the demodulated signals coming from different photodiodes.

The correction signals are computed by the global control. Signals coming from different photodiodes indicate how much each mirror has to be moved to maintain the interferometer in the longitudinal working position. The signals are then sent by digital optical links to the digital electronics of the mirror suspensions, that thus compute and produce the correction forces to be sent to the various coil–magnet actuators of the superattenuators to produce the required mirror displacement. The signal computation and transmission take place rapidly, with negligible delays for the global feedback (typically a few hundreds of microseconds are necessary for the full feedback chain), allowing a robust locking of the interferometer. Single feedback loops, based on the Pound–Drever technique (Section 3.5.6), operate on four independent combinations of the lengths mentioned above (see Virgo Collaboration, 2008b, for details).

In general, the main difficulty in cavity lock acquisition is that the photodiode error signal exists only during the crossings of the cavity resonances that take place in very small space domains. Indeed, the resonance condition occurs every half wavelength, in a domain a few nm long, depending on the finesse F of the cavity (with the line width being $2\pi/F$). An additional goal of the inertial damping system, suppressing the superattenuator mechanical resonances, is to allow the locking acquisition by reducing the mirror velocity along the beam (below 1 μm/s), in order to increase the resonance crossing time. In this way the error signal is present for a period long enough to allow the locking feedback to stop the mirror swing, and to null the error signal (which is zero at resonance). Virgo makes use of a special locking acquisition procedure, named variable finesse. This procedure (described in Virgo Collaboration, 2008b) has the significant advantage of bringing the interferometer to the working position in a predictable way, starting from the swinging mirror.

Second stage frequency stabilisation

The frequency pre-stabilisation stage taking place on the IMC (see Section 7.3), is fundamental to achieve the locking acquisition (a fluctuation $\delta\nu$ of the laser frequency in a cavity with length L having the same effect of a mirror swing $\delta L = L \cdot \delta\nu/\nu$). However, the residual frequency fluctuations in the band after pre-stabilisation (around $10^{-2} - 10^{-1}$ Hz/Hz$^{1/2}$) are much larger than the final requirements (10^{-6} Hz/Hz$^{1/2}$) for the achievement of the interferometer sensitivity. A large stabilisation of the laser frequency is achieved by the locking system, forcing the laser frequency to follow the arm 'common mode', i.e. the sum of the 3 km arm lengths (second stage frequency stabilisation); (Virgo Collaboration, 2009). The common mode is a very long and stable reference cavity in the entire Virgo band. The laser frequency is stabilised on the common mode by adding the error signal (coming from B5 photodiode, Figure 7.1) to the IMC error signal used for frequency pre-stabilisation. The common mode is not a good reference below 1 Hz, where the residual mirror motion is large due to the suspension resonances. For this reason, in the ultra-low frequency range, the common arm motion, and thus the laser frequency, is stabilised on a 30 cm long rigid optical frequency reference cavity.

Automatic alignment

Once the interferometer is locked, it is necessary to keep the mirrors aligned with each another and with respect to the beam. A good level of contrast in the interference pattern can be achieved if the alignment accuracy of the interferometer optics is close to 1 nrad (rms value). This level is not achievable by the ground-based local controls of the mirror. Indeed, the camera and the optical levers used for the mirror's local control rest on the ground, and thus the mirror angular positions are locked to the reference systems affected by seismic noise. Once the interferometer is locked and the automatic alignment is functioning, the mirror local controls are switched off.

In Virgo, the automatic alignment has been accomplished by using the Anderson technique, described in Anderson (1984). The modulation frequency is tuned in order to have the first higher-order optical mode, generated by the misalignments, resonant in the long arm cavities. This allows one to detect the alignment error signal in transmission to the long Fabry–Perot cavity, and thus to also use the terminal photodiodes for the purpose. As in the case of locking, error signals from quadrant photodiodes at different output ports (Figure 7.1) are managed by the global control system, in order to provide suitable correction signals to be sent to the coil–magnet actuators of the superattenuator marionette. The scheme of the photodiodes and actuators involved in the feedbacks is illustrated in Virgo Collaboration (2010a).

Mirror hierarchical control

In addition to the inertial damping, suppressing the mirror motion in the superattenuator resonance range (between 30 mHz and 2 Hz), a hierarchical control strategy is adopted to reduce the mirror swing, and thus the noise, coming from the finite dynamics of the DAC (see Section 7.3). The ultra-low frequency mirror displacements (below 10 mHz), induced by drifts and Earth tides, are very large (hundreds of microns). These displacements are compensated using the interferometer as a sensor, and the three coil–magnet pairs of the horizontal inertial damping as feedback actuators (Virgo Collaboration, 2004a). In this case, the huge electromechanical floor induced by the necessity of operating large actuations is suppressed by the strong filtering provided by the entire pendulum chain underneath. The residual payload displacement along the beam (only around a fraction of micron after the action of the inertial damping and tidal control) is compensated up to a few hertz from the marionette. The residual induced electromechanical floor is further suppressed down to an acceptable level by the mirror pendulum underneath. The reference mass actuators, pushing directly on the mirror above a few hertz, are used only to compensate for the residual mirror displacements, not exceed a few nanometers. Actuation noise measurements performed on the interferometer demonstrate that even the main contribution (from the marionette actuation point) does not affect the Virgo sensitivity, nor the sensitivity of the second generation antenna (Advanced Virgo).

Noise reduction

The identification of spurious mechanisms that can affect the sensitivity of the antenna involves many techniques. Coherence between the output photodiode signal and other signals monitoring parts of the interferometer subsystems is the easiest method to identify the noise source. Another technique is to excite and monitor a possible source of noise, in order to measure the transfer function from the disturbance to the output photodiode. The level of noise induced during the interferometer operations can be inferred by multiplying the measured transfer function by the 'standard' input noise (measured by the sensor of the source when no excitation is applied). This technique allows one to estimate the floor induced on the interferometer by a given mechanism, even if it is not dominant in the sensitivity curve (i.e. is masked by other noise during operations). Various actions can be necessary to reduce the impact of a given noise source, such as the remote tuning of the parameters of the interferometer control, hardware operations on external optical benches and even on the optical payload or suspended benches, with the opening of the vacuum system (with a longer break of operations).

After a few months of the scientific run VSR2, the high frequency sensitivity was dominated by the shot noise, depending on the power injected in the interferometer (during VSR2 about 17 W are available at the output of the input mode cleaner). In the low frequency range, several technical noise sources are present. Once the locking and the automatic alignment are acquired, one of the main goals is to reduce the noise induced by the control loops. The feedback gain in the low frequency range (at least below a few hertz) must be high in order to compensate for the wide swing of the mirrors and interferometer working point. As mentioned above, each feedback locks a given degree of freedom of the interferometer to a sensor (photodiode or quadrants). The sensor exhibits a given noise floor (due to diffused light, electronic noise, etc..). It is important to stress that re-injection of noise coming from the photodiode does not take place only below the unitary gain frequency of the feedback. Control stability requirements do not allow the feedback force spectrum to cross the unitary gain frequency with an infinite slope. The control roll-off (connecting the high-gain range to the low-gain one, where the loop is not necessary) has a finite slope. In other words, it is not possible to achieve high gain below 10 Hz without having significant gain in the detection band. This finite gain above the unitary gain is responsible for the reintroduction of sensor noise also in the detection band. For this reason, it is very important to reduce the sensor noise, by suppressing the diffused light impinging on the photodiode, and improving the readout noise electronics. Noise subtraction techniques to remove the control action, measured on the feedback signals from the dark fringe, are implemented with a large accuracy – as shown in Virgo Collaboration (2010c), these techniques strongly reduce the control noise. Another important mechanism affecting the low frequency sensitivity is due to the diffused light coming from the optical benches. A fraction of this spurious light recombines with the main beam and is phase modulated by the bench vibrations (seismic, acoustic, etc.). A strong mitigation of diffused light noise has been achieved by using diaphragms, or optical dampers, on the optical benches, and by means of acoustic isolation systems. An additional noise mechanism in the low frequency range is related to the coupling of the environmental electromagnetic noise with the magnets used for their position control.

Calibration

Interferometer calibration (Rolland, 2009) is a procedure to convert a signal from the main photodiode (in watts) into differential displacement of the arm length (in meters). A white noise is injected in the coil–magnet actuators, used to steer one of the end cavity mirrors. In this way, one can measure, at the output photodiode, how the locking feedback loops compensate the effects of a differential mirror strain, and thus infer what would be the mirror motion in the absence of loops.

For this purpose it is necessary to know the mirror displacement provided by the coil–magnet actuator once a given current is applied to the coils (absolute calibration). This calibration is achieved by injecting into the coils sinuisodal signals with the interferometer in a 'free-running Michelson configuration'. This means that the end cavity mirrors are kept misaligned and the interferometer is not locked. One can count the number of fringes swept because of excitation and obtain in this way an absolute measurement of the displacement induced on the mirror.

7.5 Virgo+ upgrades

Virgo+ operated very close to its design sensitivity, with a NS–NS horizon around 10 Mpc. The optical scheme of Virgo+ interferometer is very similar to the Virgo one. The main improvements were the use of a higher power laser (50 W), together with several upgrades on the injection bench, and of a thermal compensation system for the mirror deformation, induced by light absorption. Other upgrades were the use of new electronics for detection and control, and the implementation of solutions to reduce the diffused light in the interferometer. In 2010, monolithic suspensions were assembled to reduce the thermal noise floor. The new mirrors at the input of the cavity will have a higher reflectivity, in order to achieve a cavity finesse of 150. This will improve the antenna sensitivity, with the goal of extending the NS–NS horizon up to 50 Mpc. The design sensitivity curve of Virgo+ after the insertion of monolithic suspensions is displayed in Figure 7.3.

New injection bench

An amplification of the slave laser was achieved in Virgo+ by an amplifier, made using four Nd:YVO4 laser crystals, longitudinally pumped by a fiber-coupled laser diode. An output power of 50 W at 1064 nm was achieved in this way. The power presently used at the input of the ITF is about 20 W. A resonant triangular cavity placed on the external bench (pre-mode cleaner) enabled the reduction of the fluctuations of the power and of the frequency of the laser.

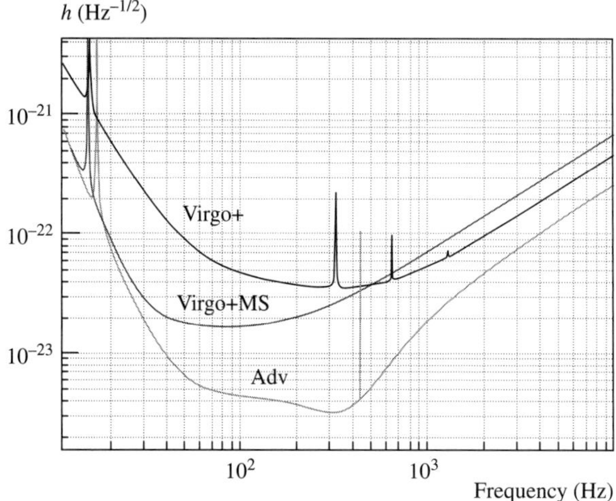

Figure 7.3 The design sensitivity of Virgo+ with monolithic suspension and high reflectivity mirror. A 25 W laser power (after the IMC, i.e. impinging on the PR mirror), a finesse of 150, and a recycling gain of 20, have been considered. The curve is compared with the previous design sensitivity of Virgo+ (shown as Virgo +MS; same input power, without monolithic suspension, finesse of 50, and recycling gain of 43) and with the nominal Advanced Virgo (AdV) sensitivity, discussed in Section 7.6.

Thermal compensation system

A temperature gradient in the mirror is induced by the small fraction of optical power absorbed by high-reflectivity coating. The dominant mechanism is the change of its refractive index (several parts per million), producing beam shape distortions larger than those induced by the geometrical deformation of the reflecting surface. The resulting thermal lensing can destroy the system's optical properties (a 10 km extra focal lensing is already an issue). This effect actually limits the amount of laser power that can be used in gravitational wave detectors. The main problems are related to the power recycling cavity. Here, other cavity modes induced by thermal lensing can resonate, as well as the fundamental one. This results in a deterioration of the sideband beam profile. As a consequence, the error signal used to read the dark fringe and to control the recycling cavities, given by the superposition between the carrier and the sideband field, becomes noisy. This induces an increase of the noise in the interferometer and renders its control difficult.

A thermal compensation system (TCS) for the input mirrors in each cavity was set up to balance the temperature distribution. The TCS is based on a pre-stabilised CO_2 laser projector (10.64 μm) that shines a heating pattern onto the high-reflectivity surface of the input mirror. A high-resolution wavefront sensor (phase camera) is used as a monitor. This measures the spatial profile (and phase) of the beam field at radio frequency sideband, at the output of the interferometer. The radio frequency sidebands are significantly more sensitive to misalignments or other spatial distortions of the power recycling cavity than the carrier field, and thus represents a good monitor of the TCS effects in recovering the correct mirror performance. Radiation pressure variations induced by power fluctuations of the CO_2 laser cause spurious displacements of the test mass. The TCS power stabilisation level ($\delta P / P$), necessary to be compliant with the Virgo+ sensitivity requirements, must be of the order of $10^{-7} Hz^{-1/2}$ at 30 Hz.

Monolithic suspensions

The monolithic suspensions (Amico *et al.*, 2002) are thin fiber mirror suspensions made out of fused silica. Laboratory tests have shown that a low dissipation pendulum with a quality factor close to 10^8 was achievable (about a factor 100 better than in Virgo). This solution is very similar to the one planned for Advanced Virgo. Monolithic suspensions were assembled on the four cavity mirrors in 2010.

7.6 Towards the next generation

The design sensitivity of Advanced Virgo, and the dominant noise contributions, are reported in Figure 7.4. The sensitivity of the antenna will be limited by the pendulum thermal noise of the mirror suspension below a few tens of hertz, by shot noise (i.e. quantum noise) above 400 Hz, and by a combination of the mirror thermal noise (dominated by coating dissipation) and shot noise in the intermediate frequency range. The corresponding NS–NS horizon is about 150 Mpc. It is important to remember that the rate of detectable events scales more or less with the volume of accessible sky, and thus approximately with the cube of the horizon. For this reason, the new interferometer is expected to detect many events

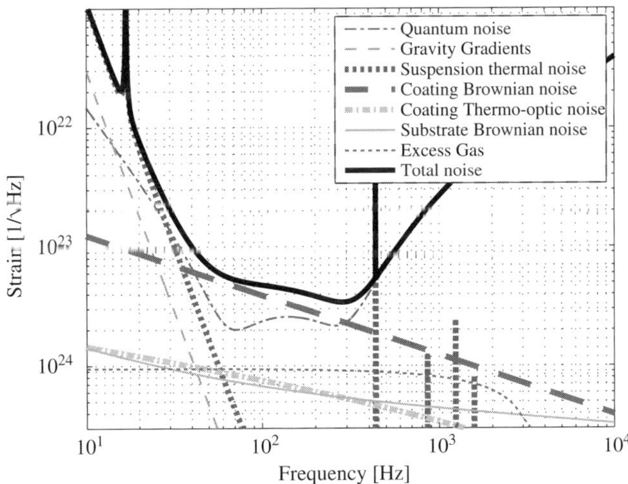

Figure 7.4 Reference Advanced Virgo sensitivity and expected noise contributions. The curves have been calculated with a transmittance of 11% of the dual recycling mirror, a dual recycling phase of 0.15 rad, 125 W of laser power entering the interferometer, a cavity finesse of 888, and a power recycling factor of 21.5. The chosen dual recycling parameters optimise the horizon for coalescing binary neutron stars. The Virgo design sensitivity is also shown for comparison.

per year, initiating the era of the gravitational wave astronomy. The presence of the dual recycling cavity (Section 3.5.2) allows the optimisation of the interferometer sensitivity for different astrophysical sources. The scientific case of Advanced Virgo is well illustrated in Chapter 4 of AdV Baseline Design (2009).

As mentioned above, measurements of the superattenuator isolation performance have shown that the residual seismic noise at the mirror level does not affect the Advanced Virgo sensitivity curve, such as the noise coming from the control of the mirror position. On the other hand, major upgrades are necessary for other subsystems. Baseline solutions are summarised here, even if some choices might be changed upon new findings. More details on Advanced Virgo subsystems can be found in AdV Baseline Design (2009).

Optical layout of Advanced Virgo

The Advanced Virgo optical layout is illustrated in Figure 7.5. In order to reduce the shot noise, the light power circulating in the arms is maximised by increasing either the finesse of the arm cavity up to around 900 (150 in Virgo+ and 50 in Virgo) and, according to the expected interferometer losses, the finesse of the power recycling cavity (around 70). In this configuration, the laser power impinging on the beam splitter is 2.7 kW and 760 kW in the arm cavities. As shown in the Section 3.5.4, the mirror thermal noise scales with the inverse of the radius of the impinging beam. A beam waist of 8.5 mm will be set close to the center of the cavity, in order to have a beam spot around 6 cm in radius on

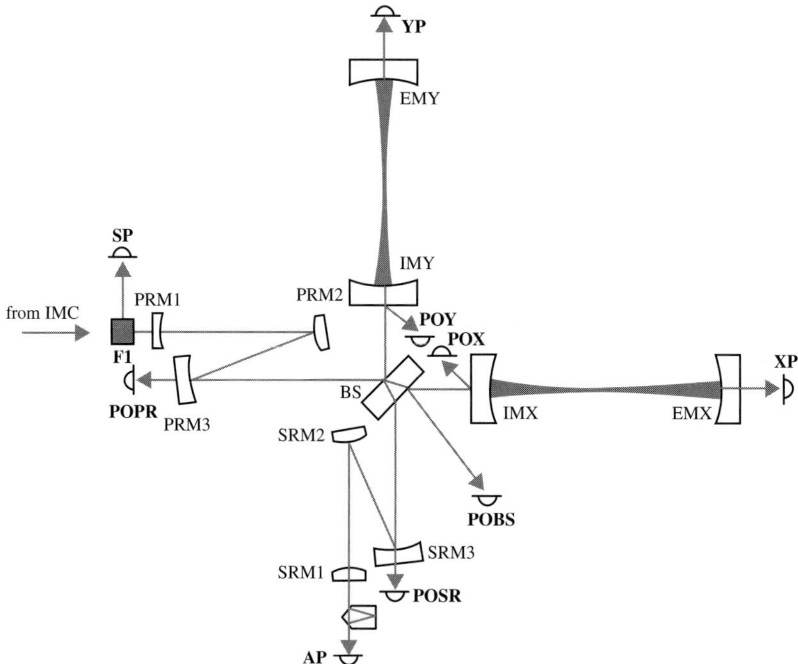

Figure 7.5 Advanced Virgo optical layout.

both cavity mirrors. These are the largest spots that can be accepted with a mirror small enough to be machined with a good optical surface. The present baseline foresees the use of non-degenerate (or stable) cavities both for recycling and dual-recycling, with two turning mirrors and a folded path (see Figure 7.5). The non-degenerate condition means that the high-order modes are not at resonance when the fundamental Gaussian mode resonates in the cavity. This is acheived by taking the spacing of the first transversal mode much larger than the linewidth of the cavity itself, and paying attention to avoid resonance of the higher-order modes. In this configuration, the interferometer is more robust against mirror thermal deformations and misalignments, responsible of high-order mode generation. The non-degeneracy is obtained increasing the length of the cavity. Due to infrastructure constraints, which prevent a significant lengthening of the central cavities, this condition will be likely achieved in Advanced Virgo by a folded path, such as that depicted in Figure 7.5. The additional mirrors of the power recycling cavity (PRM1, PRM2, PRM3) in Figure 7.5 must be suspended from the same superattenuators used for the beam splitter and for the injection bench. The same applies to the dual recycling cavity mirror (SRM1, SRM2 and SRM3). Different arrangements are also under study in order to incorporate the non-degenerate cavity design in Advanced Virgo, since the multi-mirror optical payload has still to be studied in detail, and its design and integration with suspensions could prove challenging. The response of an interferometer with dual recycling is given by a combination of dual recycling parameters, and the arm cavity finesse. We have seen in Section 3.5.2 that the dual recycling mirror transmittance changes the finesse of the dual recycling cavity, and thus the bandwidth of the detector, whereas the fine tuning of the length affects the cavity central frequency and, as a consequence, the frequency of the antenna peak sensitivity. The sensitivities achievable in Advanced Virgo for different choices of the dual recycling parameters are discussed in AdV Baseline Design (2009). It is important to stress that, once the mirror transmittance is fixed, the tuning of the sensitivity peak can be performed remotely, by adjusting the dual recycling mirror position by just a few tens of nanometers.

High power laser

The Advanced Virgo laser will have a power of about 200 W (four times that of Virgo+), with the goal of delivering at least 125 W in ITF input in the Gaussian fundamental mode, after mode filtering at the input of the injection system and the pre-mode cleaner, located on the external bench. The laser is based on the so called DPSSL technology (Diode Pump Solid State Laser) (Frede *et al.*, 2005), which foresees the use of a master laser and two stages of amplification. The possibility of using a fiber amplifier to achieve these specifications is currently under study.

Advanced Virgo injection system

The working principle of the Advanced Virgo injection system is identical to the Virgo one, with the IMC used as a reference to lock the laser frequency, and the reference cavity to stabilise the IMC length in the low frequency range. The same geometry, finesse and reference cavity of Virgo, will be used. The main effort concerns the optical properties of the cavity mirrors, which must have a coating absorption below 1 ppm, and an rms micro-roughness less than 1 nm. This will reduce thermal effects and diminish optical losses. Other stringent requirements for the Advanced Virgo injection system, such as a small intensity noise of the beam sent to the interferometer (which must be less than $3 \cdot 10^{-9} \mathrm{Hz}^{-1/2}$ at 10 Hz) are reported in AdV Baseline Design (2009). After the IMC the intensity stabilisation system, similar to the Virgo one, will provide the signal for stabilising the laser relative intensity noise, and reach the requirements.

Advanced Virgo Mirrors

The baseline foresees the use of Fabry–Perot cavity mirrors with a diameter of 35 cm and a 20 cm thickness (the same diameter as the Virgo mirrors but twice as thick and heavy). This choice will enable reduction of the pendulum thermal noise (scaling with the inverse of the square root of the mirror mass) and the radiation pressure noise (see Section 3.5.2). A mirror substrate made from a new type of fused silica, with a bulk absorption three times smaller than that used in Virgo (keeping the same quality factor, refractive index homogeneity and residual strain) has been selected. The roughness of the mirror must be reduced from the present value of a few nanometers (corresponding to 300 ppm of optical losses) down to fractions of a nanometer, to fulfil the requirement to have 75 ppm of optical losses. Regarding the mechanical losses, the dominant dissipation mechanism in the mirror is still the coating. Lower dissipations have been measured titanium-doped using a tantalum pentaoxide (Ta_2O_5) coating (Harry *et al.*, 2007) (about 1.6 to 1.8×10^{-4}, in comparison with typical values of around 2×10^{-4}).

Advanced Virgo thermal compensation system

The wavefront distortions and the deformation of the high-reflectivity mirror surface are corrected by applying an external CO_2 laser on both faces of each cavity input mirror. As for Virgo+, the major problem concerns fluctuations of the laser intensity that can induce a spurious displacement of the mirror. In Advanced Virgo the relative intensity noise of the CO_2 laser must be reduced by about a factor ten with respect to Virgo+. This requirement is considered too challenging, and thus a new transmission optic (compensation plate) will be suspended from the superattenuator in the recycling cavity, behind the input mirror. The CO_2 laser will be used to adjust the lens compensation in order to correct for the optical wavefront distortion occurring on the mirror. In this way, the CO_2 laser radiation pressure has no effect on the high-reflectivity surface of the mirror in the cavity, where the measurement of the arm differential motion (i.e. of the gravitational wave effect) takes place.

Other improvements

A total pressure of 10^{-9} mbar is necessary to ensure the noise induced by the residual gas is below the lowest point of the Advanced Virgo sensititivity curve (around $3 \times 10^{-24} \mathrm{Hz}^{-1/2}$ at 300 Hz). Cryogenic traps will be used to separate the long pipes containing the arms of the interferometer and the central area vacuum chambers, where the superattenuator and the optical payloads are located. This will allow a bake-out to be performed on the tubes, without affecting the sophisticated components around the mirrors and its suspensions. Significant work will be necessary on the Virgo infrastructure to reduce the anthropogenic noise inside the experimental buildings. In particular, quieter machines and adequate acoustically insulated rooms around the external benches will be implemented. The strategy to be exmployed to achieve a stable locking and alignment, and the corresponding control noise, is yet to be determined. High accuracies (two orders of magnitude better than Virgo) in longitudinal length control will be required in order to avoid the re-injection of noise due to beam power fluctuations. As mentioned in Section 7.4, the locking of a high-finesse cavity is a challenging operation, since the crossing time of the resonance gets smaller as the finesse increases. One technique to overcome this problem is to use an additional laser, with a different wavelength, in order to decrease mirror reflectivity (5–15%) and thus lower cavity finesse. Once the cavity is locked on the additional laser, it is set in a deterministic way to its working point, where the main laser is at resonance, and the main error signals can be used. An offset in the auxiliary laser error signal is introduced, and a smooth transition to lock the cavity on the interferometer laser can be performed.

Acknowledgements

The authors gratefully acknowledge the support of the the Italian Istituto Nazionale di Fisica Nucleare and the French Centre National de la Recherche Scientifique for the construction and operation of the Virgo detector.

References

AdV Baseline Design. 2009. *https://pub3.ego-gw.it/itf/tds/VIR-027A-09.*
Amico, P., *et al.* 2002. *Rev. Sci. Instrum.*, **73**, 3318–3323.
Anderson, D. 1984. *Appl. Optics*, **23**, 2944–2949.
Arnaud, N., *et al.* 2005. *Nucl. Instrum. and Meth. in Phys. Res. A*, **550**, 467–489.
Beauville, F., *et al.* 2003. *Rev. Sci. Instrum.*, **74**, 2564–2569.
Beauville, F., *et al.* 2006. *Class. Quant. Grav.*, **23**, 3235–3250.
Beccaria, M., *et al.* 1997. *Nucl. Instrum. and Meth. in Phys. Res. A*, **394**, 397–408.
Bernardini, A., *et al.* 1995. *Rev. Sci. Instrum.*, **70**, 3463–3472.
Bondu, F., *et al.* 2002. *Class. Quant. Grav.*, **19**, 1829–1833.
Cagnoli, G., *et al.* 1999. *Phys. Lett. A*, **255**, 230–235.
Cagnoli, G., *et al.* 2000. *Rev. Sci. Instrum.*, **71**, 2206–2210.
Dattilo, V. 2003. *Phys. Lett. A*, **318**, 192–198.
Frede, M., *et al.* 2005. *Optics Express*, **13**, 7516–7519.
Harry, G., *et al.* 2007. *Class. Quant. Grav.*, **24**, 405–415.
Losurdo, G., *et al.* 2001. *Rev. Sci. Instrum.*, **72**, 3653–3661.
Marka, S., Masserot, A. and Mours, B. Tech. rept. LIGO-T020036-00-D (2002), 1997. Available at http://www.ligo.caltech.edu/docs/T/T020036-00.pdf.
Rolland, L. 2009. In: *Proceedings of the 8th Amaldi Conference on Gravitational Waves,* Columbia University, New York.
Vinet, J.Y., *et al.* 1996. *Phys. Rev. D*, **54**, 1276–1286.

Vinet, J.Y., *et al.* 1997. *Phys. Rev. D*, **56**, 6085–6095.
Virgo Collaboration. 2004a. *Astrop. Phys.*, **20**, 629–640.
Virgo Collaboration. 2004b. *Class. Quant. Grav.*, **21**, S433–S440.
Virgo Collaboration. 2008a. *Applied Optics*, **47**, 5853–5861.
Virgo Collaboration. 2008b. *Astrop. Phys.*, **30**, 29–38.
Virgo Collaboration. 2009. *Phys. Rev. A*, **79**, 053824.
Virgo Collaboration. 2010a. *Astrop. Phys.*, **33**, 131–139.
Virgo Collaboration. 2010b. *Astrop. Phys.*, **33**, 182–189.
Virgo Collaboration. 2010c. *Astrop. Phys.*, **33**, 75–80.

8

GEO 600

H. Lück and H. Grote

The GEO 600 detector is a 600 m baseline interferometer in Hannover, Germany. We begin by discussing the history of this detector and go on to describe the techniques used to achieve initial target sensitivity. We then describe a series of upgrades and advanced techniques that will increase the sensitivity of this instrument for frequencies above 500 Hz. We finish by summarising the future plans of GEO 600.

8.1 A bit of history

GEO 600 is a British / German gravitational wave detector (see Grote for the LIGO Scientific Collaboration, 2008) located in Germany close to the city of Hannover. The scientific goal of GEO 600, beside taking data for gravitational wave detection, is the demonstration and testing of techniques for advanced detectors. GEO evolved from a collaboration between the groups working on the Garching 30 m and the Glasgow 10 m prototypes. In 1989 these groups proposed to build an underground 3 km gravitational wave detector, 'GEO', in the Harz mountains in northern Germany (Hough *et al.*, 1989) based on earlier proposals (Maischberger *et al.*, 1985; Leuchs *et al.*, 1987). Although reviewed positively, a shortage of funds on both ends, in the British Science and Engineering Research Council and after the German reunification also in the German funding bodies, made the realisation of this large project impossible. The collaborators decided to try obtaining funds for a shorter detector and compensate for the shortness by implementing more advanced techniques. A suitable stretch of land to build a 600 m instrument was found 20 km south of the city of Hannover, owned by the Universität Hannover and the state of Lower Saxony. Funding could finally be obtained from PPARC (Particle Physics and Astronomy Research Council), MPG (Max-Planck-Gesellschaft), BMFT (Bundesministerium für Forschung und Technologie), and the State of Lower Saxony. The construction of GEO 600 started on September 1995. The following years were busy with the construction of the infrastructure and the vacuum system, which was finished in 1998 just before the implementation of suspension systems and optics started. The two input mode cleaners (see Figure 8.1) of GEO 600 were operative in 1999. GEO 600 was equipped with medium quality test optics in 2001 in order to gain experience with the handling and installation of 5 kg optics suspended from wires just a few times the thickness of a hair. Two of the mirrors were already 'monolithically' suspended by thin glass fibres (see Section 8.2). At the same time the principal investigators of GEO and LIGO signed a memorandum of understanding, agreeing on the sharing and joint analysis of data acquired by the instruments. The first stable operation of the power recycled interferometer was achieved in December 2001 immediately followed by a short coincidence test run with the LIGO detectors (Willke *et al.*, 2004), testing the stability of the system and getting acquainted with data storage and exchange procedures. The first scientific data run (S1), again together with the LIGO detectors, was performed in August and September of 2002, where GEO 600 achieved a duty cycle (fraction of time taking science data) of 97%. The work on implementation of signal recycling (discussed later in this section) and the exchange of all temporary optics for the final, monolithically suspended optics ended with another joint data run in fully dual recycled mode at the end of 2003 and into 2004. As the sensitivity of all detectors improved, the collaborative data runs were extended in length, culminating in the fifth data run (S5) lasting from November 2005 to October 2007. This data run, in which the Virgo detector also participated, concluded the operation of the first generation of gravitational wave detectors.

8.2 GEO 600 techniques

To achieve a sensitivity comparable to the larger LIGO (Sigg for the LIGO Scientific Collaboration, 2008) and Virgo detectors (Acernese *et al.*, 2008), GEO 600 had to use techniques to increase the sensitivity that were not planned for the first generation of the other instruments, such as signal recycling, multiple-stage pendulums with monolithic suspension, and electrostatic actuators.

In the early 1990s when the decision on the optical layout of GEO 600 had to be made, signal recycling (Meers, 1988) and resonant sideband extraction (Mizuno *et al.*, 1993a; Mizuno, 1995) were just being developed in the Garching laboratories around the 30 m prototype (Mizuno *et al.*, 1993b; Heinzel *et al.*, 1996). Optical delay lines had shown severe problems with scattered light in the experiments (Winkler, 1983). The technique of using arm cavities in combination with dual recycling, a combination of power recycling (Schnier *et al.*, 1997) and signal recycling (Strain and Meers, 1991; Freise *et al.*, 2000) was considered to be too immature in terms of robust control schemes for all relevant degrees of freedom to be used in GEO 600. Hence the choice fell on a simpler Michelson interferometer with folded arms (without arm cavities), giving an effective arm length of 1200 m, and dual recycling (see Figure 8.1).

The lasers used in the prototypes at that time were stabilised Ar^+ lasers with a wavelength of 514 nm. These 'dinosaurs', consuming tens of kilowatts to produce a few watts of noisy laser light, were replaced in the large-scale interferometers by the just evolving solid state Nd:YAG lasers, which have a much higher efficiency (see Kane and Byer, 1985; Frede *et al.*, 2004) and are inherently much more stable. The longer wavelength of 1064 nm also relaxed the requirements on mirror surface accuracy and micro roughness. In the 1990s these lasers showed good power stability only at frequencies in the MHz range. A modulation technique, which shifted the laser power stability requirements from the detection band into the radio frequency range, was used for generating the gravitational wave (GW) signal and

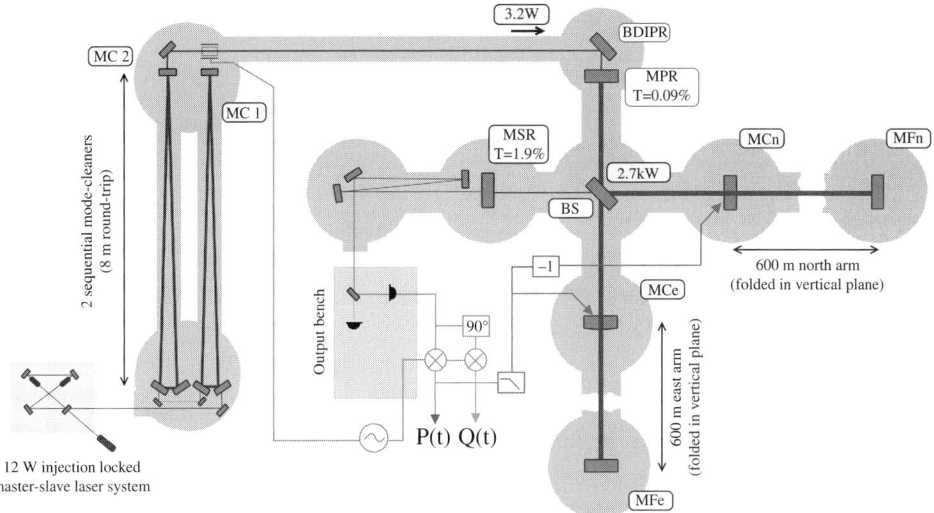

Figure 8.1 Schematic layout of GEO 600 with heterodyne read-out, as used until 2009. The light of the laser is sent through the two mode cleaners (MC1 and MC2) and enters the interferometer at the power recycling mirror (MPR) with a power of 3.2 W. The light power is enhanced in the power recycling resonator and reaches about 3 kW at the beam splitter (BS). Note that the arms are folded, i.e. the light path goes from the beam splitter via the far mirrors (MFn and MFe) to the inboard mirrors (MCn and MCe), which are suspended 25 cm above the outgoing beam, and then back via the same far mirrors to the beam splitter. The signal recycling mirror (MSR) is located to the left of BS in this figure.

with it the error signal for controlling the differential length of the interferometer arms. The GEO team decided to use Schnupp modulation (Schnupp, 1988) for this purpose. Phase modulating the light before it enters the interferometer in combination with a slight arm length difference (13.5 cm in the case of GEO 600) yields a signal at the modulation frequency. The amplitude of this signal depends on the deviation from the 'dark fringe' (the operating point of destructive interference at the output port) and hence gives an error signal for controlling the differential length degree of freedom and also yields the GW signal. The same modulation sidebands are also used to create error signals for an automatic alignment system as described in Grote *et al.* (2004). In the first generation all gravitational wave detectors used this readout scheme, which is called RF (radio frequency) or heterodyne readout.

Power recycling

In GEO 600 the light source is a master/slave 12 W Nd:YAG laser with a wavelength of 1064 nm (Zawischa *et al.*, 2002). Two subsequent optical resonators, the mode cleaners, with a finesse of about 2000 each, filter the light to lower the spatial fluctuations and reduce the higher-order transversal optical modes (Gossler *et al.*, 2003), before the beam is injected into the interferometer through modulators and Faraday isolators (see Figure 8.1). In order to optimise the signal-to-shot-noise ratio (SNR) the light power in the interferometer arms has to be as high as possible. Whereas the other large GW detectors use optical cavities in the interferometer arms to increase the light power, in conjunction with a relatively low power recycling factor, in GEO 600 the power is resonantly enhanced only in the power recycling cavity by the mirror MPR (see Figure 8.1) with a transmission of 900 ppm, with a high power recycling factor of about 1000. In this scheme, for one interferometer arm (the one in direct transmission) half of the light power has to traverse the beam splitter twice. The absorption in the beam splitter substrate leads to local heating, which results in a thermal lens (Winkler *et al.*, 1991; Hild *et al.*, 2006). This thermal lens limits the maximum light power level that can be used in GEO 600 to about 10 kW. Until 2009 the power level used at the beam splitter was about 3 kW.

Signal recycling

Similar to Power Recycling (where the carrier power being reflected by the interferometer back to the input port is resonantly enhanced) an additional recycling mirror (MSR in Figure 8.1) in the interferometer output port can resonantly enhance the signal sidebands. Bandwidth and centre frequency of the signal recycling cavity are determined by reflectivity and microscopical position of this signal recycling mirror, respectively. In GEO 600 a transmission of the mirror of 2% and a detuning from carrier resonance of about 2.2 nm was used until mid 2009. This resulted in a bandwidth (FWHM) of about 700 Hz and a detuning frequency of 530 Hz. A resulting noise curve is shown in Figure 8.2.

Seismic isolation and triple pendulum suspensions

For isolating the mirrors from ground motion GEO 600 uses a triple pendulum suspension system (Plissi *et al.*, 2000; Husman *et al.*, 2000). In places where fast actuation on the mirrors is required, an almost identical reaction pendulum is suspended next to the main pendulum chain. For the upper two stages of the pendulums, at low frequencies up to roughly 10 Hz and actuation ranges up to a few millimetres, magnet–coil actuators are used. The

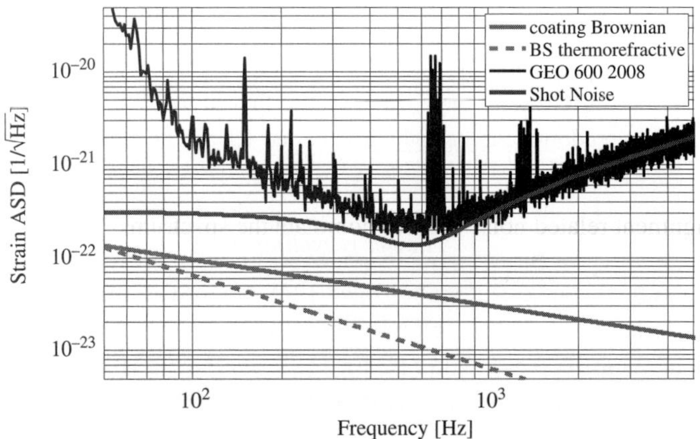

Figure 8.2 Noise contributions and composite noise of GEO 600 as of 2008

motion of the upper mass with respect to the holding frame is sensed with shadow sensors and the signal is fed back (with appropriate filtering) to the magnet–coil actuators to damp the mechanical resonances of the suspension. In the main interferometer electrostatic drive actuators are used (Hewitson *et al.*, 2007). These allow low-noise, small range (a few μm), and fast actuation directly at the mirror level without sacrificing the high mechanical quality factors of the mirrors by the use of magnets and without inducing environmental magnetic noise coupling .

The lowest stage of the GEO suspension is made in a quasi-monolithic way by suspending the mirror from the penultimate mass via four fused silica fibres of $\approx 210\,\mu$m diameter, which are welded to small fused silica plates, which in turn are bonded to the mirror sides using a hydro-catalysis bonding technique (Sneddon *et al.*, 2003). The internal modes of the fibres themselves are slightly damped by a coating of amorphous Teflon to ease the requirements on the length control loops (Gossler *et al.*, 2004).

8.3 The status in late 2009

From January 2006 to September 2009 GEO 600 collected almost 1000 days of scientific data. During the first 21 months, GEO 600 and LIGO jointly took data in the fifth scientific run S5 of the LIGO Scientific Collaboration, with Virgo joining part of the time. The last part of the data-taking period, called 'Astrowatch', which started in November 2007, covered the downtime of the LIGO (L1 and H1) and Virgo detectors during their upgrading period to the 'enhanced' versions, in order to avoid missing the detection of (unlikely) strong galactic events. Part of the time during the Astrowatch period was set aside for noise investigations and experiments preparing future upgrades. Even with these investigations, GEO 600 reached an observation time of 86.0% of the overall time. During Astrowatch GEO 600 was partly joined by the 2 km LIGO detector, LIGO-H2, at the Hanford site. The average sensitivity of GEO 600 was roughly a factor of two lower than that of the LIGO-H2 detector for frequencies above 500 Hz. Typical sensitivities of the GEO 600 and the LIGO-H2 detector during astrowatch are shown in Figure 8.3.

Throughout the whole data-taking period GEO 600 used Schnupp modulation for heterodyne readout, signal recycling detuned to a frequency of 530 Hz, a light power at the input to the power recycling cavity of about 3 W with a power recycling factor of about 1000. The peak sensitivity of GEO 600, corresponding to a minimum of the noise floor of $2.2 \times 10^{-22}\,\mathrm{Hz}^{-1/2}$, is reached around the detuning frequency of 530 Hz (see Figure 8.2). At higher frequencies the sensitivity of GEO 600 is mostly shot-noise limited. Below 100 Hz technical, alignment related noise sources dominate the spectrum.

8.4 Upgrade plans

With the beginning of the science run S6/VSR2 between the LIGO scientific collaboration and Virgo in July 2009, the upgrade program of the GEO 600 detector started. The new configuration will mainly concentrate on sensitivity improvements at frequencies above 500 Hz, where about an order of magnitude can be gained by lowering the shot noise. Besides

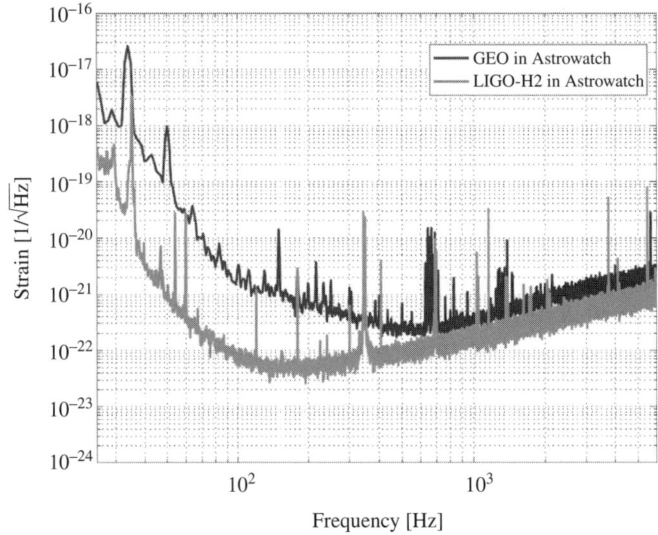

Figure 8.3 Noise performance of GEO 600 and LIGO-H2 during the 'Astrowatch' data run.

technical noise sources, thermal noise (mainly coating Brownian noise and thermorefractive noise at the beam splitter) will put a limit on the sensitivity of GEO 600 below 500 Hz The upgrades will include the following subsystem improvement steps:

- tuned signal recycling and DC readout
- implementation of an output mode cleaner
- injection of squeezed vacuum into the asymmetric port
- reduction of the signal recycling mirror reflectivity
- increased light power inside the interferometer.

All of these upgrade steps are independent from each other, as far as implementation and control is concerned. Therefore, we plan to establish full locking and analysis of the performance after each upgrade step, including data taking, to gain experience with the latest configuration.

Shot noise in an interferometric GW detector originates mainly from the vacuum fluctuations entering the output port (Caves, 1981). The contributions that come from the fluctuations on the input laser beam are small compared to those entering the output port because the input laser beam is well stabilised, filtered by the power recycling cavity, the interferometer is operated close to the dark fringe, and the Schnupp asymmetry and the imbalance in the losses of the arms of the interferometer are small. Reducing the vacuum fluctuations entering the output port in the 'phase quadrature' can therefore reduce the observed shot noise. This will be discussed later in the section on squeezing.

From RF readout to DC readout

So far GEO 600 used a heterodyne readout technique with Schnupp modulation to read out the gravitational wave signal and obtain error signals for the differential Michelson length control. In GEO 600 the modulation frequency is in the radio frequency range at about 14.9 MHz, and therefore the readout technique is also called RF readout or heterodyne readout, as mentioned in Section 8.2. Unfortunately this method increases the observed shot noise by collecting additional (vacuum) noise from twice the modulation frequency. These noise sidebands beat with the modulation sidebands at the modulation frequency, producing noise at the modulation frequency which gets demodulated into the signal frequency band (Buonanno *et al.*, 2003). This RF readout technique was chosen for all first generation GW detectors because the lasers used were shot-noise limited only in the MHz frequency range. The improved amplitude stability of (stabilised) solid state ND:YAG lasers in the detection frequency band, in addition to the filtering properties of the power recycling cavity, now allows to switch to a self–homodyne readout technique, also called DC readout. Here the operating point is slightly (in the range of about 5 pm to 50 pm) detuned from the dark fringe, which yields a dependence of the light power at the output port on the arm length difference. In the frequency space picture the laser carrier frequency serves as the local oscillator to beat with the sidebands induced by the gravitational waves. As no noise contributions are mixed into the GW frequency band from other parts of the spectrum that do not carry GW information, the shot noise for DC readout is lower than for RF readout. All large interferometric GW detectors will switch to DC readout on their way to the 'advanced' generation. The first experiences with DC readout in the GEO 600 detector are reported in Hild *et al.* (2009); Grote (2010); Degallaix (2010). DC readout is also beneficial for the injection of squeezed states: the squeezing level and the correct phasing has to be taken into account only at the laser carrier frequency and the respective signal sideband frequencies, but not at the modulation frequencies; see also the section later on squeezing.

Schnupp modulation with a reduced modulation index will still be used in GEO 600 for generating the error signals of the differential wave–front sensing for the auto–alignment system. As these reduced modulation sidebands will be removed with an output mode cleaner (see next section), RF oscillator phase noise (which was close to limiting the noise in RF readout) will no longer be a problem.

Output mode cleaner

The light at the output port of GEO 600 is not a pure Gaussian TEM_{00} mode but, owing to the imperfections of mirror surfaces and thermal effects in the optics, also contains higher-order modes. Due to different Guoy phases these higher-order modes, i.e. the non-TEM_{00} light, do not fulfil the same dark fringe interference conditions and hence do not respond to a differential length change with the same sign and amplitude as the main part of the beam, i.e. the TEM_{00} mode. Therefore the relative signal content in the higher-order modes is much lower than in the fundamental mode. The higher-order light, on the other hand, fully contributes to the shot noise, which reduces the maximally possible detector sensitivity. There are different solutions to this problem: One solution is to make the local oscillator beam stronger, such that the shot noise from higher-order mode light gets lower in proportion to the signal output and to the shot noise from the local oscillator light. This works for RF readout as well as for DC readout (where the DC offset determines the power in the laser carrier light, i.e. the local oscillator), as long as the local oscillator beam does not have too much higher-order mode content itself. Increasing the local oscillator level has technical constraints however, as more light power has to be detected, and the interferometer can become less stable in the case of DC readout with a large dark-fringe offset. If higher-order modes cannot be sufficiently reduced by thermal compensation systems (see the section on thermal compensation later), another solution is the implementation of an output mode cleaner (OMC), which can remove most of the higher-order mode light before detection. This is planned for GEO 600. The GEO 600 OMC (Degallaix, 2010) is a four-mirror cavity with a round trip length of roughly 66 cm as shown in Figure 8.4. The mirrors are mounted on a fused silica base plate with epoxy glue cured with UV light. The finesse of about 150, giving a FWHM line width of about 3 MHz, assures that the modulation sidebands (14.9 MHz) are reflected by the OMC when it is tuned to carrier resonance. With proper mode matching, the light in the TEM_{00} mode will get transmitted and the unwanted light in higher TEM modes will be reflected. The target transmission for the OMC is 98%. One of the mirrors is mounted on a piezoelectric actuator. Dithering of the OMC length with this piezo actuator allows the generation of error signals for the length control and for the auto-alignment system aligning the beam onto the OMC. More details are described in Prijatel (2010). The OMC will be put into an additional vacuum vessel attached to (but separated by a viewport from) the main GEO vacuum system. This vessel will contain some mode matching optics for the OMC, the OMC itself, a Faraday isolator for the injection of squeezed light, and one or more photo diodes. The vacuum level foreseen for operation is

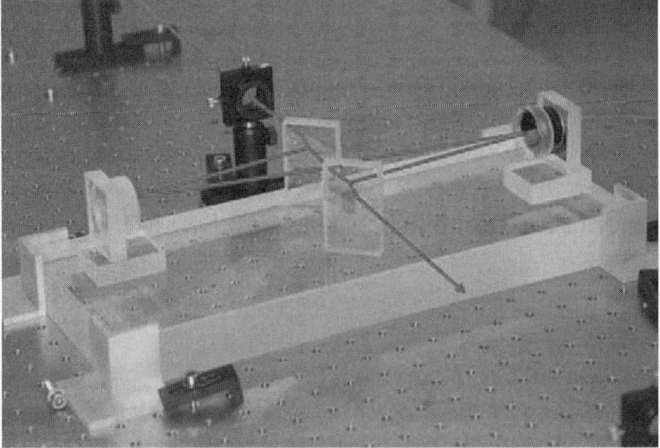

Figure 8.4 The output mode cleaner for GEO 600.

about 10^{-2} mbar. All the optical elements will be placed on a platform seismically isolated with a Minus-K® isolation system (http://www.minusk.com/).

Squeezing

At frequencies above 500 Hz shot noise is the dominant noise source in GEO 600. The upgrade plans are therefore aiming at reducing the shot noise level. The interferometer is operated close to the 'dark fringe'. Almost all light entering either the input or the output port is hence reflected back to the same port. The vacuum fluctuations entering and being reflected back to the output port are interfering with the carrier light, which is present due to the slight offset from the 'dark fringe' for DC readout, and give rise to the shot noise on the photo detector. Only the fluctuations in the same quadrature as the carrier light at the output port causes differential phase noise in the interferometer arms, hence we will call it the phase quadrature. As these fluctuations are causing the shot noise, the injection of squeezed light, where the noise in the phase quadrature is decreased, lowers the shot noise detected by the photo detector. The squeezed vacuum injected into the output port of GEO 600 will have a squeezing level of about 10 dB and with tolerable losses of about 15% can reach a shot noise reduction of about 6 dB, i.e. a factor of two in strain amplitude spectral density. The squeezing level is defined as the inverse ratio of the fluctuations in the relevant quadrature (here the phase quadrature) with repsect to the 'unsqueezed' vacuum fluctuations. The injection will be done via the Faraday isolator mentioned above. To make optimal use of the squeezing over the full detection frequency range, the squeezing ellipse (i.e. the phase of the injected squeezed light) must be optimally oriented with respect to the light leaving the output port for all frequencies within the detection bandwidth. In the case of a signal recycling cavity detuned with respect to the carrier, this can be achieved by sending the squeezed light through additional filter cavities (Harms *et al.*, 2003; Kimble *et al.*, 2001) or using twin signal recycling (Thüring *et al.*, 2007, 2009). For tuned (zero frequency detuning) signal recycling, the orientation of the squeezing ellipse is frequency independent and no filtering of the squeezed light is required as long as radiation pressure noise does not contribute to the sensitivity. The expected improvement in the GEO 600 sensitivity by using squeezing (together with a series of other changes) is shown in Figure 8.5.

Signal recycling modifications

In the past, GEO 600 used signal recycling mostly *detuned*, such that it is most sensitive at a signal frequency of 530 Hz. To improve the sensitivity over the full frequency range of interest, GEO 600 will use squeezing together with *tuned* signal recycling (Hild *et al.*, 2007). The achievable sensitivity is shown in Figure 8.5. The phase modulation caused by a gravitational wave creates two signal sidebands around the carrier light. Only one of these sidebands is resonantly enhanced in a detuned signal recycling cavity. The other sideband is far away from the resonance of the detuned signal recycling cavity. Tuned signal recycling, on the other hand, resonantly enhances both signal sidebands, providing higher optical gain on resonance. This results in lower shot noise at the resonance frequency than detuned

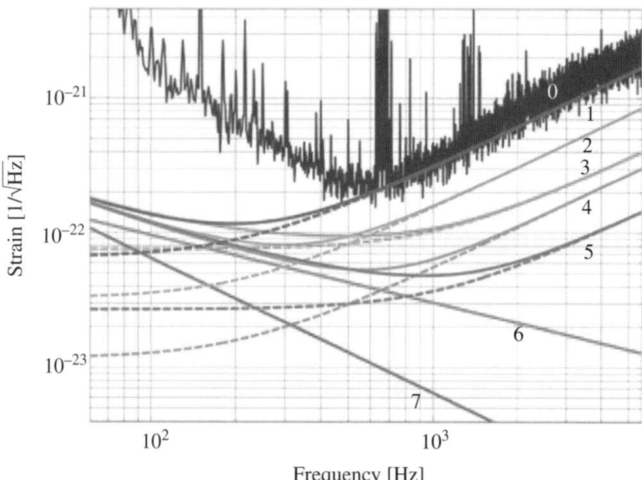

Figure 8.5 Noise contributions to GEO 600 for different configurations. Solid lines indicate total noise, dashed lines represent shot noise only. Technical noise sources are neglected here. Key: (0) GEO 600 sensitivity 2009; (1) tuned SR, $T_{SR} = 2\%$, $P = 3.2$ W; (2) tuned SR, $T_{SR} = 2\%$, $P = 3.2$ W, 6 dB squeezing; (3) tuned SR, $T_{SR} = 10\%$, $P = 3.2$ W, 6 dB squeezing; (4) tuned SR, $T_{SR} = 2\%$, $P = 20$ W, 6 dB squeezing; (5) tuned SR, $T_{SR} = 10\%$, $P = 20$ W, 6 dB squeezing; (6) coating thermal noise; (7) thermo-refractive noise of the beam splitter.

signal recycling. In the case of GEO 600 this does not influence the total sensitivity as at low frequencies the total noise is dominated by thermal noise and technical noises anyway.

Widening the bandwidth of GEO 600 by increasing the transmission of the signal recycling mirror from 2% to 10% reduces the high-frequency shot noise at the expense of the low-frequency noise. Being limited by thermal noises at frequencies below 700 Hz anyway, this only has a marginal influence on the overall noise for low frequencies. The crossover point between coating thermal noise and shot noise only moves from 700 Hz (curve 4 in Figure 8.5) to 800 Hz (curve 5 in Figure 8.5), while the high-frequency total noise drops by a factor of about two. Increasing the bandwidth further would raise the low-frequency shot noise to an undesirable level, while giving only marginal improvement at high frequencies.

Light power increase

The final step in the series of foreseen GEO 600 upgrades is an increase in light power inside the interferometer by a factor of about ten. Several changes are needed to achieve this goal, as described below.

Mode cleaner modifications

GEO 600 uses two consecutive optical resonators for removing light that is not in the fundamental Gaussian TEM_{00} mode and for reducing spatial beam fluctuations. Scattering inside these mode cleaners due to imperfections of the mirror surfaces results in a loss of light power between the laser and the interferometer of about 50%. If these losses are reduced, the throughput of the mode cleaners and hence the power available at the power recycling mirror can be increased. When bringing the mode cleaners into resonance, the light power inside the resonators and with it the radiation pressure on the mirrors suddenly changes. This sudden change of force acting on the mirrors exceeds the dynamical range of the longitudinal feedback system and needs to be compensated by an additional feed-forward control. Increasing the light power inside the mode cleaners, both by reducing the optical losses and by increasing the laser power (see below), would lead to undesirable problems with radiation pressure in the mode cleaners when keeping the same cavity finesse. Reducing the reflectivity of the coupling mirrors and with it the power build-up inside the mode cleaners, improves the throughput and reduces radiation pressure problems at the same time. The filtering properties of the initital mode cleaners were designed very conservatively (Winkler *et al.*, 2007) and will still yield a sufficiently high suppression level after the decrease in finesse. Exchanging the mode cleaner mirrors will increase the throughput by little less than a factor of two. The high-power compatibility of the subsequent auxiliary optics (electro-optic-modulator, Faraday isolators) still needs to be tested and confirmed.

Laser power increase

GEO 600 uses a 12 W master–slave Nd:YAG laser with a wavelength of 1064 nm at a power of about 6 W. With the optical losses in the mode cleaners of about 50% a laser power of about 3.2 W is available at the power recycling mirror (see Figure 8.1). The laser will be exchanged for a master–slave–amplifier system delivering about 30 W, which has been developed by the Laser Zentrum Hannover. It is the same kind of laser that is being used in Enhanced LIGO (Smith for the LIGO Scientific Collaboration, 2009). The increase in laser power by a factor of about 5, together with the increase in the throughput of the mode cleaners, will give a power increase of a factor of 10 in the power recycling cavity.

Local control changes

The mechanical resonances of the triple pendulum for seismic isolation of the main interferometer mirrors are actively damped by a local control system using shadow sensors to monitor the upper mass movement (Plissi *et al.*, 2000, Willke *et al.*, 2002). These shadow sensors are operated with a DC light source (an infrared LED) and unfortunately also detect some of the light scattered by the interferometer mirrors. This means that slight fluctuations in the circulating light power, which can originate from a slight misalignment of one of the interferometer mirrors, result in a signal at the shadow sensor which is treated as a movement of the upper mass. A 'compensation' signal is fed back to the actuator. This in turn misaligns the mirror controlled by this sensor and can lead to an instability. The coupling increases with the light power inside the interferometer. In GEO 600 this effect has already been observed at the current circulating light power level and could make the foreseen increase in laser power impossible. Operating the light source of the shadow sensor with current modulated in the kHz range and accordingly demodulating the signal of the photo-detector can make the system insensitive to low-frequency fluctuations of the light power in the interferometer. If the system fulfils the expectations, all local control systems of the main GEO optics will be converted to AC operation. Initial experiments have already shown an encouraging performance.

Thermal Compensation System

As a consequence of the missing arm cavities and the high power recycling factor of GEO 600, as described earlier in the section on power recycling, the full light power in the interferometer arms passes through the beam splitter. Although GEO uses fused silica glass (Suprasil$^{®}$ 311SV) with very low absorption of $0.5\,\text{ppm}\,\text{cm}^{-1}$, about $15\,\text{mW}$ will be absorbed inside the beam splitter with the upgraded light power of $30\,\text{kW}$ at the power recycling mirror. The local temperature increase causes a thermal lens which limits the maximum light power that can be used in the interferometer. A system shining infrared light onto the beam splitter, which is absorbed at the beam splitter surface, will be used to reduce the thermal lensing. Instead of using a CO_2 laser as in the other GW detectors (Waldman for the LIGO Science Collaboration, 2006; Acernese *et al.*, 2008), an incandescent light source will be used in GEO 600. This will minimise the fluctuations in heating power to the required level, which is beyond the current possibilities of a CO_2 laser. Due to the lack of arm cavities GEO 600 is as sensitive to beam splitter displacement as to a displacement of the mirrors in the interferometer arms. Because they have arm cavities, LIGO and Virgo are mostly sensitive to displacement of the cavity mirrors, but not to displacement of the beam splitter. On the other hand in LIGO and Virgo, due to the high light power in the interferometer arms and not at the beam splitter, the thermal effects mostly arise at the inboard cavity mirrors and are corrected using a correction plate outside of the cavity. The absorbed power in the GEO 600 beam splitter of $15\,\text{mW}$ is much less than the $0.4\,\text{W}$ of heating anticipated for the coating of the input mirrors of the arm cavities of advanced LIGO and advanced Virgo, so less heating power will be required for the GEO thermal compensation system.

8.5 In the future

As far as can be foreseen now, in the period from 2011 to 2015 GEO 600 will be the only interferometer in the world that can be online for astrophysical observations for a substantial amount of time. The LIGO interferometers and Virgo will both be upgrading to the advanced versions of their instruments. Therefore, according to current planning, GEO 600 will spend more and more time in observation mode, as all the upgrades foreseen for GEO 600 at this time will have been implemented, and the sensitivity should approach the planned goal.

Acknowledgements

The authors thank the German Centre for Quantum Engineering and Space-Time Research, QUEST, for support.

References

Acernese, F., *et al.* 2008. *Classical and Quantum Gravity*, **25**, 114045.
Buonanno, A., Chen, Y., and Mavalvala, N. 2003. *Phys. Rev. D*, **67**, 122005.
Caves, C. M. 1981. *Phys. Rev. D*, **23**, 1693–1708.

Degallaix, J. *et al.* 2010. *Commissioning of the tuned DC readout at GEO600. J. Phy. Conf. Ser.* **228** 012013.

Frede, M., *et al.* 2004. *Classical and Quantum Gravity*, **21**, S895–S901.

Freise, A., *et al.* 2000. *Physics Letters A*, **277**, 135–142.

Gossler, S., *et al.* 2003. *Review of Scientific Instruments*, **74**, 3787–3795.

Gossler, S., *et al.* 2004. *Classical and Quantum Gravity*, **21**, S923–S933.

Grote, H. and the LIGO Scientific Collaboration 2010. *The GEO 600 status. Class. Quantum Grav.* **27** 084003.

Grote, H., *et al.* 2004. *Appl. Opt.*, **43**, 1938–1945.

Grote for the LIGO Scientific Collaboration, H. 2008. *Classical and Quantum Gravity*, **25**, 114043.

Harms, J., *et al.* 2003. *Phys. Rev. D*, **68**, 042001.

Heinzel, G., *et al.* 1996. *Physics Letters A*, **217**, 305–314.

Hewitson, M., *et al.* 2007. *Classical and Quantum Gravity*, **24**, 6379–6391.

Hild, S., *et al.* 2006. *Applied Optics*, **45**, 7269–7272.

Hild, S., *et al.* 2007. *Classical and Quantum Gravity*, **24**, 1513–1523.

Hild, S., *et al.* 2009. *Classical and Quantum Gravity*, **26**, 055012.

Hough, J., *et al.* 1989. *MPQ Report*, **147**.

Husman, M. E., *et al.* 2000. *Review of Scientific Instruments*, **71**, 2546–2551.

Kane, T. J. and Byer, R. L. 1985. *Optics Letters*, **10**, 65–67.

Kimble, H. J., *et al.* 2001. *Phys. Rev. D*, **65**, 022002.

Leuchs, G., *et al.* 1987. *MPQ Report*, **129**.

Maischberger, K., *et al.* 1985. *MPI for Quantum Otpics Report No. MPQ96*.

Meers, B. J. 1988. *Phys. Rev. D*, **38**, 2317–2326.

Mizuno, J. 1995. *Comparison of optical configurations for laser-interferometric gravitational-wave detectors.* M.Phil. thesis, Leibniz Universität Hannover.

Mizuno, J., *et al.* 1993a. *Physics Letters A*, **175**, 273–276.

Mizuno, J., *et al.* 1993b. *Physics Letters A*, **175**, 273–276.

Plissi, M. V., *et al.* 2000. *Review of Scientific Instruments*, **71**, 2539–2545.

Prijatel, M., *et al.* 2010. *J. Phys. Conf Ser.* **228** 0120140.

Schnier, D., *et al.* 1997. *Physics Letters A*, **225**, 210–216.

Schnupp, L. 1988. *Presentation at European Collaboration Meeting on Interferometric Detection of Gravitational Waves.* (Sorrento, Italy, October 1988).

Sigg for the LIGO Scientific Collaboration, D. 2008. *Classical and Quantum Gravity*, **25**, 114041.

Smith for the LIGO Scientific Collaboration, I. R. 2009. *Classical and Quantum Gravity*, **26**, 114013.

Sneddon, P. H., *et al.* 2003. *Classical and Quantum Gravity*, **20**, 5025–5037.

Strain, K. A. and Meers, B. J. 1991. *Phys. Rev. Lett.*, **66**, 1391–1394.

Thüring, A., *et al.* 2007. *Opt. Lett.*, **32**, 985–987.

Thüring, A., *et al.* 2009. *Opt. Lett.*, **34**, 824–826.

Waldman for the LIGO Science Collaboration, S. J. 2006. *Classical and Quantum Gravity*, **23**, S653–S660.

Willke, B., *et al.* 2002. *Classical and Quantum Gravity*, **19**, 1377–1387.

Willke, B., *et al.* 2004. *Classical and Quantum Gravity*, **21**, S417–S423.

Winkler, W. 1983. *MPQ Report*, **74**.

Winkler, W., *et al.* 1991. *Phys. Rev. A*, **44**, 7022–7036.

Winkler, W., *et al.* 2007. *Optics Communications*, **280**, 492–499.

Zawischa, I., *et al.* 2002. *Classical and Quantum Gravity*, **19**, 1775–1781.

Part 3

Technology for advanced gravitational wave detectors

9

Lasers for high optical power interferometers

B. Willke and M. Frede

Many experiments in modern physics set highly demanding requirements on their laser light sources. Precision metrology, laser cooling experiments and the quantum engineering of atoms and molecules are some example areas in which very stable lasers are indispensable. The first generation of interferometric gravitational wave detectors have been among the laser applications with the most challenging requirements, simultaneously requiring low fluctuations in power, frequency and beam pointing as well as high power levels of 10 W. Laser sources for second generation gravitational wave detectors need to fulfill even more demanding requirements, which we will discuss in the first section of this chapter. The second section is devoted to the design of lasers for advanced detectors followed by a section in which we discuss their stabilisation. The last section covers some laser concepts for third generation gravitational wave detectors.

9.1 Requirements on the light source of a gravitational wave detector

One of the fundamental noise sources of laser interferometric gravitational wave detectors directly related to the laser light is the shot noise in the interferometer readout. The ratio of a potential gravitational wave signal to the readout shot noise is proportional to the square root of the light power in the interferometer. Hence, gravitational wave detectors need high-power lasers in combination with resonant optical cavities to achieve high circulating power levels in the interferometer. First generation detectors use lasers with approximately 10 W light power and second generation instruments will require power levels of order 200 W. In general, increasing the light power in the interferometer improves the sensitivity until the noise introduced by fluctuating radiation pressure forces on their mirrors reaches the same level as the readout shot noise. At this power level the independent sum of the two is equal to the standard quantum limit (SQL). For light power, P, at frequencies, f, above the pendulum resonances, the response of the suspended mirrors of mass, m, to fluctuating radiation pressure forces can be approximated by

$$\delta x \approx \frac{1}{2\pi^2 c} \frac{\delta P}{m f^2},$$

such that the mirror displacement caused by radiation pressure shot noise scales as $\delta x \propto \sqrt{P}/mf^2$. Hence, radiation pressure shot noise is most relevant at low frequencies, whereas readout shot noise dominates for higher frequencies. Signal recycling techniques as well as the use of squeezed light make this analysis more complicated, such that only detailed simulations can reveal the frequency dependence of the quantum noise contributions to the interferometer sensitivity. The circulating power in first generation gravitational wave detectors is too small to reach the SQL, such that radiation pressure noise is negligible at all frequencies. In advanced detectors, however, both flavors of quantum noise – the readout shot noise and the radiation pressure shot noise – have to be considered. As the radiation pressure noise scales in proportion to the circulating power P, the old paradigm 'the more circulating power the better the sensitivity' no longer applies.

For advanced gravitational wave detectors a circulating power level of almost 1 MW was chosen, which can only be achieved by a combination of laser powers of order 200 W and optical cavities. For future detectors, the required laser power might be even higher if mirrors with larger masses are used or if a reduced finesse of the optical cavities is beneficial. Totally abandoning cavities in future detectors is unlikely, however, as they not only provide a power build-up, but also filter the temporal and spatial fluctuations of the beam.

As all advanced gravitational wave detectors rely on the use of several non-degenerated optical resonators, their lasers need to provide single-frequency radiation in a well defined and stable spatial mode. Furthermore, linearly polarised radiation is required to achieve a good interference contrast in the presence of non-symmetrical birefringence effects in the beam splitter or the interferometer arms. Power fluctuations of the laser beam can limit the sensitivity of detectors as well. The interferometers are stabilised to the dark-fringe operation point at which the coupling of laser power fluctuations into the gravitational wave channel is much smaller than at the mid-fringe point. The DC readout method employed by advanced detectors, however, requires a small deviation from the dark-fringe such that some carrier light is detected on the main photodiode. Fluctuations of this light, as well as of the light fields that produce radiation pressure forces on the suspended mirror, can limit the sensitivity. Frequency noise can couple into the interferometer output signal due to an intentional or accidental asymmetry of the storage time in the interferometer arms or via scattered light fields. Hence, sophisticated techniques are required to stabilise the laser frequency and power by means of feedback control loops or passive filtering. Due to the complicated coupling paths of laser fluctuations into the gravitational wave channel, detailed simulations are required to deduce the exact frequency-dependent stability requirements for the pre-stabilised laser light (see for example Somiya *et al.*, 2006, 2007).

For advanced gravitational wave detectors, simulations reveal that the frequency noise in the interferometer needs to be as low as several $\mu\text{Hz}/\sqrt{\text{Hz}}$, and a relative power noise in the lower $10^{-9}/\sqrt{\text{Hz}}$ range is required. Such extreme stability levels can only be achieved by starting with a low-noise laser system which is then further stabilised by means of nested control loops with sophisticated sensing methods. The laser stabilisation scheme employed for the Advanced LIGO detector is one possible way to meet such demanding requirements. Before we discuss the key elements of this stabilisation scheme in the third section of this chapter, we will first review methods to generate high-power single-frequency single-mode laser light.

9.2 Lasers for advanced gravitational wave detectors

One big challenge in the development of laser systems for the use in gravitational wave detectors arises from the fact that the high-power radiation has to have a diffraction-limited spatial profile and must be linearly polarised. Furthermore, the laser has to operate in a single-frequency mode. As already mentioned above, low fluctuations in power, frequency and beam pointing are required, which can only be achieved with a combination of a low-noise laser design and additional feedback noise control. For the latter, the laser design has to incorporate wide-range, high-bandwidth actuators. The duty cycle of gravitational wave detectors is strongly coupled to the reliability and long-term stability of their lasers. Drifts in the laser parameters have to be small enough not to exceed the range of interferometer control systems. Furthermore, the laser should be easy to maintain. To meet all these requirements several choices and trade-offs need to be made in the laser design process.

The tasks of generating high laser power and achieving the single-frequency requirement are usually split by the use of a master-oscillator power-amplifier (MOPA) design or by injection locking a high-power oscillator to a single-frequency master laser. Due to its high intrinsic stability a Nd:YAG non-planar ring oscillator (NPRO) is often used as the master laser. This laser type was invented by Kane *et al.* (1985) and NPROs are commercially available with output powers of up to 2 W (see InnoLight, 2009). A single Nd:YAG crystal forms a non-planar resonator, which in combination with an internal Faraday rotation provides all the features required for a unidirectional ring laser. Both the Faraday effect, achieved by placing a permanent magnet close to the crystal, and the non-planar design, cause a rotation of the polarisation state of the circulating light field. These two effects, however, add for one traveling direction of the beam and subtract for the opposite. This fact, together with the polarisation-dependent reflection coefficient of the front facet, cause different round trip losses for the two directions and the system responds with single-directional operation.

The monolithic design ensures small mechanical fluctuations of the cavity length, the dominant source of frequency noise for NPROs, are according to Willke *et al.* (2000), small changes in the crystal temperature caused by fluctuations in the pump power. These temperature fluctuations couple via the temperature dependence of the index of refraction and, to a smaller extent, through the thermal expansion of the crystal to the optical length of the resonator and thus to the laser frequency. For active frequency control a piezoelectric element mounted on top of the laser crystal can introduce a stress-induced change in the index of refraction in the laser resonator and can hence be used as a tuning element for the laser frequency. This actuator has a response of approximately 1 MHz/V and a useful bandwidth of about 100 kHz. For slow, wide-range actuation of the laser frequency the well-controlled crystal temperature can be changed with a tuning coefficient of approximately 3 GHz/K. The mode-hop-free tuning range of the laser is typically more than 10 GHz.

The relative power noise (RPN) of commercially available NPROs in the frequency band from 10 Hz to 10 kHz is RPN $\leq 1 \times 10^{-6}/\sqrt{\text{Hz}}$. Each NPRO comes with an internal noise-eater that suppresses the 30 dB high relaxation oscillation peak at around 1 MHz and reduces the low frequency RPN by a factor of approximately five.

In addition, NPROs have, due to mode selective pumping, a good spatial beam profile with in most cases less than 3% of the power in higher-order modes. They are slightly elliptically polarised and their polarisation extinction ratio can, by means of retarding plates, be improved from 1/6 to 1/100. A detailed characterisation of NPROs can be found in Kwee and Willke (2008), which provides information on the unit-by-unit fluctuations of eight NPROs, as well as day-by-day fluctuations of one unit over 3.5 months. The performance described above makes NPROs well suited as light sources for gravitational wave detectors. However, thermal-optical effects in the gain medium at the time of writing limit the output power to 2 W, and further power amplification stages are required.

One of the main objectives in high-power generation is to handle the thermal load in the laser crystals. Pump light is absorbed in the crystals and, due to the quantum defect, a relevant fraction of the pump power (e.g. approximately 24% in Nd:YAG systems; see Brown, 1998) is converted into heat. Furthermore, non-radiative decay from the pump to the ground level and absorption processes starting from the upper laser state produce heat. In a steady-state situation this heat has to be removed via a cooling interface which produces a temperature gradient within the crystal. The resulting stress sets an upper limit to the acceptable pump power, as this stress needs to stay below the stress-fracture limit of the crystal material. Furthermore, a thermal lens is generated by the temperature gradient in combination with the temperature dependence of the refractive index of the crystal, the thermal expansion of the crystal material and the photoelastic effect (see Koechner, 1970). To keep this thermal lens as small as possible is one of the main challenges in power scaling of fundamental-mode solid-state lasers. One solution, described by Kane *et al.* (1983, 1985) and Eggleston *et al.* (1984), for example, is to choose a zig-zag path of the laser mode in the direction of the thermal gradient and hence average over the thermal effects. Two laser systems for gravitational wave detectors based on slabs have been designed following this concept. One such system, proposed by Mudge *et al.* (2000), uses a side-pumped slab in a stable–unstable resonator configuration to allow power scaling by extending the slab in the unstable direction. Power levels up to 80 W were demonstrated by Mudge *et al.* (2002) before thermal lens effects caused a loss of mode control. In a second laser design, an end-pumped slab was used by Sridharan *et al.* (2006) in an amplifier configuration. A careful choice of the crystal angles and of the surface roughness of the slab was found to be essential to suppress parasitic oscillations, which cause degradation of the spatial beam quality and limit the achievable output power.

Another way to reduce the thermal lens is to align the laser mode parallel to the thermal gradient and choose the crystal aspect ratio such that the mode radius is large compared with the crystal thickness. This concept has been applied in the thin-disc lasers described in Giesen *et al.* (1994). The thin disc (typically 100–200 μm thick) allows excellent heat dissipation but requires highly doped laser materials and multi-pass arrangements to compensate for the small absorption length. Hence, the thin-disc design can only be used in combination with laser materials with high doping capabilities, such as Yb:YAG.

In fiber lasers (see, for example, Jeong *et al.*, 2004a), the thermal lens problem is reduced by extending the length of the active medium to produce a more beneficial surface-to-volume ratio. The laser mode is defined by the fiber core and for small cores ($d \leq 10\,\mu$m) is diffraction limited. On the down side, small-core diameters correspond to high laser intensities at which non-linear effects such as Raman and Brillouin scattering limit the laser performance. A more detailed description of fiber lasers for use in gravitational wave detectors is given in the last section of this chapter.

The concept chosen for the Advanced LIGO laser is an end-pumped rod laser design, which has some advantages in comparison to the above-mentioned laser concepts. In the end-pumped rod design the pump light can be matched to the circular beam shape of the Gaussian laser mode such that a very good spatial overlap between the gain medium and the laser mode is achieved. This leads to a high efficiency and a good beam quality of the laser system. The thermal effects in rod lasers are strong compared to the other designs but, as will be discussed in the following, can either be compensated or used advantageously to achieve stable fundamental-mode operation.

One limiting effect in rod lasers is the depolarisation caused by stress-induced birefringence, as described by Koechner (1970) and Tidwell *et al.* (1992). The temperature gradient causes stress in the radial rod direction, causing the principal axes of the birefringence to be in the radial and tangential directions. Hence, a linearly polarised laser mode senses different refractive indices at different locations: horizontally polarised light, for example, senses n_{radial} at the left and right side of the crystal, whereas at the top and bottom of the crystal it sees $n_{\mathrm{tangential}}$. At other locations the laser beam senses a combination of n_{radial} and $n_{\mathrm{tangential}}$. Hence, the rod acts as a retardation plate, the orientation and retardation of which depends on the location in the rod. In addition the spatial distribution of n_{radial} and $n_{\mathrm{tangential}}$ forms polarisation-dependent thermal lenses. This effect is called bi-focusing. Both depolarisation and bi-focusing can limit the achievable output power of linearly polarised lasers.

Depolarisation and bi-focusing can both be reduced to a large extent by choosing a naturally birefringent laser material such as neodymium doped yttrium vanadate ($Nd:YVO_4$), as was done by Frede *et al.* (2007) for the intermediate power stage of the Advanced LIGO laser. In this laser system the NPRO light passes four $Nd:YVO_4$ crystals in an amplifier configuration (see Figure 9.1). Each crystal is end-pumped by a $400\,\mu$m diameter, NA:0.22 fiber-coupled laser diode with a nominal power of 45 W. To increase the lifetime of these diodes the pump power is set to 33 W during normal operation. The laser crystals with a $3\,$mm $\times\,3\,$mm cross-section consist of a 2 mm undoped and an 8 mm long-doped $Nd:YVO_4$ region. The doping concentration is 0.3 at.%. The undoped region was bonded to the crystal to reduce the thermal stress at the plane where the pump beam enters the doped region. To separate the pump light from the laser light, dichroic 45° mirrors with an antireflection coating for the pump wavelength of 808 nm and a high-reflection coating for the laser wavelength of 1064 nm were implemented. Efficient cooling of the laser crystals was realised by wrapping them with $500\,\mu$m thick indium foil and by mounting this assembly in water-cooled copper blocks. After passing the four laser heads the 2 W NPRO beam is amplified

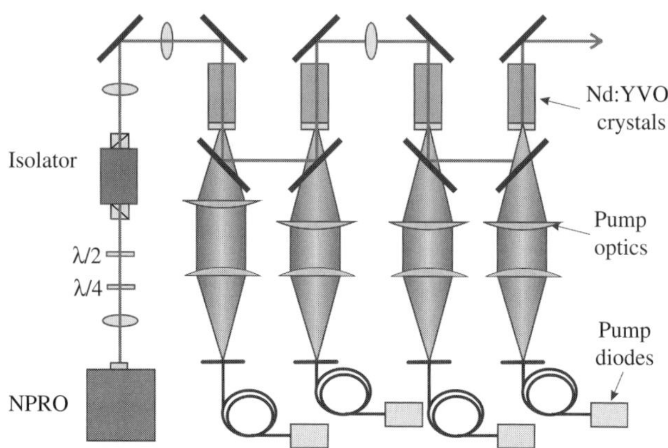

Figure 9.1 Schematic layout of the 35 W front-end laser system. The output of an NPRO is amplified by four Nd:YVO$_4$ crystals each pumped with a fiber-coupled laser diode.

to 35 W with retained good beam quality. In a power scaling experiment by Frede *et al.* (2007), output power levels of up to 65 W for a seed power of 20 W were achieved. The small difference in the center wavelength of the emission profiles of Nd:YAG and Nd:YVO$_4$ can be compensated by temperature tuning of the NPRO. Due to the approximately 0.9 nm (FWHM) broad emission profile of Nd:YVO$_4$, an efficient amplification was achieved over a large NPRO temperature range from 25°C to 40°C (see Frede *et al.*, 2007). The pump spot diameter in each crystal was optimised to achieve a high extraction efficiency without degrading the spatial beam profile. The spatial beam quality of the NPRO (97% TEM$_{00}$ mode content) is only slightly degraded by the amplification process, and 95% of the power of the 35 W laser was measured to be in the TEM$_{00}$ mode. This small degradation can be attributed to the third and fourth amplification stage, as after the first two stages a TEM$_{00}$ mode content of 97% was measured. The RPN of the 35 W laser is identical to the NPRO RPN at frequencies below 50 Hz and exceeds it by a factor of up to 5 for frequencies in the audio band. Lasers of this type were installed in the enhanced LIGO detectors and in the Virgo+ detector and showed stable and reliable performance. The combined system of the NPRO and the vanadate amplifier stages will be called the front-end laser in the following.

The advantages of the naturally birefringent front-end laser material cannot be utilised at much higher power levels due to the low fracture limit of Nd:YVO$_4$. Hence, a crystal material such as Nd:YAG and a depolarisation compensation scheme must be used.

The high-power stage for the Advanced LIGO laser (see Figure 9.2) is an injection-locked ring oscillator with four laser heads and two birefringence compensation units each formed by an imaging system and a quartz rotator. Each laser head consists of a water-cooled Nd:YAG rod and a pump unit. The rods are 3 mm in diameter and have a 40 mm long 0.1 at.% doped region. Two undoped, 7 mm long end-caps are bonded to the doped crystal to reduce the heat load at the rod faces and to allow for a uniform cooling of the active region by supporting and sealing the rods at these undoped ends. The pump units consist of seven fiber-coupled 808 nm laser diodes, each with a nominal output power of 45 W. As in the

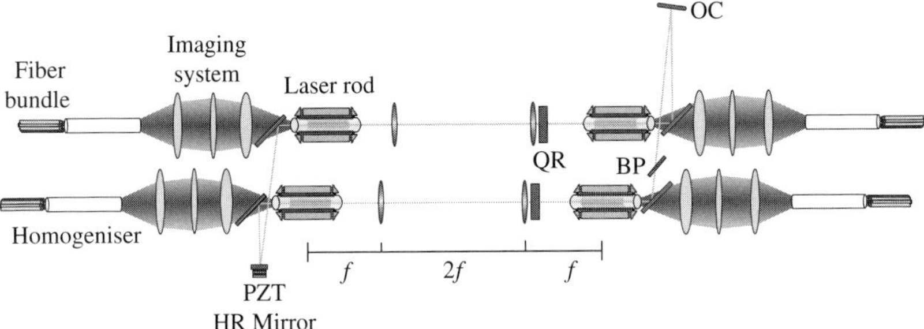

Figure 9.2 Schematic of the 200 W high-power stage. The ring laser consists of four laser rods grouped in two pairs, each with a depolarisation unit: two lenses and a quartz rotator (QR). Due to a Brewster plate (BP) the laser produces linearly polarised light which is partly transmitted through the output coupler (OC). A mirror mounted on a piezoelectric actuator (PZT) can be used to injection lock this laser to the 35 W front-end laser.

case of the front-end laser, the laser diodes are operated at reduced output power to increase their lifetime. The light leaving the fiber bundle is coupled into a 2 mm diameter, 100 mm long fused silica rod to spatially mix the pump light of the different fibers and to produce a uniform intensity profile which avoids hot-spots in the laser crystal. An additional benefit of this homogeniser is the fact that the spatial pump-light distribution does not change when one laser diode fails. In that case the reduced power can be compensated by the remaining six diodes and the faulty diode can be replaced at a convenient time. A telescope focuses the pump light into the laser rod where the pump light is guided by total internal reflection, due to the water-cooled rod surface acting as a waveguide. A highly reflective coating at 808 nm on the end-face of the rod leads to a double pass configuration for the pump light. This has the advantage of a more uniform pump-light absorption along the crystal axis than a single-pass arrangement would provide (see Frede *et al.*, 2004).

The above-mentioned bi-focusing and birefringence problem could be almost completely suppressed by the following compensation scheme (already described by Lü *et al.*, 1996): The four rods are grouped in two pairs, each with a lens combination that images one rod of a given pair into the second one. In addition, a quartz rotator changes the polarisation state of the beam between the two rods of a pair by 90°. Hence, the light has sensed approximately an equal distance of n_radial and $n_\mathrm{tangential}$ after traveling through a laser rod, the rotator and a second laser rod. This scheme works almost perfectly for a two-head pair, as described in Frede *et al.* (2004). Equally good results are achieved in the four-head Advanced LIGO laser. Less than 3% of the light power can be found in a characteristic depolarisation pattern in the beam reflected by the Brewster plate that serves as a polariser in the resonator. In an un-compensated setup depolarisation losses of up to 25 – 30% could be found. An aperture in the resonator, or a sophisticated design of the stability ranges of the high-power resonator as described in Frede (2007), ensure a fundamental-mode operation of the laser. The latter makes use of the aberrations of the thermal lens, which cause a dependence of its refractive power on the laser mode size. In fact higher-order spatial modes sense a reduced refraction in the outer lens areas which do not influence the fundamental mode. In the resonator's stability range this effect shows up by a shifted stability zone of the higher-order modes compared to the fundamental mode. Based on this difference, a specific pump power together with a specific resonator design can be chosen, such that only the fundamental mode of the resonator is stable.

To form the full Advanced LIGO laser system, as shown in Figure 9.2, the high-power ring laser is injection locked to the 35 W front-end laser. The two lasers are kept within the injection-locking range by means of a feedback control system with automatic lock acquisition functionality. The error signal for this feedback system is generated by utilising the Pound–Drever–Hall scheme.

A prototype version of this laser system has an output power of 220 W and it takes typically less than 10 ms to acquire the injection-locked state (Winkelmann *et al.*, 2010). Once injection-locked, a control loop feeding back to the length of the high-power oscillator cavity via a fast piezoelectric actuator (PZT) and a slow long-range actuator keeps the system in lock for several days until an external disturbance occurs. Even though particularly high attention in the laser design was paid to choose low-noise techniques, the temporal and spatial fluctuations of the laser beam are still too high to meet the demanding requirements of gravitational wave detectors. Hence, additional laser stabilisation is required, which will be discussed in the following section.

9.3 Laser stabilisation

The light sources of advanced gravitational wave detectors have to meet the following stability requirements: The frequency noise in the interferometer needs to be below several μHz$/\sqrt{\text{Hz}}$, which corresponds to a relative stability of $10^{-20}/\sqrt{\text{Hz}}$. The relative power noise, especially at low Fourier frequencies where the radiation pressure coupling is most significant, has to be in the low $10^{-9}/\sqrt{\text{Hz}}$ range, and the pointing of the beam injected into the interferometer has to be below corresponding angular fluctuations of $10^{-11}\,\text{rad}/\sqrt{\text{Hz}}$ or lateral shifts of $10^{-10}\,\text{m}/\sqrt{\text{Hz}}$. Furthermore, these three laser parameters have to be stable in the control band of the detectors (0.1 Hz–10 Hz) to avoid non-linear noise up-conversion into the gravitational wave band and saturation in length or alignment control loops of the detector.

Such extreme stability levels can only be achieved by nested control loops with sophisticated sensing schemes. The laser stabilisation employed for the Advanced LIGO detector is one possible way to meet such demanding requirements and will serve as an example to discuss the key elements of laser stabilisation.

Figure 9.3 shows a schematic layout of the Advanced LIGO pre-stabilised laser system (PSL), including a pre-mode-cleaner (PMC) and the frequency stabilisation path. The PMC is a rigid-spacer ring-cavity with a round-trip length of 2 m. One of the tasks of the PMC is to reduce the higher-order mode content and pointing of the laser beam. Only a modest pre-filtering of the beam is required within the PSL, as the largest fraction of the spatial filtering of the laser is provided further downstream inside the instrument's vacuum system by the suspended mode cleaners. In the Advanced LIGO case, the PMC has to filter higher-order modes of the laser beam to a level of $\leq 5\%$ and needs to reduce the beam pointing

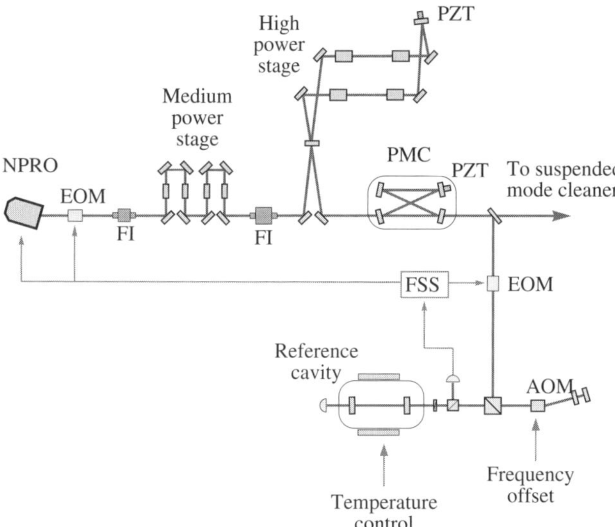

Figure 9.3 Layout of the Advanced LIGO pre-stabilised laser system including a schematic of the nested frequency control scheme. EOM, electro-optical modulator; AOM, acousto-optical modulator; FI, Faraday isolator; PMC, pre-mode-cleaner resonator; PZT, piezoelectric transducer; FSS, frequency stabilisation electronics.

by a frequency-dependent factor of up to several hundred. Both tasks are accomplished by a PMC with a finesse of $F \approx 130$. The PMC furthermore provides filtering of technical RPN (as opposed to quantum RPN or shot noise) in the MHz frequency range, where the interferometer sensing and control requires that the technical RPN stays below the shot noise of 100 mW. The RPN filtering transfer function of an optical resonator can be described as a low-pass with a corner frequency equal to the half-width at half-maximum linewidth of the resonator. In the case of the advanced LIGO PMC, the corner frequency is approximately 580 kHz such that a power noise filtering by a factor of 11 at 9 MHz is achieved. The length of the PMC is controlled via a PZT to keep the PMC resonant with the injected light. A Pound–Drever–Hall sensing scheme, which employs the same phase modulation sidebands as the injection-locking control, provides the error signal for this feedback control loop.

Concerning the frequency stability, the goal for the PSL of gravitational wave detectors is to achieve an intermediate level in the $100 \, \mathrm{mHz}/\sqrt{\mathrm{Hz}}$ range and to provide high-bandwidth actuators for control loops that use the long gravitational wave detector cavities as frequency references. In the Advanced LIGO case (see Figure 9.3) the laser is pre-stabilised to a rigid-spacer high-finesse reference cavity that is suspended in a thermally controlled vacuum chamber on the laser table. For this purpose a small fraction of the power is picked off the main laser beam and is injected into the reference cavity after a set of phase modulation sidebands are imprinted on the beam with an electro-optical modulator (EOM). The Pound–Drever–Hall method is used to determine the deviation of the laser frequency from one of the resonant frequencies of the reference cavity and this information is used in a control loop to minimise this deviation. The control signal is fed back to the PZT frequency actuator of the NPRO and to a second EOM placed between the NPRO and the 35 W amplifier, such that the EOM can act as a fast phase corrector.

The limit of the achievable frequency stability in such a scheme is set by thermal noise in the coatings of the reference cavity mirrors, as was shown by Numata *et al.* (2004). This limit is several orders of magnitude higher than the stability required inside the interferometer, and the PSL has to provide actuators for further frequency control. A double-passed acousto-optical modulator (AOM) is placed between the pick-off point and the reference cavity such that it can effectively change the frequency of the main laser beam with respect to the beam that is stabilised to the reference cavity. This AOM, together with the temperature control of the reference cavity, serve as the actuators for feedback control systems that stabilise the laser to the suspended mode cleaner and the interferometer. The main coupling path of laser frequency fluctuations into the gravitational wave detector's output signal is given by an asymmetry in the interferometer arms which distinguish the system from a frequency-noise insensitive white-light interferometer. For the advanced detectors these asymmetries are expected to be at the same level as they were for the initial detectors, and hence the frequency stability requirements for the PSL stay almost the same. Therefore, the Advanced LIGO PSL will use the same frequency stabilisation scheme as initial LIGO.

Various mechanisms couple laser power fluctuations into the gravitational wave channel. These coupling paths depend strongly on the interferometer design and can be divided into two main classes. The first class couples directly via fluctuations in the light power detected on the dark port photodiode. In a real interferometer there is always at least a small amount of light on this photodiode – for example due to an intentional or accidental deviation from the dark-fringe operation point of the gravitational wave detector. In this case, a change of the light power injected into the detector would cause a change in the power on the output photodiode and would be interpreted as a gravitational wave signal. The second class couples indirectly via radiation pressure fluctuations into the position of the suspended interferometer mirrors. Even though the beam splitter ideally causes the technical power fluctuations on the laser beam to be split symmetrically into the two interferometer arms, any asymmetry, for example in the optical losses in the arms, can break this balance. This would result in an unequal power in the two arms and hence in a difference in the radiation pressure force on the suspended mirrors. Correspondingly a differential mirror displacement would occur which is then falsely interpreted as a gravitational wave.

Experimental investigations on existing gravitational wave detectors by Smith *et al.* (2008) as well as simulations for advanced detectors by Somiya *et al.* (2006, 2007) show complex features in the power noise coupling to the interferometer output. Depending on the origin of the carrier field in the interferometer output signal (differential loss in the interferometer arms or a dark-fringe offset), the RPN sidebands get a different weighting in the homodyne or heterodyne detection at the interferometer readout port. Furthermore, the tuning of the signal recycling cavity can strongly influence the effect of RPN on the interferometer sensitivity. Coupling of power fluctuations into length and alignment control loops makes the situation even more complex such that only a detailed simulation of the interferometer behavior can reveal the frequency-dependent RPN requirement for advanced instruments.

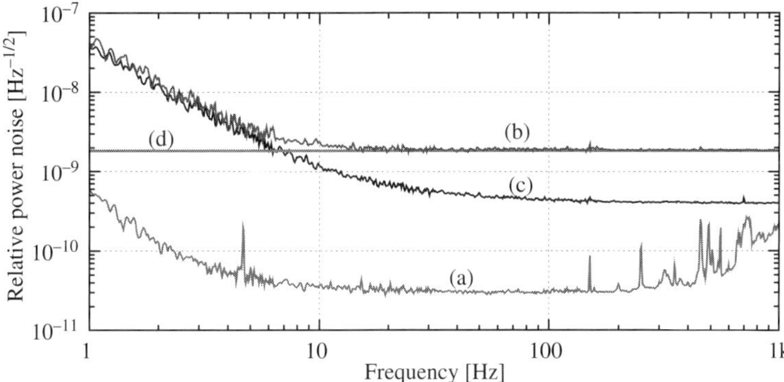

Figure 9.4 Linear spectral density of relative power noise measured in an NPRO power stabilisation experiment: (a) in-loop noise measurement; (b) out-of-loop result; (c) electronic noise; (d) estimated shot noise.

The most stringent requirement in the Advanced LIGO PSL was calculated to be RPN $\leq 2 \times 10^{-9}/\sqrt{\text{Hz}}$ at 10 Hz. As the main challenge of power stabilisation experiments is the sensing of the power fluctuations, initial stabilisation experiments were performed with easy-to-use and commonly available low-power lasers. Several such experiments were limited by unexplained excess noise. At the time of writing (2010) the required stability level has been demonstrated (see Kwee *et al.*, 2009). In this experiment extreme care was taken in the design of the electronics and in the reduction of scattered light in the optical setup. The laser beam was spatially filtered before it was detected with a multi-photodiode detector and the experiment was performed in a dust-free environment.

Figure 9.4 shows the result of this experiment. Curve (a) shows the noise spectral density measured with an in-loop detector when the control loop is closed. This curve represents the free-running noise divided by the loop gain. Curve (b) is the out-of-loop noise measured with an independent photodetector. Curve (c) shows the projected electronic noise (incoherent sum of noise in the two photodetectors), and the expected shot noise is indicated by curve (d). Curve (b) shows an RPN level of better than RPN $\leq 2.4 \times 10^{-9}/\sqrt{\text{Hz}}$ for frequencies between 10 Hz and 1 kHz. Under the assumption of equal-size but independent noise sources in the in-loop and out-of-loop photodetector, this noise level needs to be divided by $\sqrt{2}$ to estimate the noise on a laser beam available for downstream experiments such as a gravitational wave detector.

For a more detailed description of the laser stabilisation for advanced gravitational wave detectors we refer the reader to a review article by Willke (2010).

9.4 Lasers for third generation interferometers

The laser sources required for third generation laser interferometric gravitational wave detectors are strongly dependent on optical design. All reflective interferometers, possibly with heavy silicon mirrors, can handle much higher power levels than initial or advanced instruments. Detectors based on Sagnac interferometers can operate with a laser source even with low temporal coherence, which has advantages concerning the coupling of stray light into the gravitational wave channel. Layouts with optical cavities require single-frequency lasers with a high frequency stability. Coating thermal noise considerations might require a shorter wavelength of the laser light, whereas at longer wavelength silicon has a high transmission and can be used as the material for transmissive optics. The preferred spatial beam profile might not be a fundamental Gaussian beam; close to a flat-top profile or a higher-order Laguerre Gaussian mode might be preferable. Even though thermal loading of the interferometer might limit the useful power level in the interferometer, an increase in the laser power might allow one to abandon the power recycling mirror or to reduce its reflectivity.

Hence, the interferometer layout as well as advances in laser technology will strongly influence the laser design of third generation gravitational wave detectors. Possible topologies are master-laser power-amplifiers (MOPA), injection-locked systems, or a mixture of the two. Different types of gain-medium geometries can be accommodated, for example slabs, rods, fibers or discs, which could be made from crystalline material, glasses or ceramics. The variety of possible requirements makes it impossible to make general predictions as to the laser source for third generation detectors. We will only highlight some design considerations in this section.

At the development time of second generation gravitational wave detectors, neodymium-doped yttrium aluminum garnet (Nd:YAG) was the best choice as the gain material for 100 W class lasers. However, in the future, should kilowatt-class lasers become necessary, ytterbium-doped YAG, which lases at 1030 nm, could replace the Nd dopant, as Yb:YAG has a lower quantum defect and a better thermal management. Furthermore, the 940 nm pump diodes used have potentially longer lifetimes. Its main disadvantage is that Yb:YAG is a quasi-3-level system and thus more sensitive to increased temperatures within the gain medium.

Different design concepts are proposed to produce lasers with power levels of several 100 W and to amplify these systems into the kW region. The main concerns are the thermal management in the gain material and the reduction of beam aberrations. One way to reduce the thermal effects is to use a zig-zag beam path to average over the thermal gradient in the laser crystal (as already discussed in Section 9.2). Edge-pumped slab geometries can be combined with conduction-cooling techniques which avoid vibrations introduced by cooling fluids in conventional layouts. However, one of the main challenges in using slabs is to avoid parasitic oscillations within the high-gain regions.

An efficient birefringence compensation can reduce problems caused by depolarisation and by defocusing. However, rather than compensating effects caused by the thermal gradients, it is better to reduce these gradients by the use of multi-segmented laser rods. The maximum peak temperature of an end-pumped laser-rod or slab can be reduced, as shown in the work by Wilhelm *et al.* (2009). To decrease the overall head load in a Nd:YAG laser medium the pump wavelength can be changed from 807 nm to 885 nm, which reduces the quantum defect and therefore the overall heat load by more than 30% (see Lavi and Jackel, 2000; Frede *et al.*, 2006).

Core-doped rods can be used to achieve an easier and more stable fundamental mode operation, as shown in the work by Kracht *et al.* (2006). These rods are comparable to a double clad fiber, as described by Bedö *et al.* (1993), where only the inner core of the rod is doped and the outer core is used as a waveguide for the pump light. As the gain is only present in the doped inner core of the rod, this concept is similar to mode-selective pumping, but has the advantage that no high brightness pump source is required.

Optical fiber amplifiers have a high potential to offer single-frequency output at higher efficiencies and at lower cost than solid-state amplifiers at similar power levels (see, for example, the overview paper by Limpert *et al.*, 2007). Until several years ago diode-pumped fiber amplifiers were limited to power levels of several watts due to the unavailability of high brightness pump diodes and due to non-linear effects in the fiber such as stimulated Raman scattering and stimulated Brillouin scattering (SBS). The invention of large mode-area (LMA) fibers and of photonic crystal fibers (PCF) has enabled output powers of single-mode fiber lasers to exceed 1 kW while retaining excellent efficiencies (see, for example, Jeong *et al.*, 2004b). The large effective core diameter of these fibers decreases the average intensity of the light at the laser wavelength in the fiber and thereby increases the threshold of non-linear processes. The large inner cladding of the double-clad LMA fibers allows high-power multi-mode pumps to be coupled into the fiber. Bending losses can be used to ensure that the output remains single-mode, despite the large diameter of the core.

The limiting factor for narrow-linewidth high-power fiber lasers for the use in gravitational wave detectors is the onset of SBS. A state-of-the art fiber amplifier system with 150 W of output power with a good output beam profile (92% in TEM_{00}) is described in Hildebrandt *et al.* (2006). The optical-to-optical efficiency of this system with respect to incident pump power is 78% for a 195 W pump source. A good polarisation ratio of about 1 : 100 was achieved. Novel ideas to increase the SBS threshold are under investigation. A promising concept is to shift the Brillouin frequency along the fiber to lower the effective Brillouin gain for each frequency component. This could be achieved by temperature or strain gradients, or by varying doping concentrations along the fiber.

In addition to the reduction of the non-linear scattering effect, the reliability and noise performance of high-power fiber lasers need to be further analysed and possibly improved to meet the requirements of third generation gravitational wave detectors. In particular thermal effects and contamination at the air–glass interface have to be considered. The main problem is the large light intensity at these interfaces, which could be reduced by un-doped beam expansion sections at the fiber ends or by all-fiber solutions for the pump-light coupling. One big advantage of fiber lasers is that they are compact and simple compared to the complex solid-state laser systems. Furthermore, modern splice techniques allow one to produce an all-fiber system including the master oscillator, the high-power stage and possibly even a mode-cleaning fiber if required.

Reducing the laser wavelength allows one to reduce the thickness of the coating layers, and subsequently reduces coating thermal noise. Several gas lasers and dye lasers operate at wavelengths in the visible spectrum but their efficiency, reliability, controllability and noise performance rule them out as suitable lasers for gravitational wave detectors. In case the interferometer design requires a tunable light source or several closely spaced wavelengths,

Ti–Sa (titanium-sapphire) lasers could be chosen either at their fundamental wavelength (650 – 1070 nm) or in a frequency-doubled layout. Frequency doubling or even tripling of high-power near-infrared lasers are the most promising options to provide a single-frequency high-power light sources at shorter wavelength. An attractive approach is the second-harmonic-generation (SHG) in quasi-phase-matched ferroelectric materials such as magnesium-doped periodically poled $LiNbO_3$ (MgO:PPLN), MgO-doped periodically poled stoichiometric $LiTaO_3$ (MgO:sPPLT) and periodically poled $KTiOPO_4$ (PPKTP). Green power levels of 16 W have been demonstrated by the conversion of a solid-state laser by Tovstonog *et al.* (2008) and almost 10 W were achieved in a SHG experiment using an infrared fiber laser by Samanta *et al.* (2009). In a recent experiment Meier *et al.* (2010) were able to demonstrate 130 W of green light by frequency doubling an Advanced LIGO laser in a LBO crystal.

Erbium-doped fiber lasers emit around $1.56\,\mu m$ whereas the absorption in silicon is small compared to the initially used silica at $1\,\mu m$ wavelength. For an efficient design with low non-linear effects in single-frequency operation, the erbium-doped fiber should have high pump absorption and should be as short as possible. Pumping at 1480 nm is sufficient but the currently available pump power is limited to several watts. To overcome this limitation Yb co-doping and pumping at 980 nm can be used. This allows high pump absorption but also implicates a second gain band at the Yb wavelength around $1\,\mu m$. Nevertheless current experimental results show sufficient suppression of the $1\,\mu m$ oscillation in an Er–Yb fiber amplifier resulting in more than 8 W of single-frequency output power at 1556 nm (see Kuhn *et al.*, 2009).

As many different applications drive laser development worldwide, many laser concepts at different wavelengths and power levels are already available and advances in several fields are to be expected. Even though the large variety of optical layouts and topology options for gravitational wave detectors require a similarly large range of different laser parameters, we expect that at least one, if not more, laser designs will allow one to build a laser source with the required power, wavelength and spatial profile. However, there is currently no application for a high-power laser that has requirements on the temporal and spatial stability as stringent as those for gravitational wave detectors. Hence, a specific laser development program for third generation detectors will be required to design and build a reliable laser with sufficiently low free-running noise, an appropriate spatial beam profile and good controllability.

Acknowledgements

The authors gratefully acknowledge the support of the German Volkswagen Stiftung and the QUEST cluster of excellence at the Leibniz University Hannover.

References

Bedö, S., Lüthy, W. and Weber, H. P. 1993. *Optics Communications*, **99**, 331 – 335.
Brown, D. 1998. *Quantum Electronics, IEEE Journal of*, **34**, 560–572.
Eggleston, J., *et al.* 1984. *Quantum Electronics, IEEE Journal of*, **20**, 289–301.

Frede, M. 2007. *Einfrequentes Laserlicht höchster Brillianz.* Ph.D. thesis, Leibniz Universität Hannover, Germany.

Frede, M., Wilhelm, R. and Kracht, D. 2006. *Opt. Lett.*, **31**, 3618–3619.

Frede, M., *et al.* 2004. *Opt. Express*, **12**, 3581–3589.

Frede, M., *et al.* 2007. *Opt. Express*, **15**, 459–465.

Giesen, A., *et al.* 1994. *Applied Physics B: Lasers and Optics*, **58**, 365–372.

Hildebrandt, M., *et al.* 2006. *Optics Express*, **14**, 11071–11076.

InnoLight. 2009. Mephisto Product Line, http://www.innolight.de/.

Jeong, Y., *et al.* 2004a. *Conference on Lasers and Electro-Optics, Technical Digest*, **2**, 1065–1066.

Jeong, Y., *et al.* 2004b. *Opt. Express*, **12**, 6088–6092.

Kane, T., Eckardt, R. and Byer, R. 1983. *Quantum Electronics, IEEE Journal of*, **19**, 1351–1354.

Kane, T., Eggleston, J. and Byer, R. 1985. *Quantum Electronics, IEEE Journal of*, **21**, 1195–1210.

Koechner, W. 1970. *Appl. Opt.*, **9**, 2548–2553.

Kracht, D., *et al.* 2006. *Opt. Express*, **14**, 2690–2694.

Kuhn, V., *et al.* 2009. *Opt. Express*, **17**, 18304–18311.

Kwee, P. and Willke, B. 2008. *Appl. Opt.*, **47**, 6022–6032.

Kwee, P., Willke, B. and Danzmann, K. 2009. *Appl. Opt.*, **48**, 5423–5431.

Lavi, R. and Jackel, S. 2000. *Appl. Opt.*, **39**, 3093–3098.

Limpert, J., *et al.* 2007. *Selected Topics in Quantum Electronics, IEEE Journal of*, **13**, 537–545.

Lü, Q., *et al.* 1996. *Optical and Quantum Electronics*, **28**, 57–69.

Meier, T., Willke, B. and Danzmann, K. 2010. *Opt. Lett.*, **35**, 3742–3744.

Mudge, D., *et al.* 2000. *Selected Topics in Quantum Electronics, IEEE Journal of*, **6**, 643–649.

Mudge, D., *et al.* 2002. *Classical and Quantum Gravity*, **19**, 1783–1792.

Numata, K., Kemery, A. and Camp, J. 2004. *Phys. Rev. Lett.*, **93**, 250602.

Samanta, G. K., Kumar, S. C. and Ebrahim-Zadeh, M. 2009. *Opt. Lett.*, **34**, 1561–1563.

Smith, J. R., *et al.* 2008. *Classical and Quantum Gravity*, **25**, 035003.

Somiya, K., *et al.* 2006. *Phys. Rev. D*, **73**, 122005.

Somiya, K., *et al.* 2007. *Phys. Rev. D*, **75**, 049905.

Sridharan, A. K., *et al.* 2006. *Appl. Opt.*, **45**, 3340–3351.

Tidwell, S., *et al.* 1992. *Quantum Electronics, IEEE Journal of*, **28**, 997–1009.

Tovstonog, S. V., *et al.* 2008. *Opt. Express*, **16**, 11294–11299.

Wilhelm, R., *et al.*, 2009. *Opt. Express*, **17**, 8229–8236.

Willke, B. 2010. *Laser & Photonics Reviews*, **102**, 529–538.

Willke, B., *et al.* 2000. *Optics Letters*, **25**, 1019–1021.

Winkelmann, L. *et al.* 2010. *App. Phys. B*, **102**, 529–538.

10

Thermal noise, suspensions and test masses

L. Ju, G. Harry and B. Lee

In this chapter we discuss how thermal noise can reduce the sensitivity of gravitational wave interferometric detectors and how these losses can be suppressed. We begin with a brief introduction to the general theory of thermal noise. Then we discuss the techniques and choices of materials that can minimise thermal noise losses in three key areas: suspensions, test mass substrates and test mass coatings.

10.1 Introduction

A mechanical system that is in equilibrium with a thermal reservoir will have mechanical modes, each with average energy $k_B T$, with k_B being the Boltzmann constant and T the temperature. Randomly fluctuating thermal forces from this bath drive the mechanical system and, through its mechanical impedance, a random displacement results. It is this thermally induced displacement noise that is known as thermal noise. For a low-loss system, most of the thermally induced energy exists within a small frequency band centered around the mechanical mode frequency. The off-peak broad thermal noise power in other frequency regions is referred to as thermal noise floor. According to the fluctuation dissipation theorem (Callen and Welton, 1951), the general power spectrum of the minimal fluctuation force is F_{th}

$$F_{th}^2 = 4\, k_B\, T\, R(\omega) \tag{10.1}$$

where $R(\omega)$ is the real part of the impedance Z of the system. Using the force–impedance relationship $Z = F/v = F/(\omega\, x)$, the above equation can be expressed as

$$x_{th}^2 = \frac{4\, k_B\, T\, \sigma(\omega)}{\omega^2} \tag{10.2}$$

with $\sigma(\omega)$ the real part of the admittance $Y(\omega) = 1/Z(\omega) = \omega\, x/F$.

For a simple harmonic motion the thermal noise power spectrum of the oscillating object is given by (Saulson, 1990):

$$x^2(\omega) = \frac{4 k_B T}{m} \frac{\omega_0^2 \phi(\omega)}{\omega\big((\omega_0^2 - \omega^2)^2 + \omega_0^4 \phi^2(\omega)\big)} \tag{10.3}$$

Here ω_0 is the resonant frequency of the oscillator. It is through this equation that we could design an oscillator with minimum thermal noise:

1. Minimise the temperature T of the oscillator.
2. Maximise m to a reasonable value.
3. Minimise the loss, $\phi(\omega)$.

In an interferometric gravitational wave detector thermal noise can be reduced by lowering the temperature of the test masses while keeping all other parameters unchanged. This is the motivation behind the LCGT project (Large scale Cryogenic Gravitational wave Telescope) (Kuroda *et al.*, 2003), where the final stages of the suspension system are cooled to below 20 K. For LCGT, it is proposed that sapphire will be used as the test mass and the suspension fibre material. Although a reduction in temperature to 20 K will only reduce the thermal noise power directly by about one order of magnitude, the reduction in temperature has been shown to reduce the mechanical losses in several materials such as sapphire (Braginskii *et al.*, 1985), resulting in further thermal noise reduction. Operating the interferometer at cryogenic temperatures also has other advantages, such as a reduction in thermal lensing (Winkler *et al.*, 1991). However, operating cryogenic interferometer will add complexity to the system. The cryogenic part of the system has to be integrated with the room temperature vibration isolation system. There are also issues such as vibration from the cryopumps and potential contamination of the cold mirrors that need to be carefully addressed.

Larger masses imply more inertia, and thus more resistance to spurious forces, including thermal ones and radiation pressure noise. However, larger masses also require stronger suspension elements and control forces, placing a practical limit on the maximum test mass size. Other considerations such as availability and optical properties often take precedence when selecting test mass sizes. The test masses planned to be used by Advanced LIGO and Advanced Virgo are 40 kg. Future detectors such as the ET (Einstein Telescope) (EGO, 2010) that are made especially for low-frequency operation may consider the use of test masses in the order of 100 kg.

The largest proportion of thermal noise research has been devoted to the minimisation of the loss, $\phi(\omega)$. The choice of suspension and test mass materials, optical coating to the test masses, as well as the attachment of the suspension, all have a large influence on the level of thermal noise.

In laser interferometer gravitational wave detector research, thermal noise is mainly divided into three classes: the suspension thermal noise, the test mass substrate thermal noise and the mirror coating thermal noise. These are discussed below.

10.2 Suspension thermal noise

Suspension thermal noise is due to mechanical loss in the test mass pendulum suspensions. The primary suspension mode is the pendulum mode with frequency close to 1 Hz, which is lower than the observation frequencies. The suspension thermal noise discussed here is the off-peak noise floor.

In a typical pendulum suspension, the restoring force is due to two effects: the gravitational effect and the elastic effect from the bending of the suspension elements. The effective pendulum loss angle, ϕ_{pend} can be expressed as:

$$\phi_{\text{pend}} = \phi_{\text{mat}} \frac{E_{\text{el}}}{E_{\text{grav}} + E_{\text{el}}} \approx \phi_{\text{mat}} \frac{E_{\text{el}}}{E_{\text{grav}}} \tag{10.4}$$

Here ϕ_{mat} is the loss angle of the suspension material, E_{el} is the maximum amount of stored elastic energy and E_{grav} is the maximum amount of stored gravitational energy. Since the lossless gravitational restoring force can be made much larger than the lossy elastic restoring force, $E_{\text{el}}/E_{\text{grav}} = k_{\text{el}}/k_{\text{grav}} << 1$, it can be seen that the loss of the pendulum could be much lower than that of the loss of the suspension material loss. This effect is called dissipation dilution. The effective pendulum loss can also be expressed in terms of physical and geometric quantities as (Saulson, 1990):

$$\phi_{\text{pend}}(\omega) = \phi(\omega)_{\text{mat}} \frac{N\sqrt{T_0 Y_0 I}}{2mgL} \tag{10.5}$$

Here N is the number of suspension elements, m the suspended mass, g the acceleration due to gravity, L the suspension length, T_0 the tension in the suspension elements, Y_0 the Young's modulus of the suspension material and I the cross-section moment of inertia of the suspension element.

The simple pendulum suspension system will have a set of violin string modes from the suspension wire. The frequencies of the violin string modes depend on the length, thickness and tension of the suspension element. With a 1 Hz pendulum length, it is difficult to get materials strong enough to allow the suspension wires to be thin enough for the violin mode frequencies to be more than a few hundred hertz, which is still within the detection band, but the sensitivity is expected to be shot-noise limited. The high Q-factors of the violin string modes mean that the violin string thermal noise will be observed as high narrow spikes piercing the shot-noise floor. A readout scheme has been proposed to monitor the suspension thermal noise and then remove the peaks from the data (Santamore and Levin, 2001). This will have negligible effect on the signal-to-noise ratio (SNR) of the detector. The method however requires accurate sensing for each suspension fiber and a sensor noise lower than $2 \times 10^{-11}\,\text{m}/\sqrt{\text{Hz}}$ to apply to advanced detectors. Alternative suspension with a cross-section geometry that varies along the suspension element length (Lee *et al.*, 2005, 2006) may have the potential to provide a smaller number of violin modes within the detection frequency band.

When considering suspension loss, the material loss ϕ_{mat} is in fact a combination of several losses: bulk material loss ϕ_{bulk}, surface loss of the suspension elements ϕ_{surf} and the thermoelastic loss of the suspension $\phi_{\text{thermelas}}$.

$$\phi_{\text{mat}} = \phi_{\text{bulk}} + \phi_{\text{thermelas}} + \phi_{\text{surf}} \tag{10.6}$$

The bulk loss and the thermoelastic loss are material related and are frequency dependent, while the surface loss depends on the geometry and the quality of the surface of the suspension elements.

Suspension materials

There are several desirable properties when considering a material for test mass suspension. Firstly, the internal loss factor of the bulk material, ϕ_{bulk} should be as low as possible. Secondly, a high tensile strength of the suspension elements is required. This will result in higher dissipation delusion and thus lower thermal noise. Furthermore, for a thin suspension element with negligible bending stiffness compared to the tensile restoring force, the violin mode frequencies are given by:

$$f_n = \frac{n}{2L}\sqrt{\frac{T_0}{\rho_L}} \tag{10.7}$$

where f_n is the nth violin mode frequency, L is the suspension length, T_0 is the tension and ρ_L is the linear mass density of the suspension element. It can be seen that by maximising the tension to density ratio, T_0/ρ_L, a higher first mode and more widely spaced violin modes will exist.

Table 10.1 compares five different materials: C85 steel, niobium, fused silica, crystalline silicon and sapphire. The material that stands out is fused silica, having the highest tensile strength, lowest volume density and very low loss angle. However, silica fibres are brittle and are prone to moisture damage, should micro-cracks exist on the surface. The tensile strength is strongly dependent on the preparation and handling of the fibre; the tensile strength of fused silica given in the table is the average value.

Thin steel wire suspension was used in initial detectors (LIGO, Virgo and TAMA). This is simply because steel is strong, and easy and cheap to obtain. But steel has relatively high loss ϕ_{bulk} ($\sim 10^{-3}$). The GEO project pioneered fused silica suspensions (Plissi *et al.*, 1998; Willke *et al.*, 2002) and the advanced detectors such as Advanced LIGO and Advanced Virgo are going to use fused silica fibre suspension. However, silica suspension will not be suitable for cryogenic detectors due to poor thermal conductivity and reduced Q-factor at low temperature. There has been research into the possibility of other suspensions constructed from silicon (Amico *et al.*, 2004; Reid *et al.*, 2006), sapphire (Uchiyama *et al.*, 2000; Tomaru, 2002) and niobium (Ju *et al.*, 1999). Monocrystalline silicon and sapphire show promise, particularly at cryogenic temperatures, whilst niobium exhibits the best properties of all metals. Changing the aspect ratio of the suspension element by using ribbons (Gretarsson *et al.*, 2000; Lee *et al.*, 2005) rather than fibres can also result in an effective reduction in ϕ_{pend} through an improvement in the level of dissipation dilution, although trade-off is complicated by the exact details of the geometry (Cumming *et al.*, 2009).

Surface loss

The loss angle of a bulk material is not sufficient to indicate the loss angle of long, thin suspension filaments since they have significant surface area compared to their volume. Surface loss is one phenomena that must be considered. Gretarsson and Harry (1999) have studied the surface-induced loss of fused silica fibres, characterising the loss as follows:

$$\phi_{\text{surf}} = \phi_{\text{bulk}}\left(1 + \mu\frac{d_s}{V/S}\right) \tag{10.8}$$

Table 10.1. *Suspension material properties. The sources of the values are referenced. The loss angles given are bulk losses except for steel, for which the internal loss for suspension element size samples is given*

Material	Loss angle ϕ	Tensile strength (Gpa)	Density ρ_v (kg/m^3)
C85 steel	$\phi \simeq 2 \times 10^{-4}$[*a]	3[b]	7800
Niobium	$\phi_{\text{bulk}} \simeq 3 \times 10^{-7}$[c]	0.4[d]	8570
Fused silica	$\phi_{\text{bulk}} \simeq 6 \times 10^{-8}$[e]	4[b]	2200
Silicon (crystalline)	$\phi_{\text{bulk}} \simeq 5 \times 10^{-9}$[f]	1[f]	2330
Sapphire	$\phi_{\text{bulk}} \simeq 3 \times 10^{-9}$[g]	2[h]	4000

[b]Amico *et al.* (2001) [a]Cagnoli *et al.* (1999) [c]Veitch *et al.* (1987)
[d]Goodfellow (2006) [e]Penn *et al.* (2006) [f]Amico *et al.* (2004)
[g]Braginskii *et al.* (1985) [h]MolTech (2010)

Here ϕ_{bulk} is the loss angle of the bulk material, d_s the dissipation depth, which is a characteristic distance beyond which the loss effect is negligible, V/S the volume to surface ratio and μ a geometrical factor relating to the geometry of the suspension element and the mode of oscillation. For transverse oscillations, μ is given by:

$$\mu = \begin{cases} 2 & \text{fibres} \\ \frac{(3+a)}{(1+a)} & \text{ribbons} \end{cases} \tag{10.9}$$

where a is the aspect ratio (thickness to width) of the combined suspension elements. It can be shown that the surface loss becomes significant when the dimension of the suspension elements approaches d_s. It has been found (Gretarsson *et al.*, 2000) that d_s for fused silica is approximately 150 μm. Given that mirror masses are $\sim$ 40 kg, fused silica suspensions will likely have dimensions in the order of several hundred micrometres, and thus the relevant loss angle of the material may be several tens of times higher than the bulk loss.

A similar study for thin single crystal silicon cantilevers was conducted by Yasumura *et al.* (2000), resulting in a similar characterisation of surface loss. Thus it is likely that for small dimension silicon suspensions, an increase in the material loss angle of at least a factor of ten can be expected, when compared to the bulk loss value of 3×10^{-8}. Although the bulk loss for niobium is in the order of 3×10^{-7} at room temperature (Veitch *et al.*, 1987), thin niobium suspensions in the order of 100 μm, with correct treatment, would have a loss angle of $\phi = 10^{-5}$ (Ju *et al.*, 1999). The geometry-dependent Q-factor of niobium is given as $Q_{\text{Nb}} \simeq 10^7 (V/S)^{1/2}$ (Ju *et al.*, 1999), which is different from that of fused silica. The surface loss properties for sapphire are currently not as well known as for fused silica, but similar levels of degradation may be expected.

Thermoelastic loss

Another contribution to material loss ϕ is the frequency-dependent phenomenon known as thermoelastic loss (Zener, 1937, 1938). This loss results from the coupling of strain to

Table 10.2. *Suspension material thermophysical properties at 300K. The thermoelastic loss peak frequencies were obtained for ribbon thicknesses of 50 μm in all except niobium, where a thickness of 100 μm was used, due to its lower tensile strength*

Material	α (K^{-1})	κ (W m^{-1} K^{-1})	C_v (J kg^{-1} K^{-1})	Thermoelastic loss peak frequency (Hz)
Steel	1.2×10^{-5}	49	486	8122
Niobium	7.3×10^{-6}	53.7	265	3714
Fused silica	5.1×10^{-7}	1.38	772	511
Silicon	2.6×10^{-6}	149	710	56592
Sapphire	5.4×10^{-6}	46	770	9384

temperature due to the non-zero thermal expansion coefficient of the suspension material. As the suspension element bends due to pendulum mode or violin mode oscillation, one side of the suspension element is slightly compressed whilst the other is slightly stretched. Consequently, the compressed side is heated while the stretched side is cooled and the resulting temperature gradient causes heat to flow, causing energy loss. The time for the heat flow to reduce the temperature gradient depends on the oscillation frequency. This effect is observed as a frequency-dependent mechanical loss.

The thermoelastic loss can be characterised as follows:

$$\phi_\mathrm{th}(\omega) = \Delta \frac{\omega\tau}{1 + (\omega\tau)^2}$$

(10.10)

where Δ is related to the Young's modulus Y_0, thermal expansion coefficient α, temperature T, volume density ρ_v and volume specific heat C_v by:

$$\Delta = \frac{Y_0 \alpha^2 T}{\rho_\mathrm{v} C_\mathrm{v}}$$

(10.11)

The value τ is related to the geometry of the suspension by:

$$\tau = \begin{cases} \frac{\rho_\mathrm{v} C_\mathrm{v} r^2}{1.08\pi\kappa} & \text{fibres} \\ \frac{\rho_\mathrm{v} C_\mathrm{v} t^2}{\pi^2\kappa} & \text{ribbons} \end{cases}$$

(10.12)

where r and t are the radius and thickness of the suspension element, respectively, and κ is the thermal conductivity.

From the above equations, it can be seen that the thermoelastic loss peak occurs when $\omega\tau = 1$. Thus the peak frequency increases with larger values of thermal conductivity κ or smaller suspension element dimension. Generally, it is desirable to locate the peak as far from the detection band as possible.

A table of thermophysical properties at room temperature, along with the thermoelastic peak frequency, is given in Table 10.2. It is observed that the largest variation between materials occurs for the value of thermal conductivity, κ. Consequently, it is the high thermal conductivity of silicon and the low thermal conductivity of fused silica that accounts for the highest and lowest thermoelastic peak frequency in silicon and fused silica respectively. The low thermal expansion coefficient, α, of fused silica will also result in a lower level of thermoelastic loss.

Suspension attachments

The attachment of the suspension element to the test mass is a potential source of mechanical loss. First, it can introduce noise through poorly defined boundary conditions for the suspension. Second, it could increase the loss of the test mass as a whole and thus increase the test mass thermal noise level. Care must be taken that it is not a dominant source.

Steel wire suspensions have been used in all the first generation interferometric gravitational wave detectors except for GEO 600. In these detectors the steel wire is arranged such that a semicircular loop is formed at the bottom of the suspension, within which the test mass is held. The upper ends of the steel wire suspension elements are clamped to the preceding suspension stage. The advantage of such a suspension method is its simplicity, as well as the flexibility it provides by allowing the test masses to be removed easily for replacement, upgrade, repair or cleaning. This arrangement has the disadvantage of added thermal noise due to friction between the steel wire and the circumference of the test mass. This is a particular problem at the point close to the equator of the test mass where the steel wire begins to deviate. Stand-offs have been used to better define the boundary conditions and reduce the friction.

Fused silica fibre suspensions are proposed for the final suspension stage in room temperature advanced detectors. Fused silica fibres cannot be used in the same way as steel wire suspensions. The suspension fibres are instead bonded permanently to both the final test mass and the preceding suspension stage. This is achieved by bonding fused silica 'ears' at or near the equatorial position around the test mass circumference using silicate bonding techniques (Rowan *et al.*, 1998; Amico *et al.*, 2002). The ears provide a small platform from which the fused silica wire is welded to the silica ear to form the suspension. Since all interface points are via permanent bonds, the amount of thermal noise introduced through slip–stick phenomena is minimised. Very low loss has been reported using this silicating bonding method (Rowan *et al.*, 1998; Smith *et al.*, 2003).

The monolithic suspension means that replacement of the test masses or the suspension element would be somewhat complicated should the need arise. The use of removable modular suspensions was explored (Lee *et al.*, 2005; Lee, 2007). With high pressure contact points of the modular and the test mass, the loss due to slip–stick at the contact points could be minimised (Taniwaki *et al.*, 1998). One concern of such method is the increase in test mass thermal noise contributed by the loss of the holes in the test mass for accommodating the suspension module. This issue has been studied in depth by Gras *et. al.* (2004) using Levin's method (Levin, 1998a). It has been found that for typical laser beam diameters, the equatorial losses, ϕ_{hole}, can be as high as 4×10^{-3} without the test mass thermal noise increasing by more than 10%. The losses at the surface of the equatorial holes should be lower than this value with the proper polishing treatment. The modular suspension investigated so far has used niobium ribbon, which has higher loss than fused silica. It maybe possible to to use fused silica or silicon material to form a modular suspension.

10.3 Test mass thermal noise

Test mass thermal noise in a broad sense depends on the mechanical dissipation in the mirror substrate itself as well as on the reflective coatings. Thermally induced motion at the surface of the mirror is integrated over the laser spot, and it is this integrated motion that is referred to as test mass thermal noise. Below the lowest internal mode of the test mass (typically around 10 kHz), the mirror thermal noise power scales as $1/f$ within the detection band. Here we concentrate on the test mass substrate contribution to the thermal noise, while the coating thermal noise will be discussed in Section 10.4

In the case of inhomogeneously distributed loss, the loss angle ϕ is a function of position $\phi(r)$. In an interferometer, the thermal noise is sensed by the laser beam at the middle of the test mass. Levin (1998a) developed a direct approach method to calculate the thermal noise. This method relates the energy dissipated in the test mass W_{diss} in response to the laser beam pressure force F_0 on the test mass, to the admittance Y associated with the test mass internal resonance:

$$Y(f) = \frac{2W_{\mathrm{diss}}}{F_0^2}. \tag{10.13}$$

Using equation (10.2), this leads to the power spectrum of the thermal noise of the test mass

$$S_x(f) = \frac{2k_{\mathrm{B}}T}{\pi^2 f^2} \frac{W_{\mathrm{diss}}}{F_0^2}. \tag{10.14}$$

with

$$W_{\mathrm{diss}} = 2\pi \int_V \varepsilon(\vec{r})\phi(\vec{r}, f)\mathrm{d}V. \tag{10.15}$$

Here $\phi(\vec{r}, f)$ is the loss angle of the test mass, which can be position and frequency dependent, and $\varepsilon(\vec{r})$ is the energy density of elastic deformation when the test mass is maximally deformed under the applied laser pressure.

Thus to calculate the thermal noise of the test mass we could simply apply a virtual pressure

$$P(\vec{r}, t) = F_0 \cos(2\pi ft)f(\vec{r}), \tag{10.16}$$

to the face of the test mass according to the laser intensity and profile $f(\vec{r})$, find the average power W_{diss} dissipated in the test mass under this pressure and use equation (10.14) to obtain $S_x(f)$. With the help of finite element modelling, it is relatively easy to calculate W_{diss} for a test mass with certain intrinsic loss plus different surface coating loss as well as any other localised loss due to suspension attachment and damping elements. One result from this analysis is that any loss/defect near the laser beam would have a larger effect on the thermal noise than those located far away from the laser beam. Thus the contribution of the loss due to the attachment of the suspension element, which is usually at the edge of the test mass, has a smaller contribution to the overall thermal noise. It also can be seen that the effect of the mirror coating on thermal noise is then important. This will be discussed in detail in Section 10.4. Levin's approach to thermal noise calculations is also very important in other analysis such as designing of the damping scheme for parametric instability control (see Chapter 13) so that thermal noise performance is diminished minimally. It is worth to point out that before Levin's approach developed, the test mass thermal noise was calculated using mode expansion method, by treating the thermal motion of the test mass as the sum of all the internal resonance modes motion. This method would only give a good approximation of the thermal noise of a homogeneous free test mass.

Test mass materials

Current room temperature detectors chose to use fused silica test masses. The loss of silica test masses were observed to be of the order of $\phi \sim 10^{-6}$ (Traeger *et al.*, 1997; Taniwaki *et al.*, 1998) with a lower value approaching $\sim 10^{-8}$ in some resonant modes (Penn *et al.*, 2001; Braginsky *et al.*, 1992). Fused silica also has very low optic absorption of < 1ppm/cm (Loriette and Boccara, 2003; Stefan Hild, 2006) as well as low bulk scattering ~ 0.7ppm/cm (Chen *et al.*, 2011).

Other test mass materials such as sapphire have been proposed (Ju *et al.*, 1993). The excellent thermal and mechanical properties of sapphire makes it a promising material for use in test masses, especially in cryogenic detectors (Kuroda *et al.*, 2003). It would be an excellent candidate in terms of thermal noise, thermal lensing compensation (Hello, 2001; Mansell *et al.*, 2001) and parametric instability control (Ju *et al.*, 2006). However, the high thermal expansion coefficient of sapphire means that the thermoelastic noise (Braginsky *et al.*, 1999) would exceed the substrate thermal noise for room temperature detectors (Liu and Thorne, 2000). Besides, there are challenges to producing large sapphire samples with excellent optical (Yan *et al.*, 2006) and mechanical properties compared with fused silica.

Silicon has also been considered for advanced detectors (Rowan *et al.*, 2003). It has very low mechanical loss as well as desirable thermal properties. It however, requires a different laser frequency ($\lambda = 1.55\,\mu$m) to use silicon as a test mass in the gravitational wave detector, or in an all reflective configuration.

10.4 Coating loss

Theory

Coating thermal noise is best understood using Yuri Levin's theorems (Levin, 1998b, 2008) which give thermal noise from the losses experienced by an applied 'force' in the shape of the readout. This 'force' is a virtual pressure applied to the face of an optic with a shape the same as the laser beam used to detect the optic's motion. Using a virtual entropy rather than a pressure allows for the prediction of thermo-optic noise (Evans *et al.*, 2008).

Brownian thermal noise

Applying Levin's theorem with a virtual pressure to a coated test mass, infinite in radial extent, readout at the top surface with a Gaussian beam yields a prediction for Brownian thermal noise power spectral density of (Harry *et al.*, 2002)

$$S_x\left(f\right) = 2k_{\mathrm{B}}T\phi_{\mathrm{eff}}\left(1-\sigma^2\right) / \left(\pi^{3/2}fwY_0\right),\tag{10.17}$$

where k_{B} is Boltzmann's constant, T is the temperature, σ is the Poisson's ratio of the test mass substrate material, f is the frequency, w is the Gaussian beam width, and ϕ_{eff} is the effective loss angle of the mirror, given by

$$\phi_{\mathrm{eff}} = \phi + d/\left(\sqrt{\pi}wY_\perp\right)\tag{10.18}$$
$$\left\{\left(Y/\left(1-\sigma^2\right) - 2\sigma_\perp^2 YY_{||}/\left[Y_\perp\left(1-\sigma^2\right)\left(1-\sigma_{||}\right)\right]\right)\right\}\phi_\perp$$
$$+Y_{||}\sigma_\perp\left(1-2\sigma\right)/\left[\left(1-\sigma_{||}\right)\left(1-\sigma\right)\right]\left(\phi_{||}-\phi_\perp\right)$$
$$\left.+Y_{||}Y_\perp\left(1+\sigma\right)\left(1-2\sigma\right)^2/\left[Y\left(1-\sigma_{||}^2\right)\left(1-\sigma\right)\right]\phi_{||}\right).$$

Here d is the coating thickness, and Y, σ and ϕ are the Young's moduli, Poisson's ratios and loss angles of the substrate (no subscript) and the coating for stresses perpendicular ($\perp$) and parallel ($_{||}$) to the optic face. The coating may be anisotropic like this if it is made up of alternating layers of different materials. This derivation assumes the loss angles of the coating and substrate materials for shear stress are the same as those for tensile stress. This is likely not the case, although they may be similar. A complete derivation of effective ϕ without this assumption has not been completed, although there is work ongoing (Hong *et al.*, 2009).

A real coating is not read out entirely at the surface, but rather some of the laser power penetrates into the layers. A full analysis where the thermal noise contribution of each layer is weighted by the laser power at that layer to obtain a more complete version of equations (10.17) and (10.18) has not been done. However, an estimate from an approximation technique gives a correction that reduces the thermal noise by about 5% from the result of equation (10.17) for a coating on a second generation detector mirror.

Similarly, real test masses will not be infinite in radial extent. As long as the ratio of test mass diameter to beam width is above about 5, and the mirror thickness is more than the diameter, the above equations are good to a few per cent (Somiya and Yamamoto, 2009). For the case of Advanced LIGO, with a diameter of 34 cm, beam width of 6.2 cm, and thickness of 20 cm, the infinite case formula gives a results 2.6% less than the finite case (Somiya and Yamamoto, 2009). There is also an approximation for when the mirror thickness is much less than the diameter, (Somiya and Yamamoto, 2009).

If all Poisson ratios are small in equation (10.18), it can be simplified to

$$\phi_{\mathrm{eff}} \approx \phi + d/\left(\sqrt{\pi}w\right)\left(Y/Y_\perp\phi_\perp + Y_{||}/Y\phi_{||}\right).\tag{10.19}$$

Further assuming alternating layers of two coating materials (a high index and a low index), the Young's moduli and loss angles of the coating can be estimated from bulk values of the individual coating materials;

$$Y_\perp = (d_1 + d_2) / (d_1/Y_1 + d_2/Y_2) \tag{10.20}$$

$$Y_\| = (Y_1 d_1 + Y_2 d_2) / (d_1 + d_2) \tag{10.21}$$

$$\phi_\perp = Y_\perp / (d_1 + d_2) \, (\phi_1 d_1 / Y_1 + \phi_2 d_2 / Y_2) \tag{10.22}$$

$$\phi_\| = (Y_1 \phi_1 d_1 + Y_2 \phi_2 d_2) / \left(Y_\| (d_1 + d_2) \right), \tag{10.23}$$

where subscripts 1 and 2 refer to the two materials.

Thermo-optic noise

Using entropy with the same volume density as the readout for the applied 'force' in Levin's theorem yields a prediction for thermo-optic noise (Evans *et al.*, 2008). This is noise that arises from temperature fluctuations in the coating which couple to the position readout through both thermal expansion and changes in index of refraction. An optic of infinite radial extent readout with a Gaussian laser beam that penetrates the multilayer dielectric coating has a predicted thermo-optic noise of

$$S_x(f) = \frac{2\sqrt{2}}{\pi} \frac{k_\mathrm{B} T^2}{w^2 \sqrt{2\pi \kappa C_\mathrm{s} f}} \left(\alpha_\mathrm{c} d - \beta_\mathrm{c} \lambda - \alpha_\mathrm{s} d \frac{C_\mathrm{c}}{C_\mathrm{s}} \right)^2, \tag{10.24}$$

where κ is the thermal conductivity of the substrate, C_s is the heat capacity per unit volume of the substrate, α_c is the coating volume averaged thermal expansion coefficient $(1/L)\mathrm{d}L/\mathrm{d}T$, β_c is the effective coating thermorefractive coefficient $\mathrm{d}n/\mathrm{d}T$, λ is the wavelength of light, α_s is the thermal expansion coefficient for the substrate material, C_c is the coating volume averaged heat capacity per unit volume, and d is the total thickness of the coating. One important point in equation (10.24) is that the sign between the first and second term is negative, so if both the thermal expansion and thermal diffractive coefficient are the same sign these two terms can partially cancel.

Materials

First generation interferometric gravitational wave detectors used alternating quarter wave layers of high index tantala and low index silica (Abbott and The LIGO Scientific Collaboration, 2003). These materials were chosen for their low optical absorption, which is important for control of thermal lensing in the mirrors. Second generation detectors have to contend with coating thermal noise, and it was found (Harry *et al.*, 2002) that adding a titania dopant to the tantala reduces the mechanical loss ϕ of the high index material and thereby the overall coating thermal noise. This is all done while actually slightly improving the optical absorption and index of refraction (important for keeping the layer thickness and therefore d in equation (10.18) low) and leaving the Young's modulus and Poisson ratio essentially unchanged. A list of mechanical, thermal, and optical parameters for silica, tantala, and 25% titania doped tantala are shown in Table 10.3.

The thermal expansion α and thermal refraction β in equation (10.24) are important in predicting thermo-optic noise and are also shown in Table 10.3. These are not as well known

Table 10.3. *Important parameters for coating materials used in first and second generation detectors. Numbers with * by them indicate where the value has not been measured for titania-doped tantala and are taken to be the same as for tantala*

Parameter	Silica	Tantala	Titania–tantala
$\phi \left(\times 10^{-5} \right)$	4	44	23
Y (GPa)	73	140	140*
σ	0.17	0.23	0.23*
$\mathrm{d}n/\mathrm{d}T \left(\times 10^{-6}\ \mathrm{K}^{-1} \right)$	8	14	14
$(1/L)\,\mathrm{d}L/\mathrm{d}T \left(\times 10^{-6}\ \mathrm{K}^{-1} \right)$	0.51	3.6	3.6*
$C_v \left(\times 10^6\ \mathrm{J} - \mathrm{m}^{-3}\mathrm{K}^{-1} \right)$	1.6	2.1	2.1*
$\kappa \left(\mathrm{Wm}^{-1}\mathrm{K}^{-1} \right)$	1.38	33	33*
Index	1.45	2.07	2.09

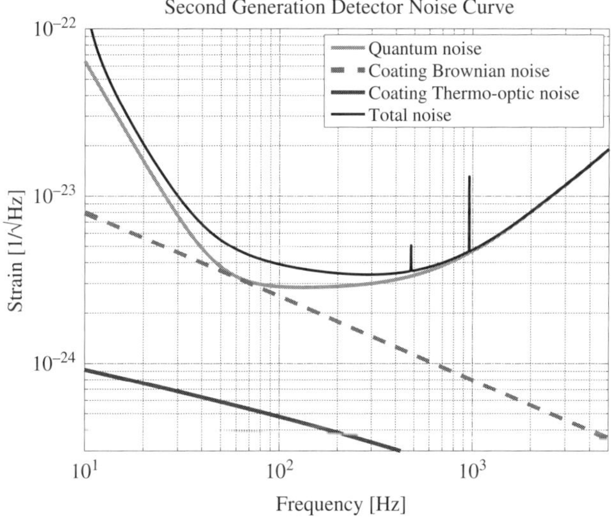

Figure 10.1 Noise levels from Brownian and thermo-optic coating noise compared to quantum noise in a typical second generation detector. This uses coating values from Table 10.3 and Advanced LIGO parameters.

for the thin film materials in use, especially the β parameter, but extensive work has gone into measuring these for better prediction of second generation noise. A typical noise curve for a second generation detector is shown in Figure 10.1.

Other materials have been studied as possible coating materials in interferometric gravitational wave detectors. These include high index materials such as niobia, hafnia, zirconia and titania; the low index material alumina; and dopants such as lutetium, tungsten, cobalt and silica. Most of these gave worse mechanical loss [tungsten-doped tantala, cobalt-doped tantala, titania-doped zirconia (Flaminio *et al.*, 2010)], represent little or no improvement over existing materials (niobia, hafnia, alumina, silica-doped tantala), and/or were unacceptable for optical reasons (zirconia, cobalt-doped tantala, tungsten-doped tantala, titania-doped zirconia, tungsten-doped titania (Flaminio *et al.*, 2010), lutetium-doped tantala). Only silica-doped titania showed improvement in mechanical loss without any offsetting problems (Gretarsson *et al.*, 2007). There is not as large an improvement as with titania-doped tantala, but silica-doped titania has not been studied as extensively.

Changes to the coating process, including using different coating vendors, has also been investigated. All vendors used ion beam deposition as the only coating technique currently available to give sub-ppm optical absorption. Most changes in process showed little or no change to mechanical loss, including changes to the substrate polish, the use of xenon instead of argon as the bombardment ion and use of a secondary ion beam of oxygen. Some changes made the ϕ noticeably worse, such as a reduction in the oxygen content. Although there may be some benefit to better understanding why some of these techniques made for worse coatings, none are planned to be used for second generation detectors.

Possible improvements in thermal noise

There are possible ways of making improvement in most of the parameters in equations (10.17) and (10.18). The ϕ and Y values can be improved by understanding better the makeup of coating materials and possibly finding better materials. Research on this is continuing (Martin *et al.*, 2009). The effective thickness d can be reduced by either finding materials with a larger difference in index and/or optimising the coating design to reduce the amount of the higher mechanical loss material (Agresti *et al.*, 2006b). It may also be possible to reduce the effective thickness by using optical cavities instead of single mirrors (Khalili, 2005), by using waveguide mirrors (Brückner *et al.*, 2010), and even to get d to zero by using coatingless mirrors (Goßler *et al.*, 2007). The effective beam width w can be increased by changing from Gaussian to other beam shapes, (Vinet, 2009). Finally, the temperature T can be reduced directly (Yamamoto *et al.*, 2006), although it is important to be aware of temperature dependencies in other parameters as some might act to increase thermal noise.

Thickness optimisation takes advantage of the fact that the mechanical loss and therefore the contribution to coating thermal noise is not the same for the high index and low index layers in titania-doped tantala/silica coatings. It is possible to get the reflectivity needed for the interferometer's optical performance using other coating designs than just alternating quarter-wave stacks of high index and low index materials. Knowing the material parameters from Table 10.3, including the thermo-optic ones, allows for an optimisation to be done. Typically there is about a 10% improvement in coating thermal noise power from optimisation. This technique is planned for some (Advanced LIGO) second generation detectors.

Using beam shapes other than the usual Gaussian can be a way to effectively increase the beam width parameter w and thus reduce the thermal noise. Care must be taken, however, to preserve the optical losses, especially from clipping, to allow for angular stability especially when used with high optical powers, and to obtain a low enough contrast defect. One shape is the 'mesa' beam (Agresti *et al.*, 2006a), which flattens the beam in the centre and makes the transition between high optical power in the centre and lower power near the mirror edges more abrupt. Mesa beams require the mirrors to have a different shape from the usual spherical. Higher order Gauss–Laguerre beams (Chelkowski *et al.*, 2009) can also be used, but care must be taken that only one mode resonates in the cavity. These alternate beam shapes are under consideration for upgrades to second generation detectors.

Finally, cooling the mirror to directly reduce all thermal noise can seem attractive. However, the thermal noise power is proportional to T while material properties can get worse with temperature more quickly. Care must be taken to choose the right materials. Cryogenic temperatures, however, can also allow for materials that don't have high performance at room temperature, such as sapphire as a substrate and suspension material. Silica, tantala, and titania-doped tantala do show higher mechanical loss at most reduced temperatures. There are temperatures where there are modest gains in both T and ϕ or lower temperatures where the T reduction still shows thermal noise improvements even though the ϕ is worse (Martin *et al.*, 2009). The evidence is not clear cut, however (Yamamoto *et al.*, 2006), and research is continuing both to better understand the current coating materials at low temperatures and to find materials with significant improvement in thermal noise at cryogenic temperatures. Cryogenics is planned for the second generation Large-scale Cryogenic Gravitational-wave Telescope (Uchiyama and The LCGT Collaboration, 2004) and will be discussed in more detail later in Chapter 15.

Acknowledgements

L. Ju and B. Lee wish to acknowledge financial support from the West Australian Government Centres of Excellence scheme and the Australian Research Council. G. Harry acknowledges the support of the United States National Science Foundation for the construction and operation of the LIGO Laboratory.

References

Abbott, B. and The LIGO Scientific Collaboration. 2003. *Nuclear Instruments and Methods in Physics Research: A*, **517**, 154.
Agresti, J., *et al.* 2006a. *Journal of Physics Conference Series*, **32**, 301.
Agresti, J., *et al.* 2006b. Page 628608 of: *Proceedings of SPIE*, vol. 6286.
Amico, P., *et al.* 2001. *Nucl. Instrum. and Meth. A*, **461**, 297.
Amico, P., *et al.* 2002. *Rev. Sci. Instrum.*, **73**, 3318.
Amico, P., *et al.* 2004. *Nucl. Instrum. and Meth. A*, **518**, 240.
Braginskii, V. B., Mitrofanov, V. P., and Panov, V. I. 1985. *Systems with Small dissipation*. University of Chicago Press.
Braginsky, V. B., Mitorfanov, V. P., and Okhrimenko, O. A. 1992. *JETP Lett.*, **55**, 432.
Braginsky, V. B., Gorodetsky, M. L., and Vyatchanin, S. P. 1999. *Physics Letters A*, **264**, 1–10.
Brückner, F., *et al.* 2010. *Physical Review Letters*, **104**, 163903.

Cagnoli, G., *et al.* 1999. *Phys. Lett. A*, **255**, 230.

Callen, H. B. and Welton, T. A. 1951. *Phys. Rev.*, **83**, 34.

Chelkowski, S., Hild, S., and Freise, A. 2009. *Physical Review D*, **79**, 122002.

Chen, X., Ju. L, Zhao, C., Flaminio, R., Lück, H., Blair, D.G. 2011. *Optics Comm.*, **284**, 4732–4737.

Cumming, A., *et al.* 2009. *Classical and Quantum Gravity*, **26**, 215012.

EGO. 2010, http://www.et-gw.eu/.

Evans, M., *et al.* 2008. *Physical Review D*, **78**, 102003.

Flaminio, R., *et al.* 2010. *Classical and Quantum Gravity*.

Goodfellow. 2006. *Material Properties, Niobium.* http://www.goodfellow.com.

Goßler, S., *et al.* 2007. *Physical Review D*, **76**, 053810.

Gras, S., Blair, D. G., and Ju, L. 2004. *Phys. Lett. A*, **333**, 1.

Gretarsson, A. M. and Harry, G. M. 1999. *Rev. Sci. Instrum*, **70**, 4081.

Gretarsson, A. M., *et al.* 2000. *Phys. Lett. A*, **270**, 108.

Gretarsson, A. M., *et al.* 2007. in *Optical Interference Coatings*, OSA Technical Digest (CD) (Optical Society of America, 2007), paper ThC10.

Harry, G. M., *et al.* 2002. *Classical and Quantum Gravity*, **19**, 897.

Hello, P. 2001. *The European Physical Journal D - Atomic, Molecular, Optical and Plasma Physics*, **15**, 373–383.

Hong, T., Yang, H., and Chen, Y. 2009. *Anatomy of Brownian thermal noise in the coated mirrors: 3-D consideration.* LIGO Document G0900906-v1.

Ju, L., Blair, D. G., and Notcutt, M. 1993. *Rev. Sci. Instrum.*, **64**, 1899.

Ju, L., *et al.* 1999. *Phys. Lett. A*, **254**, 239.

Ju, L., *et al.* 2006. *Physics Letters A*, **355**, 419 – 426.

Khalili, F. Y. 2005. *Physics Letters A*, **334**, 67–72.

Kuroda, K., *et al.* 2003. *Class. Quantum Grav.*, **20**, S871.

Lee, B. H. 2007. *PhD Thesis, University of Western Australia,* http://theses.library. uwa.edu.au/adt-WU2007.0187/.

Lee, B. H., *et al.* 2005. *Phys. Lett. A*, **339**, 217.

Lee, B. H., *et al.* 2006. *Phys. Lett. A*, **350**, 319.

Levin, Y. 1998a. *Phys. Rev. D*, **57**, 659–663.

Levin, Y. 1998b. *Physical Review D*, **57**, 659–663.

Levin, Y. 2008. *Physics Letters A*, **372**, 1941 1944.

Liu, Y. T. and Thorne, K. S. 2000. *Phys. Rev. D*, **62**, 122002.

Loriette, V. and Boccara, C. 2003. *Appl. Opt.*, **42**, 649–656.

Mansell, J. D., *et al.* 2001. *Appl. Opt.*, **40**, 366–374.

Martin, I. W., *et al.* 2009. *Classical and Quantum Gravity*, **26**, 155012.

MolTech. 2010.

Penn, S. D., *et al.* 2001. *Rev. Sci. Instrum.*, **72**, 3670.

Penn, S. D., *et al.* 2006. *Physics Letters A*, **352**, 3.

Plissi, M., *et al.* 1998. *Rev. Sci. Instrum.*, **69**, 3055–3061.

Reid, S., *et al.* 2006. *Phys. Lett. A*, **351**, 205.

Rowan, S., *et al.* 1998. *Phys. Lett. A*, **246**, 471.

Rowan, S., *et al.* 2003. Pages 292–297 of: Mike Cruise, P. S. (ed.), *Gravitational-Wave Detection, Proceedings of SPIE*, vol. 4856, doi:10.1117/12.459090.

Santamore, D. H. and Levin, Y. 2001. *Phys. Rev. D*, **46**, 042002.

Saulson, P. R. 1990. *Phys. Rev. D*, **42**, 2437.

Smith, J., *et al.* 2003. *Class. Quant. Grav.*, **20**, 5039–5047.

Somiya, K. and Yamamoto, K. 2009. *Physical Review D*, **79**, 102004.

S. Hild *et al.* 2006. *Appl. Opt.*, **45**, 7269–7272.

Taniwaki, M., *et al.* 1998. *Phys Lett. A*, **246**, 273.

Tomaru, T. 2002. *Phys. Lett. A*, **301**, 215.

Traeger, S., Willke, B., and Danzmann, K. 1997. *Phys. Lett. A*, **225**, 39.

Uchiyama, T. and The LCGT Collaboration. 2004. *Classical and Quantum Gravity*, **21**, S1161.

Uchiyama, T., *et al.* 2000. *Phys. Lett. A*, **273**, 310.

Veitch, P., *et al.* 1987. *Cryogenics*, **27**, 586 – 588.

Vinet, J.-Y. 2009. *Living Reviews in Relativity*, **12**, 5.

Willke, B., *et al.* 2002. *Class. Quant. Grav.*, **19**, 1377–1388.

Winkler, W., *et al.* 1991. *Phys. Rev. A*, **44**, 7022.

Yamamoto, K., *et al.* 2006. *Physical Review D*, **74**, 022002.

Yan, Z., *et al.* 2006. *Appl. Opt.*, **45**, 2631–2637.

Yasumura, K. Y., *et al.* 2000. *IEEE trans*, **Electron Devices 9**, 119.

Zener, C. 1937. *Phys. Rev.*, **52**, 230.

Zener, C. 1938. *Phys. Rev.*, **53**, 90.

11

Vibration isolation

The suppression of seismic ground motion is a significant challenge for ground based gravitational wave interferometric detectors. It is generally achieved through different combinations of active isolation and passive isolation. This chapter will look at two different vibration isolation systems. Part 1 will describe the mostly active vibration isolation system used by LIGO. Part 2 will discuss the mostly passive isolation systems developed at the University of Western Australia, which have features in common with isolation systems developed for many other detectors.

Part 1 : Seismic isolation for Advanced LIGO

B. Lantz for the LIGO Scientific Collaboration

One of the most significant improvements for Advanced LIGO will be to move the lower edge of the detection band from 40 Hz down to 10 Hz. This will allow us to start tracking the final inspiral of compact binary systems earlier in their evolution, and will also allow us to observe the inspirals not only of pairs of neutron stars, but also of binary systems containing more massive objects, such as black holes with 10 to 30 times the mass of the Sun, as described in Chapter 4. In order to measure gravitational waves at 10 Hz, the requirements for the motion of the Advanced LIGO mirrors are very strict. The Advanced LIGO interferometer cannot distinguish between differential length changes in the arms caused by the spacetime distortion of a passing gravitational wave and differential length changes caused by mirror motion. For an interesting discussion about interferometer configurations where this might not be true, see recent work by Kawamura and Chen (2004) and others.

The inability to distinguish between mirror vibrations (e.g. from seismic noise) and gravitational waves means that Advanced LIGO must use a high performance isolation and alignment system to support and isolate the test mass mirrors (Robertson *et al.*, 2004). To meet the Advanced LIGO sensitivity goals, the system must isolate the mirrors from ground vibrations and provide quiet actuation and alignment for the mirrors so that the resonance conditions for the LIGO cavities can be acquired and maintained. At 10 Hz, the motion of the test mass mirrors must be less than 1×10^{-19} m/$\sqrt{\text{Hz}}$, approximately 9 orders of magnitude less than the typical ground motion at the observatories at that frequency (Fritschel *et al.*, 2001).

11.1 Planned isolation platforms for Advanced LIGO

Not only must the optics be kept still in the gravitational wave detection band from 10 Hz up to several kHz, but it is also important to minimise their relative motion below 10 Hz. Since the interferometer comprises a set of coupled resonant optical cavities, there is non-linear coupling of mirror motion to signal in the gravitational wave detection readout signal as the interferometer moves away from the correct operating point.

To hold the coupling of various noise sources to an acceptable level, the length fluctuations of the 4 km arms must be below 1×10^{-14} m rms, despite, for example, diurnal tidal distortions of the ground of approximately 200 μm. The proposed seismic isolation and suspension system for the Advanced LIGO test masses uses a set of seven layered stages: an external hydraulic stage (1 layer); the BSC Internal Seismic Isolation (BSC-ISI) platform, which is a two-stage in-vacuum isolation system (two layers); and a quadruple pendulum (four layers), which supports the 40 kg test mass. Every layer provides additional isolation from both ground motion and the previous layer. LIGO has divided this system into two subsystems: *seismic isolation*, consisting of the external hydraulic stage (HEPI) (Hardham *et al.*, 2004, 2006) and the the active isolation platform (Abbott *et al.*, 2004); and *suspensions*, consisting of the multiple pendulum stages (Robertson *et al.*, 2002).

In addition to the main optics, which are each located in a large vacuum chamber called a BSC chamber (Figure 11.1), there are a considerable number of auxiliary optics which must also be isolated from ground motion. These include, for example, the power and signal recycling mirrors, the optics which prepare the light for injection into the main portion of the interferometer, and the optical filters and photodiodes for the light containing the gravitational wave information. Most of the auxiliary optics are placed in smaller vacuum chambers called HAM chambers (Figure 11.1). A simpler seismic isolation system is used for these optics. This consists of a HEPI system external to the chamber, the HAM Internal Seismic Isolation (HAM-ISI) platform, which is a single layer active platform within the chamber, and one-, two-, or three-stage pendulums for the various optics, depending on the required isolation performance.

Requirements for the BSC and HAM systems

The goal of the seismic isolation subsystem is to provide a quiet, well-controlled platform to support the various optical components. The noise requirements for the platforms are shown below in Figure 11.2. The chambers are mounted on a concrete slab, commonly called the technical slab, which is nominally separate from the slab for the building. Since the motion of the technical slab at 1 Hz (particularly the Livingston observatory) can be 1×10^{-8} m/$\sqrt{\text{Hz}}$, the HAM platform must provide at least a factor of 25 isolation from floor motion at 1 Hz, and the BSC platform should provide a factor of 1000 isolation at that frequency (Fritschel *et al.*, 2001).

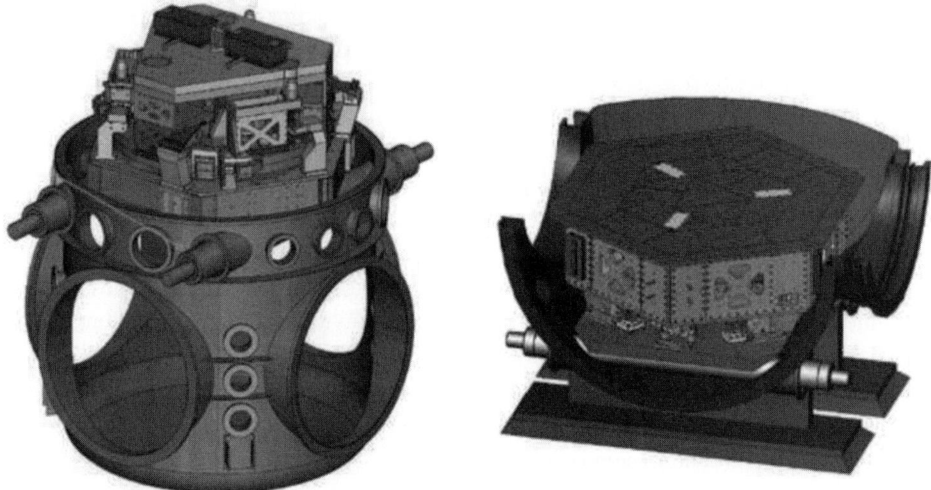

Figure 11.1 CAD renderings of the isolation platforms for the BSC chamber (left) and HAM chamber (right). The top of the BSC chamber has been rendered transparent to show the isolation platform. The doors and external supports of the BSC chamber are not shown. The HAM platform is shown in a sectioned HAM chamber.

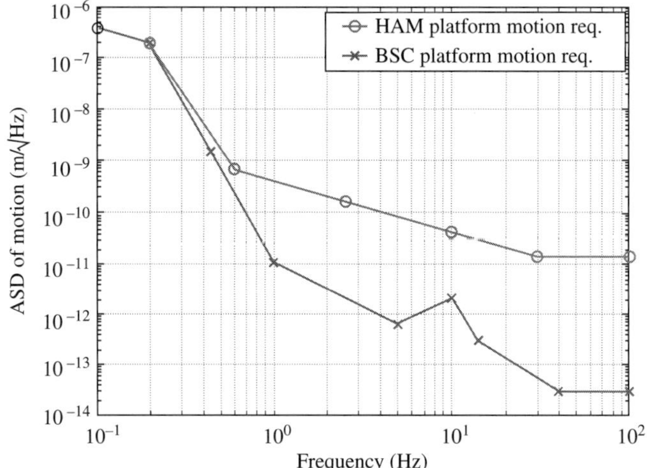

Figure 11.2 Performance requirements for the Advanced LIGO BSC and HAM isolation systems.

The platforms also provide positioning capability in six degrees of freedom (DOF) so that the interferometer can be held at the correct working point over long time periods. The hydraulically actuated outer stage (HEPI) can quietly actuate the payload by 1 mm. To accomplish the low frequency positioning, displacement sensors are used to measure the relative position of the tables with respect to their supports in all six DOFs, and actuators are used to move the table to the appropriate location. Positioning can be done at all of the stages of the seismic subsystem, but large static offsets are typically maintained by the laminar-flow quiet hydraulic actuators of the HEPI stage, which are outside the vacuum system. This minimises the heat load of the in-vacuum electromagnetic actuators on the internal stages and on the pendulum systems.

The seismic isolation system is designed to provide seismic isolation at frequencies above 0.1 Hz. This frequency is chosen for two reasons. First, it is important to provide isolation from the ground at about 0.15 Hz. This is the frequency of a large feature in the ground motion called the microseismic peak (Webb, 2002). There can be significant ground motions around this frequency caused by ocean waves, and resulting motion is typically a few µm per second at the Lousiana Observatory (Daw *et al.*, 2004). The 4 km baseline of the interferometer results in substantial differential-mode motions of the optics at these frequencies, and we have seen that it is extremely difficult to acquire lock with the interferometer if there is more than about 1 µm s^{-1} of differential velocity between the mirrors. Thus, it is important to provide seismic isolation above 0.1 Hz.

Below 0.1 Hz, it becomes increasingly difficult to provide seismic isolation in the horizontal direction because of the coupling of ground tilt into the isolation systems. This *tilt–horizontal coupling* is one of the most difficult design challenges for systems providing isolation below about 0.5 Hz, and many of the design choices for the Advanced LIGO seismic isolation system are made to minimise the impact of this coupling (Lantz *et al.*, 2009).

Tilt–horizontal coupling

Tilt–horizontal coupling arises from the inability to distinguish between a horizontal acceleration on an object and the change in lateral force on that object caused by tilting it in Earth's gravitational field, a demonstration of Einstein's equivalence principle. In practice, improving the the low frequency horizontal isolation of the system (active or passive) tends to make it more sensitive to low frequency ground tilts. This can be demonstrated with a simple example system: a horizontal mass–spring system on a tilting floor. The mass might be a stage of a simple isolation system, or it might be the proof mass of a horizontal scismometer. This system has mass, m, spring stiffness, k, and damping b. The mass is free to slide on the floor. The mass location is x_m, the attachment point to the floor is x_f. The floor is allowed to tilt with respect to local gravity, g by an angle θ. We ignore external forces on the mass, and since the motions are all small, we ignore centrifugal forces, etc. The basic equation describing this system is

$$m\ddot{x}_m = -k(x_m - x_f) - b(\dot{x}_m - \dot{x}_f) + m\,g\,\sin(\theta). \tag{11.1}$$

We define x_d as the differential motion $x_m - x_f$ between the mass and the floor. By measuring x_d, we could use this device as a seismometer to measure floor motion. We let θ be small, and we take the fourier transform so that $\dot{x}_m$ becomes $i\omega x_m$, etc. We can then rewrite equation

(11.1) as

$$x_{\rm d} = \frac{\omega^2/\omega_0^2}{-\omega^2/\omega_0^2 + i\omega/(\omega_0 Q) + 1}\left(x_f - \frac{g}{\omega^2}\theta\right). \tag{11.2}$$

where

$$\omega_0 \equiv \sqrt{k/m}, \text{ and } Q \equiv \frac{\omega_0}{b/m} \tag{11.3}$$

The differential motion is sensitive to both floor motion and floor tilt. The dynamics of the particular device are contained in the first term, and the relative sensitivity to translation and tilt are contained in the second term. This is a specific example of the general result that, for a horizontal seismometer, the ratio of the sensitivity to rotation (seismometer signal per radian of angle) to the sensitivity to horizontal motion (seismometer signal per meter of translation) at a particular frequency ω (in radians/s) is g/ω^2. Alternatively, for an isolation stage, one could solve for the motion of the mass, so equation (11.1) becomes

$$x_{\rm m} = \frac{i\omega/(\omega_0 Q) + 1}{-\omega^2/\omega_0^2 + i\omega/(\omega_0 Q) + 1}\left(x_f + \frac{g\omega_0^2}{i\omega/(\omega_0 Q) + 1}\theta\right). \tag{11.4}$$

We see that as we lower the natural frequency, the relative sensitivity to tilt increases. Thus, for a horizontal isolation system with good low frequency performance, one can use active techniques and try to compensate for the sensor's excess response to tilt at very low frequencies, or one can make a passive system with a very low natural frequency, and try to compensate for the platform's excess response to tilt.

11.2 Achieving isolation

The seismic isolation of the Advanced LIGO platforms is achieved by using both active and passive techniques. Each platform is isolated in all six DOF. To reach the desired level of performance of Advanced LIGO, we use several successive stages of isolation. Each successive stage has less motion than the stage supporting it. The slow, large-force, long-displacement actuators are located on HEPI, the outermost stage.

The basic design concept for all of the stages of the seismic isolation subsystem are similar. Each stage is supported by springs at its nominal location. The springs (four sets for HEPI and three sets for the internal isolation stages) provide passive isolation, support the weight of the stage, and set the nominal stage location in all six DOF. Each stage is instrumented with displacement sensors and inertial sensors, to measure the location and the motion of the stage in six DOF. Each stage has actuators to apply forces in all degrees of freedom.

HEPI – The Hydraulic External Pre-Isolator

The outermost stage of the system is a hydraulically driven pre-isolation stage. The laminar flow hydraulic actuators are described in detail elsewhere (Hardham *et al.*, 2004; Hardham, 2005). They provide a millimeter of throw with several hundred pounds of force, and because they are well impedance-matched to the load, they can hold that offset indefinitely with only about 10 W of power dissipation. The hydraulic actuators are also damped by an internal damping network. The acuators control the motion of the support structure that holds all the internal seismic isolation systems, which in turn supports the optical payloads. Thus, the support structure for the internal stages is stiff and well damped, providing a good base on which to do control. It also provides large throw capability so the electromagnetic actuators on the internal stages can be run with minimal static offset, maximising their typically available range and minimising their typical power dissipation.

In contrast with the internal seismic isolation system, the external hydraulic actuators are stiff, so most of the performance of this stage is achieved actively. The stage provides isolation from 0.1 Hz to about 10 Hz, and it provides positioning authority (for alignment and global offload) from zero frequency to about 10 Hz. The HEPI system was implemented for Initial LIGO at the Livingston Observatory, and has been operational there for many years. It routinely achieves a factor of 10 reduction in the microseismic peak in all three translational degrees of freedom (Hardham *et al.*, 2006) using a combination of feedback and feedforward control. The feedback control signal is derived from blending displacement and inertial sensors, as described below in Section 11.2. The isolation at the microseismic peak comes exclusively from *sensor correction* (Hua, 2005; Hua *et al.*, 2004b), where the measured ground motion (measured by a STS-2 seismometer) is fed forward to the displacement sensor measuring the relative motion between the ground and the isolated stage. When the displacement sensor signal is *corrected* by subtracting the ground motion signal, the feedback loop around that displacement sensor holds the platform still in inertial space (Hua *et al.*, 2004a).

Passive isolation

Most of the isolation performance for the BSC-ISI and HAM-ISI in the gravitational wave band (i.e. above 10 Hz) comes from passive isolation, using the basic techniques described in the previous chapter. The passive isolation of the stages is achieved by suspending the internal stages from springs and flexures. The vertical springs are triangular blade springs made from maraging steel, based on the designs used by the pendulum suspensions in the GEO project (Torrie, 2001). The horizontal compliance is achieved by suspending the stage from a 'wire' attached to the tip of each spring. Because of the weights of the stages, the 'wire' is a maraging steel rod several millimeters in diameter. Each stage is suspended from a set of three springs arranged in an equilateral triangle (Matichard *et al.*, 2009). The natural frequencies for the HAM system lie between 0.8 and 1.8 Hz (Lantz *et al.*, 2008), and the coupled frequencies of the two-stage BSC system lie between about 1.3 and 7.0 Hz. The spring and flexure arrangements are simple and work at low stress levels (the maximium stress is about 35% of yield), making them easy to use and very robust. The dimensions are chosen to minimise the spring size for a given stress level and stiffness, while minimising the tilt generated by horizontal translations. These calculations are described in detail elsewhere (Smith *et al.*, 2004; Hollander *et al.*, 2007).

Active control

The active control of the platforms both augments and complements the passive isolation. Active control is used to provide isolation between 0.1 Hz and about 20 Hz. Active control is also used to provide stable positioning control down to zero frequency. Active isolation and alignment techniques have long been used to complement passive techniques; a few interesting examples can be found in the following references: Robertson *et al.* (1982), Teague *et al.* (1998), Kramar *et al.* (1999), or Mercadal *et al.* (2001). Active control is an attractive complement to passive isolation – it offers a path to excellent low frequency performance and affords considerable flexibility in the operation of the system. The ultimate performance of active isolation system is bounded by the noise of the sensors and the stability conditions of the controls; the noise of the Advanced LIGO sensors is described below in Section 11.2. The real-time control also adds complexity to the isolation system. However, for Advanced LIGO, real-time control is already required to actively position the platforms, and real-time control is used throughout the interferometer. With actuators, position sensors, and real-time computing infrastructure in place, implementing active seismic isolation requires adding inertial sensors and increasing the bandwidth of the control loops. The active control and inertial sensing also enable various diagnostics: continuous monitoring of platform motion, calibrated motion inputs to test the the pendulum suspensions, and long range modulation of mirror positions to study interferometer noise coupling issues are a few of the diagnostics which will use the active controls in the Advanced LIGO interferometers.

The control loops are implemented as single-input, single-output (SISO) control loops. The design and performance on the prototypes and the first observatory units is described in detail in Hua (2005) and Kissel (2010). SISO control is possible because the mechanical system (the 'plant') is designed with controllability as a high priority. We combine information from displacement sensors and inertial sensors to form the feedback control signals. We also use ground seismometers as feedforward sensors for HEPI, and inertial sensors on the support structure as feedforward sensors for the HAM system. The upper unity gain frequency of the isolation loops is typically 25 to 30 Hz.

Control is performed in the 'Cartesian basis' rather than by using the signal from a single sensor to control a single loop. Each set of sensors has a fixed-value matrix which converts the signals from the six individual sensors into a set of six signals in the Cartesian basis. The coordinate system we choose is aligned with the global coordinate system for the observatory. Because of the careful mechanical design of the platforms, the coupling amongst the degrees of freedom is small enough that it can be ignored for frequencies below about 100 Hz.

The isolation platforms for Advanced LIGO support a large number and wide variety of payloads in addition to the main mirror suspensions, and require that 30 isolation platforms be commissioned. Therefore, having a simple, well defined interface significantly simplifies the commissioning and testing of the whole system. Performing control in the Cartesian basis also simplifies the interface with the other components of the detector. Most importantly, the cartesian basis simplifies the problem of tilt–horizontal coupling because it is aligned with the local vertical. By separating the horizontal, vertical, and tilt control designs, one can more easily implement the methods necessary to deal with tilt–horizontal coupling.

Blended sensors

Since there are two sets of sensors for each stage of the isolation system, relative displacement sensors for the low frequency positioning and inertial sensors for mid-band isolation, a method to combine the information from both sets of sensors is required to make the control system perform correctly. The technique we use is sensor blending (Hua, 2005). For each DOF, the sensor signals are combined by using a pair of 'complementary filters' (Sims, 1973). In general, a pair of complementary filters are two filters, L and H, which have the property that their Laplace transforms, $L(s)$ and $H(s)$, sum to 1, i.e.

$$L(s) + H(s) = 1 \tag{11.5}$$

A simple example of a complementary pair of filters is:

$$L_{\text{example}}(s) = \frac{1}{s+1}, \text{ and } H_{\text{example}}(s) = \frac{s}{s+1} \tag{11.6}$$

Here, L is a lowpass filter and H is a highpass filter. We employ filter pairs where L is a lowpass filter applied to the normalised displacement sensor signal and H is a highpass filter applied to the normalised inertial sensor signal. The filters are typically more complicated than the example above so as to achieve particular shaping of sensor noise and ground motion. The filters can also be set for use with different interferometer working conditions (e.g. acquisition versus lock), different ground motion conditions (e.g. typical operation may benefit from different filters than those applied during an earthquake or the passing of trains), or different testing setups of the platforms.

The signals from the two sensors are each normalised to be in common units before the complementary filtering, and then the two normalised filtered signals are added together to form a *supersensor*. The sensor signals are normalised by digitally filtering the sensor output by the inverse of the sensor and electronics transfer function, and scaled to a convenient level for use in the digital control.

It is important to evaluate how the noise for the various sensors couples into the system. For convenience, we can specify the noise n of the sensors and associated electronics in terms of input sensor noise, i.e. equivalent platform motion, as n_{inertial} and n_{disp}. We further define R_{disp} and R_{inertial} to be the readout chain (gain, whitening, analog to digital conversion, etc.) of the displacement sensor and the inertial sensor. Here, we ignore the complications presented by the analog to digitial conversion process, which contributes quantisation in both time and signal level. Details of the analog filtering can be seen in Lantz *et al.* (2010). A digital filter is created which is the inverse of the sensor response and and filtering chain times a scaling factor, A, from the physical unit to the digital signal. The digital supersensor signal is, therefore,

$$d = L\,A\,(x_{\text{platform}} - x_{\text{support}} + n_{\text{disp}}) \text{ (displacement sensor)}$$
$$+ H\,A\,(x_{\text{platform}} + n_{\text{inertial}}) \quad \text{(inertial sensor)} \tag{11.7}$$

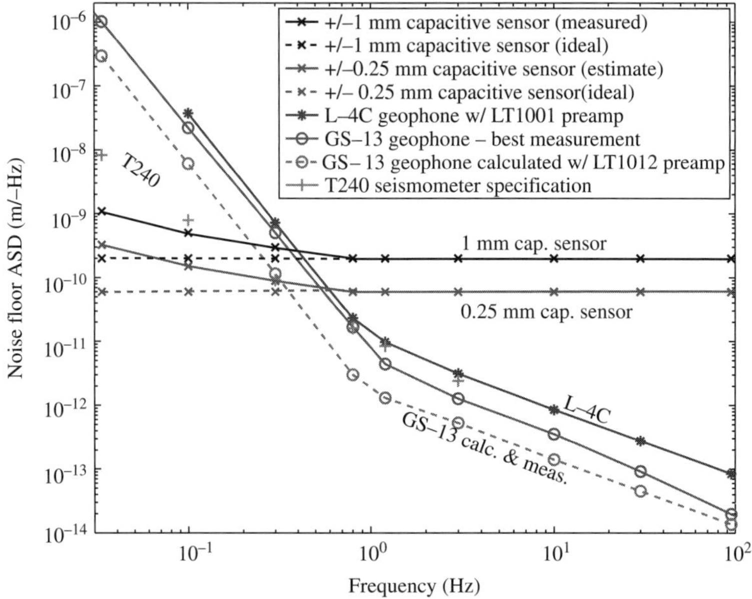

Figure 11.3 Noise estimates for the various SEI sensors (Lantz, 2009).

since L and H are complementary, this simplifies to

$$d = Ax_{\text{platform}} + A(-Lx_{\text{support}} + Ln_{\text{disp}} + Hn_{\text{inertial}}) \tag{11.8}$$

Thus, for each degree of freedom, our supersensor signal is the inertial motion of the platform plus three noise terms: (1) the lowpass filtered motion of the support structure, (2) the lowpass filtered noise of the displacement sensors, and (3) the highpass filtered noise of the inertial sensor. For horizontal sensors, there is a fourth term from the tilt-horizontal coupling. This term appears as $Ag/\omega^2 \cdot tilt_{\text{platform}}$ and so for the two horizontal translation directions, equation (11.8) becomes

$$d = Ax_{\text{platform}}$$
$$+ A(-Lx_{\text{support}} + Ln_{\text{disp}} + Hn_{\text{inertial}} + H\frac{g}{\omega^2}tilt_{\text{platform}}) \tag{11.9}$$

The limits of feedback control for active platforms are set by these four terms and the loop gain of the control.

Sensor selection

There are many sensors used by the Advanced LIGO Seismic isolation system. HEPI uses Streckeisen STS-2 low frequency seismometers on the floor for sensor correction and feedforward, Kaman DIT-5200 differential eddy current sensors and Sercel L-4C seismometers for feedback. The HAM-ISI uses L-4C seismometers mounted on the in-vaccum support structure for feedforward. The feedback sensors are Microsense 8800 capacitive sensors

(2 mm total range) mounted between the support structure and stage 1, and modified Geotech GS-13s are mounted on stage 1.

The sensors used for the BSC-ISI are: Microsense 8800 capacitive sensors (2 mm total range) mounted between the support structure and stage 1 and Microsense 8800 capacitive sensors (0.5 mm total range) mounted between the stage 1 and stage 2. The inertial sensors for stage 1 are Nanometrics Trillium-240s (used below about 2 Hz) and Sercel L-4Cs (used above about 2 Hz). The inertial sensors for stage 2 (the stage with the optical table) are modified Geotech GS-13s. The noise of the various sensors is shown in Figure 11.3. The inertial sensors in the vacuum are housed in small UHV compatible vessels so they do not contaminate the LIGO vacuum system.

11.3 Conclusions

By combining active and passive techniques, one can build systems with high performance and a range of useful capabilities. The systems for Advanced LIGO demonstrate excellent seismic isolation performance (Kissel, 2010; Matichard *et al.*, 2009) and provide a range of related capabilities including calibrated motion inputs, motion monitors, adjustable control modes, and control offload for the pendulum suspension.

Acknowledgements

This work was supported by the National Science Foundation under grant 0757896. The authors gratefully acknowledge the support of the United States National Science Foundation for the construction and operation of the LIGO Laboratory and the Science and Technology Facilities Council of the United Kingdom, the Max-Planck-Society, and the State of Niedersachsen, Germany, for support of the construction and operation of the GEO 600 detector. The authors also gratefully acknowledge the support of the research by these agencies and by the Australian Research Council, the International Science Linkages program of the Commonwealth of Australia, the Council of Scientific and Industrial Research of India, the Istituto Nazionale di Fisica Nucleare of Italy, the Spanish Ministerio de Educación y Ciencia, the Conselleria d'Economia, Hisenda i Innovació of the Govern de les Illes Balears, the Royal Society, the Scottish Funding Council, the Scottish Universities Physics Alliance, The National Aeronautics and Space Administration, the Carnegie Trust, the Leverhulme Trust, the David and Lucile Packard Foundation, the Research Corporation, and the Alfred P. Sloan Foundation.

Part 2: Passive isolation

J.-C. Dumas

11.4 Design goals and philosophy

The design goals for the isolator described here were to combine several novel approaches to isolation stages that were conceptualised at the University of Western Australia, and realise them into a full vibration isolation system with a performance targeting the require-ments of next generation gravitational wave detectors. These performance goals are to isolate test masses from seismic noise below the fundamental floor of the gravity gradient noise at $\sim$10Hz, and to achieve low residual motion at low frequencies to facilitate control requirements and cavity locking.

The design follows a passive philosophy to produce a 'soft' isolator. It is conceptually sim-ilar to the Virgo design, but it is much more compact and has an extra stage of pre-isolation in both the vertical axis and the horizontal plane. A compact system is not only practical but also reduces problems caused by internal modes. The design also has the advantage of using relatively inexpensive components, without any high quality seismometers or other expensive sensors that would be required for active isolation.

11.5 Cascade stages
Pre-isolation components

The vibration isolation system consists of multiple cascaded stages as shown in Figure 11.4 applying several different techniques to attenuate seismic noise. At the top of the suspension four short inverse pendulums legs supporting a square table on top provide the first stage of pre-isolation. This allows for horizontal translation, while being rigid to tilt. The resonant frequency of these pendulums can be tuned to ultra low frequency, providing very effective pre-isolation. Its cube shape allows for the integration of a spring system that provides vertical pre-isolation. The LaCoste linkage consists of diagonally attached springs between each of the four legs on two structures (the inverse pendulum and the LaCoste supporting frame). The four lower pivot arms were fitted with counter-weights in order to provide center of percussion tuning for all eight of them. Pre-tensed springs were used in order to obtain the 'zero-length' requirement of the LaCoste geometry (LaCoste, 1934). Horizontal springs were added and stretched more than the separation between the pivoting points, creating an inverse pendulum effect. This effect produces a negative spring constant that counteracts the significant spring constant of the flexure pivots. By making the width adjustable one can tune the spring constant in order to obtain ultra low frequencies and a large dynamic range. Therefore both pre-isolation stages have resonant frequencies below 100 mHz. Both the inverse pendulum and the LaCoste stages have a cubic geometry, which allows for their combination into a single 3-D structure, as illustrated in Figure 11.5.

The rigidity of this structure allows for the suspension of a Roberts linkage stage nested within the two pre-isolators. It is a relatively simple design consisting of a cube frame sus-pended by four wires hung off the LaCoste stage, as illustrated in Figure 11.6. Its geometry is tuned to restrict the suspension point of the load to an almost flat horizontal plane, thus making the gravitational potential energy almost independent of displacement and minimis-ing the restoring force, resulting in a low resonance frequency (Garoi *et al.*, 2003; Dumas *et al.*, 2009b). At only 1 m height the whole top section is very compact, including also the topmost stage of the multi-stage pendulum in the same volume. The pre-isolation stages are mounted on top of a rigid frame so as to have enough height to suspend the isolation chain from the top of the Roberts linkage, as seen in Figure 11.4. The interweaving of

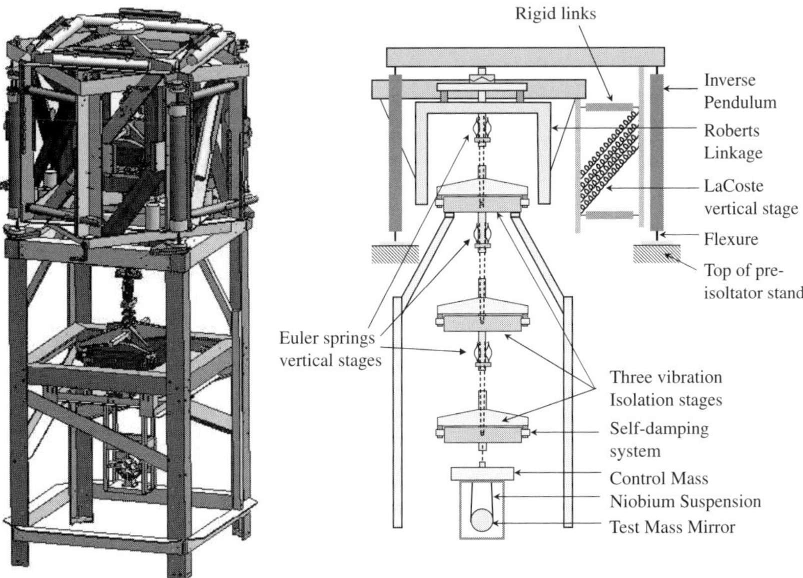

Figure 11.4 Full vibration isolator system and schematic that show the different stages of pre-isolation and the multi-pendulum stage with a test mass at the bottom of the chain.

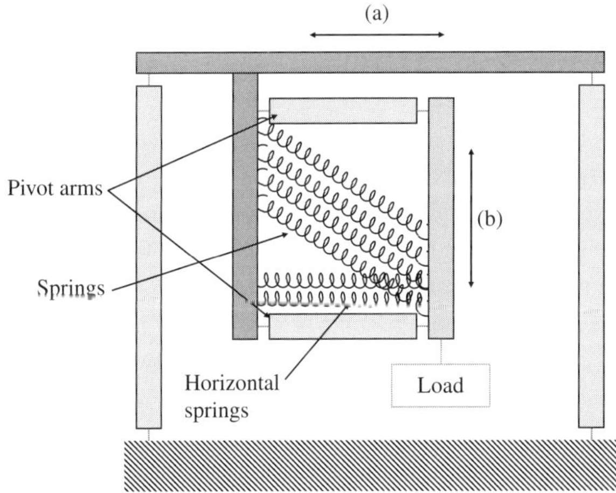

Figure 11.5 First stage horizontal and vertical pre-isolation. The pre-isolator combines two ultra low frequency stages: (a) The horizontal inverse pendulum and (b) the vertical LaCoste linkage. The concept of the anti-spring for flexure spring constant cancellation is shown.

the pre-isolation stages together with the use of large dynamic range sensors and actuators allows for an active feedback control of the pre-isolator stages, enabling performance close to the limit set by seismic tilt coupling (Dumas *et al.*, 2009a).

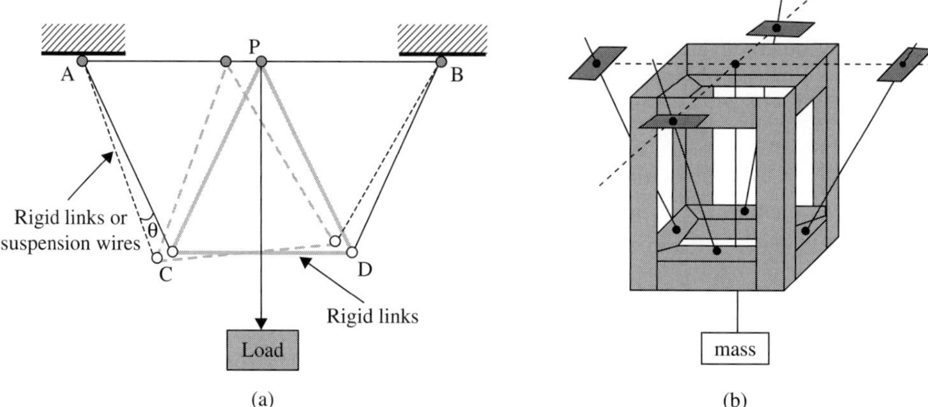

Figure 11.6 The Roberts linkage. (a) A one-dimensional diagram of a Roberts linkage with a suspended load from point P, which stays in the same plane for variations in the position of C and D. (b) The cube shaped design used in the AIGO suspension.

Isolation stages

Intermediate masses of 40 kg are suspended to form a self-damped pendulum arrangement, as illustrated in Figure 11.7. The self-damping concept consists of viscously coupling different degrees of freedom of the pendulum mass, as shown in Figure 11.8. Each intermediate mass is pivoted at its center of mass, with a light x-shaped frame fixed to the pendulum link to provide a reference against tilt (Figure 11.7). Neodymium iron boride magnets in a comb-like distribution are mounted to the frame. These are paired with intermeshing copper plates attached to each corner of the square rocker mass to create a viscous damping through eddy current coupling, reducing the Q-factor of the pendulum normal modes (Dumas *et al.*, 2004). An aluminium arm is attached on each side of the top rocker mass of the pendulum chain. The purpose is to increase the moment of inertia to further reduce the Q-factor of the lower resonant mode of the multi-stage pendulum.

Each intermediate mass in the multi-stage pendulum is attached to the next using Euler springs tuned for low frequency vertical isolation effectively attenuating the vertical component of the seismic noise. Euler spring stages can be tuned with anti-spring geometries to achieve good low frequency performance within a very compact design (Winterflood *et al.* 2002; Chin *et al.* 2004, 2005). This involves loading a vertical elastic column just beyond the Euler buckling instability, where it becomes a well behaved spring with frequency equal to $1/\sqrt{2}$ of the frequency of a pendulum of the same length. Methods were found of adding geometric anti-spring elements to this system to create even lower resonant frequencies. The Euler spring approach is advantageous because the stored elastic energy is reduced by a factor of (working range/effective length)^2, enabling the spring to be of much lower mass, and with much higher internal mode frequencies. Figure 11.9 shows a diagram of an intermediate mass with Euler spring attachment, while Figure 11.10 illustrates integration of the rocker mass with the Euler spring stage.

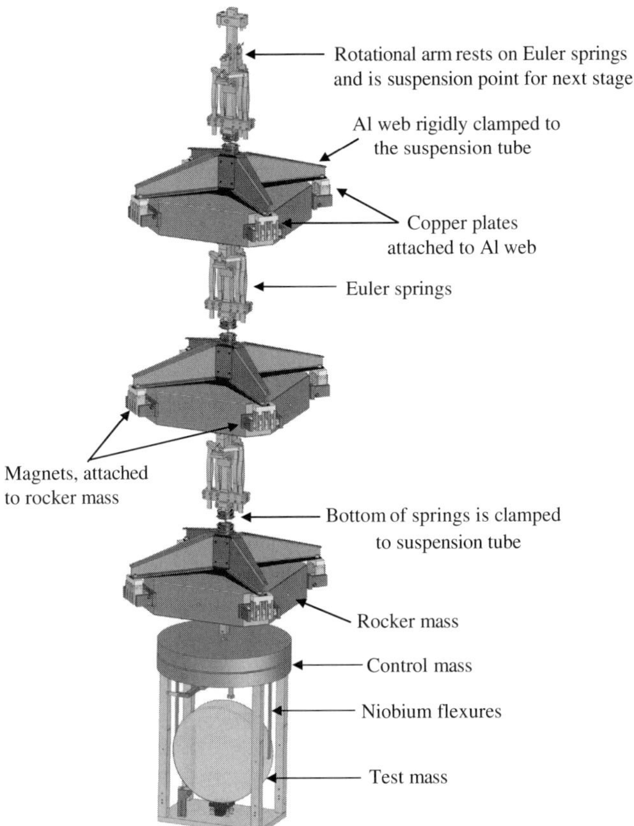

Figure 11.7 The multi-stage pendulum including three intermediate masses showing the rigid section, rocker mass, eddy current damping and Euler springs for vertical isolation. At the bottom of the chain the control mass stage provides sensing and control for a test mass suspended with niobium flexures.

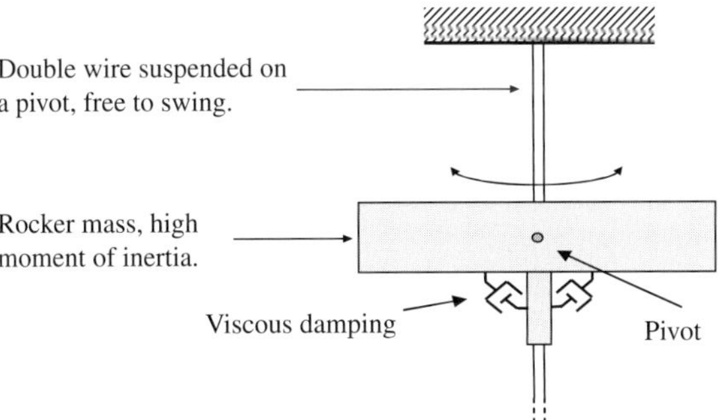

Figure 11.8 Diagram of one self-damped pendulum stage. Magnets that generate eddy currents on copper plates create the viscous damping for a high moment of inertia rocker mass.

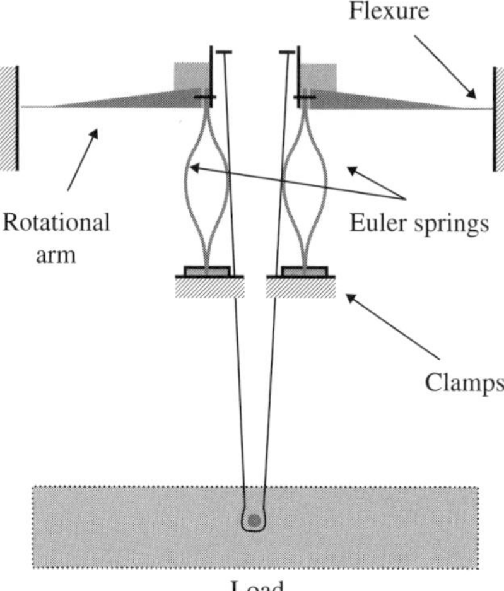

Figure 11.9 Schematic of the intermediate mass showing the Euler spring vertical stage and the attachment to the rocker mass (Chin *et al.*, 2005).

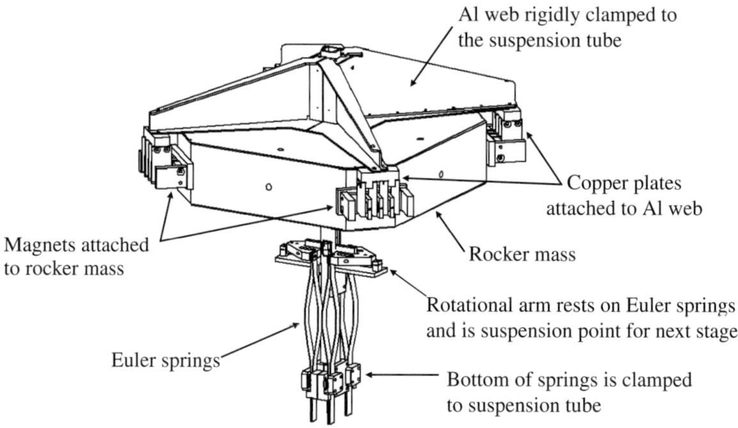

Figure 11.10 Intermediate mass showing the integration of the high moment of inertia rocker mass with the Euler spring vertical stage. The intermediate mass has a hollow tube in the center to allow for the suspension wire to go through all the stages. The figure also shows the $\times$-shaped frame that hold the copper plates on top of the rocker mass. Attached to the rocker mass are the magnets that create the damping through eddy current generation on the copper plates (Dumas *et al.*, 2004).

Control mass and test mass suspension

At the bottom of the multi-stage pendulum chain shown in Figure 11.7, a $\sim$30 kg control mass stage provides the interface to the test mass. The test mass consists of a 50 mm, 30 mm thick fused silica mirror supported in an aluminium and stainless steel cylinder. The purpose is to closely replicate a 'real' test mass and allows for the characterisation of the high performance vibration isolation chain developed at the University of Western Australia using an optical cavity. Four niobium ribbons, each 25 μm thick, 3 mm wide and 300 mm long, are used to suspend the test mass from the control mass stage. These are designed to minimise internal modes while providing a high Q-factor (Lee *et al.*, 2005).

For the initial suspension we have used temporary brass pins clamped to the suspension ribbons through a high pressure contact tooth instead of high pressure contact pins bonded to the end of the ribbons (Lee *et al.*, 2006). The ribbon clamping mechanism, as opposed to a permanent bond, is not ideal. However, the current design provides high enough contact pressure (approaching the yield strength of niobium) to minimise slip–stick friction, whilst not weakening the suspension ribbon at the point of clamping. Even though the same pin design has been used for the initial brass pins extra thermal noise may be induced. Currently, priority lies in testing the performance of the vibration isolation system. Therefore a temporary brass/niobium suspension at the expense of a possible increase in thermal noise is acceptable.

Figure 11.11 illustrates the control mass stage. A cage attached to the control mass provides both mechanical safety stops for the test mass, and a low noise reference from which actuation and local sensing of the test mass can be performed. The control mass is suspended from the vibration isolation system using a single wire. It also contains actuators

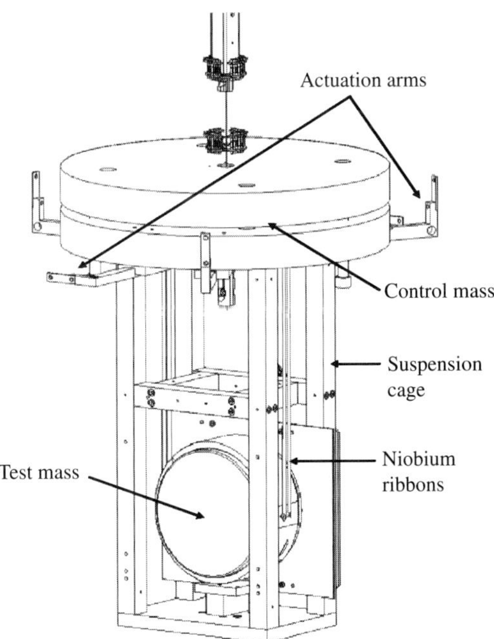

Figure 11.11 Control mass stage with test mass suspended. Actuation arms holding permanent magnets are attached to the control mass. From the control mass a suspension cage is attached and a test mass is suspended by four niobium ribbons.

and sensors from which five degrees of freedom (DOF) (translation in all three dimensions, yaw and pitch) can be accessed (Dumas *et al.*, 2009a).

11.6 Control hardware

Shadow sensors

The position of several stages of the isolation stack is monitored by the local control system through optical shadow sensors. A shadow sensor is a simple device consisting of an LED shining an infrared beam onto two photodiodes. A long, thin flag of the same width is attached to the stage to be monitored, and as the flag is displaced across the sensor the difference of the two photodiode signals forms a linear response. Each photodiode signal is amplified and converted to a signal in the ±10V range for the ADC module of the DSP board to be read by the digital control system. The shadow sensor has a relatively large dynamic range of ±5mm, and typical sensitivity of 10^{-10}m/$\sqrt{Hz}$ (Winterflood, 2001).

Magnet–coil actuator

Magnetic–coil actuators are used to control several stages, as described in Section 11.6. Each actuator consists of a pair of coils assembled together. Two designs of magnetic actuators are used in the isolator. Large actuators are used on the pre-isolation stage, for position control, drift correction, and damping ultra low frequency (ULF) normal modes; a smaller actuator is used to control the mass. The magnetic field within the actuator coils is nearly uniform within 1% in the central 10 mm of their range, allowing a large dynamic range in the control system.

Wire heating

In addition to the magnetic actuators, some stages are controlled by passing a current through particular suspension elements. The elements warm up and lengthen through thermal expansion. This control method is relatively effective when the system is under vacuum, as heat does not dissipate through convection, but only by the relatively slow processes of radiation and conduction. The advantage of this method is that it removes the complication of added parts, and in some cases provides much greater dynamic range. This is used to correct for large drifts caused by daily and seasonal ambient temperature changes. This actuation method responds with a quadratic relationship, but it is linearised in the digital feedback loop. The horizontal control of the Roberts linkage, and the vertical control of the LaCoste linkage, both employ this strategy.

The four suspension wires of the Roberts linkage are individually wired to current power supplies, allowing the length of each to be controlled by thermal expansion as they warm up. Since the Roberts linkage is very sensitive to any change of tension in any of the wires due to its carefully tuned folded configuration, a relatively small change of length is enough to control the stage through its entire dynamic range $\sim$10 mm. The position of the Roberts linkage is controlled via the circulating current, which in turn is controlled by the local control system through integral feedback to the current power supplies.

The LaCoste linkage has a large dynamic range and can be controlled through its entirety by magnetic actuators at a fixed ambient temperature. However, daily and seasonal temperature fluctuation cause drifts that would far exceed the capacity of the actuators. A $1°$ C change will offset the balance point of the LaCoste linkage by its entire range of 10 mm. For this reason, wire heating is an essential part of the LaCoste control loop. The coil springs of the LaCoste linkage are electrically connected in series, and can be heated to change the spring constant of the springs, which greatly affects the vertical position or balance point of the stage. Low frequency control (DC – 10 mHz) of this stage is achieved by regulating this current, while the magnetic actuators are used at higher frequencies ($\sim$100 milliHertz or 0.1 Hz).

Control implementation and degrees of freedom

The inverse pendulum can be monitored and actuated in three DOF, two in the horizontal plane (X and Y) and one angular (yaw ϕ), i.e. the rotation about the vertical axis. These are sensed and actuated through four shadow sensors and four actuators that are co-located on the inverse pendulum frame, as shown in Figure 11.12. The four signals are diagonalised into the three DOF X, Y and ϕ, and each is controlled independently as a single-input, single-output (SISO) feedback loop.

The LaCoste stage is a purely one-dimensional vertical stage (DOF: Z). Two actuators are mounted on opposing sides as illustrated in Figure 11.12 and Figure 11.13. They are used to damp the ULF normal mode of the stage. In addition, the LaCoste linkage can be controlled by heating the coil springs on all four sides of the stage, which are all connected to a high-current power supply. A shadow sensor mounted on the side of the pre-isolation structure monitors the vertical position of the LaCoste linkage.

The Roberts linkage in Figure 11.14 is controlled in two DOF, X and Y, by passing a current through its suspension wires. Each of the four suspension wires is electrically isolated from the rest of the vibration isolator structure and independently connected to a high-current power supply. By controlling the circulating current on each wire it is possible to control its length, and therefore the position. Since the suspension system is under vacuum

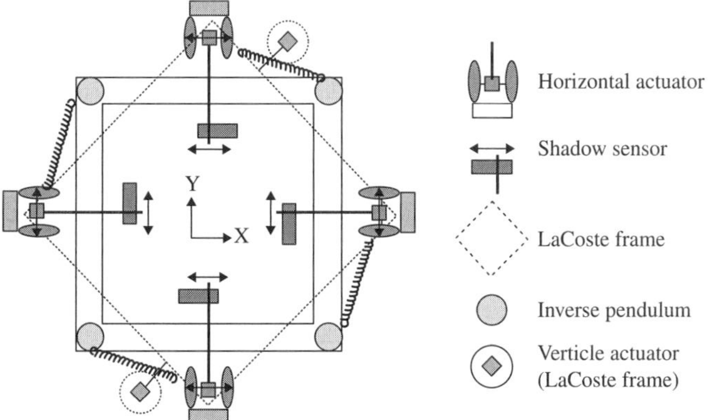

Figure 11.12 The inverse pendulum is controlled through shadow sensors and magnetic actuators.

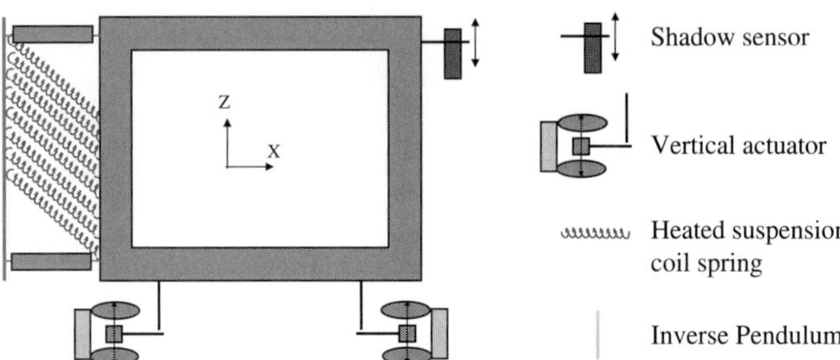

Figure 11.13 The LaCoste stage is controlled through a shadow sensor and magnetic actuator as well as through the heating of the suspension coil spring.

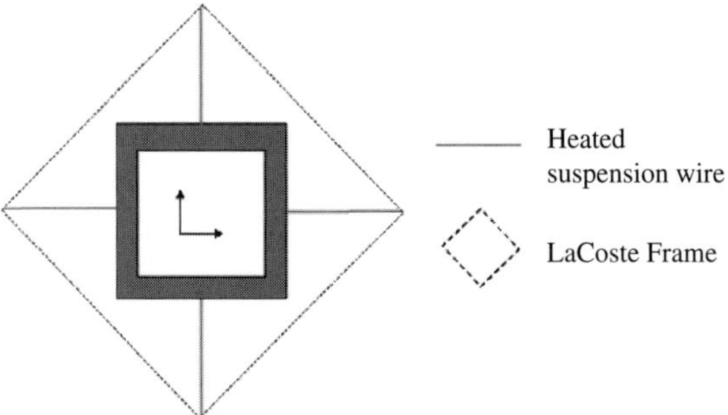

Figure 11.14 The Roberts Linkage is controlled through shadow sensors and the heating of the four suspension wires.

the heat loss by convection is insignificant. The X and Y signals from the control mass shadow sensors are used to feed back to this Roberts linkage actuation method. This control system provides a low frequency correction of any drift in the Roberts linkage and ultimately of the multi-stage pendulum and the test mass.

The control mass can be controlled in five DOF: three orthogonal translations X, Y, Z; the rotation about the vertical axis, yaw (ϕ), as shown in Figure 11.15; and the rotation about the horizontal axis perpendicular to the laser axis, pitch (θ), shown in Figure 11.16. Three horizontal actuators and shadow sensors are co-located in a 120° arrangement on the horizontal plane as seen in Figure 11.15, while two vertical shadow sensors and actuators are co-located on opposing sides of control mass along the laser axis as in Figure 11.16. The signals are digitalised by a sensing matrix into five orthogonal DOF: X, Y, Z, ϕ and θ. These are treated by five separate control loops as independent SISO systems, before the signals are recombined by a driving matrix into the appropriate actuator signals.

Optical lever

Due to poor coupling of the mirror suspension angular modes to the control mass, it is necessary to have a direct readout of the mirror angular orientation. This was achieved by a simple optical lever as illustrated in Figure 11.17. A laser outside the vacuum envelope

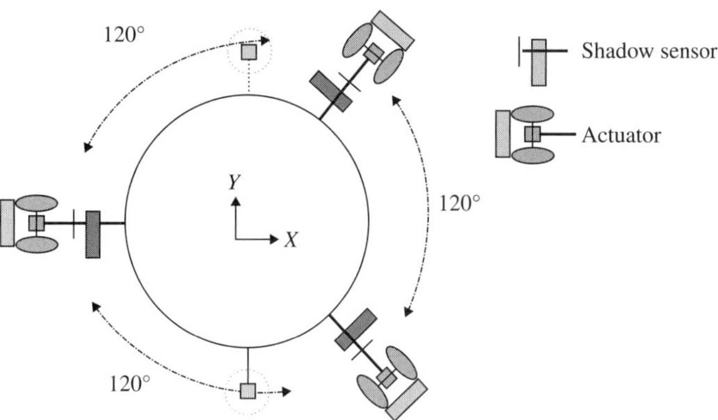

Figure 11.15 The control mass has three actuators and shadow sensors collocated on the horizontal plane, in a 120° arrangement. These three signals are converted to an orthogonal reference frame X, Y, Z and ϕ by a sensing matrix.

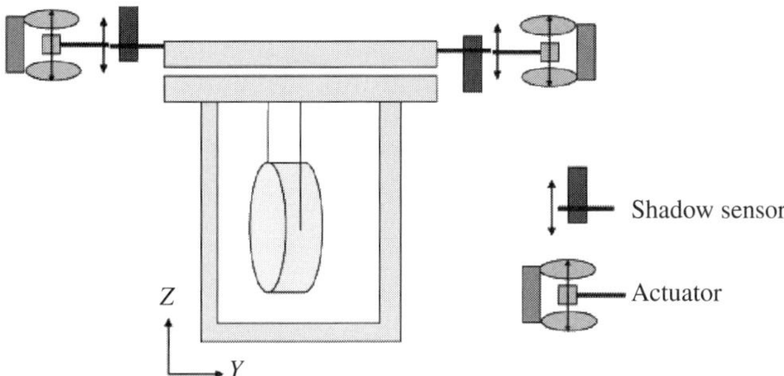

Figure 11.16 The pitch of the control mass is actuated by two vertical magnetic actuators.

is reflected off the test mass and is measured by a quadrant photodiode, also outside the vacuum envelope. In addition to being a direct measurement from the mirror surface, the optical lever provides better sensitivity to angular motion as it is placed further away from the center of rotation of the mirror, such that the same angular rotation corresponds to a much larger arc-length measured by the quadrant photodiode $\sim$5m away from the mirror, than the shadow sensors which are only 200 mm away. The drawback is the limited dynamic range: it provides $\sim$1 mrad, which is greatly exceeded by the test mass suspension oscillation when it is excited. Therefore the shadow sensor feedback is used for initial damping of the angular modes, before the optical lever signal can be used for feedback, as discussed in Section 11.7.

The digital controller

The control system is hosted by a Sheldon Instrument DSP board (www.sheldoninst.com), forming a flexible multidimensional digital control platform. The board is a SI-C33DSP on a PCI bus, based on a 150 MHz Texas Instruments TMS320VC33 DSP using a mezzanine board SI-MOD6800 to provide 32 input channels (16 bit ADC), 16 output channels (16 bit

 Vibration isolation

Table 11.1. *Input/Output channel allocation usage of DSP*

Stage	Inputs	Outputs
Inverse pendulum	8	4
LaCoste stage	2	2
Roberts linkage	4	4
Control mass	10	5
Optical lever	2	0
Auxiliary	2	1
Total:	28	16

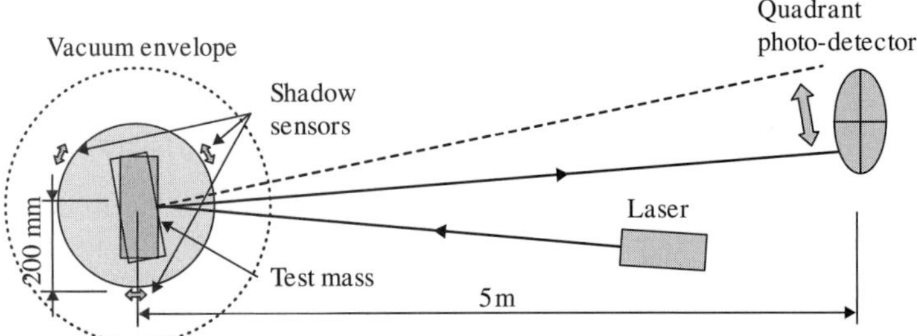

Figure 11.17 The optical lever setup, using a quadrant photodiode placed outside the vacuum envelope.

DAC), and digital input/outputs. The input channels are used as described in Table 11.1. Most input channels are used for the shadow sensors, which require two inputs each, one per photodiode. Two more inputs are used for the vertical and horizontal axis readout of the quadrant photodiode used with the optical lever, and two more are wired to auxiliary connectors to inject any arbitrary analog signal. The output channels are used for the control signals of actuators and current power supplies.

An intermediate analog system amplifies and filters the input and output channels between the DSP and the control components (shadow sensors, wire heating and actuators) with the exception of the quadrant photodetector used in the optical lever, which is integrated on a board with pre-amplification. The analog electronics consists of 13 boards in a standard 6U rack; each board contains a dual photodetector circuit for the pair of photodiodes in one shadow sensor, and a control signal circuit to drive an actuator. The dual photodetector circuit contains a transimpedance amplifier, anti-aliasing filters and an amplifier. The signal of both photodiodes is then distributed to two inputs on the DSP board. The control signal is distributed from one DSP output to the corresponding channel on the control circuit, which contains anti-aliasing filters and a high speed current amplifier, before distribution to the actuator coils. An additional board in the 6U rack contains five filter circuits for the wire heating control signals. These five signals are then distributed to five external voltage-controlled current supplies.

The control scheme, algorithms, and the user interface, are written and operated in Labview®. The built-in libraries provided by the DSP manufacturer allow for the control loops to run on the DSP board in real time including ADC and DAC at a 100 Hz sampling rate. The user interface is run on a host PC to monitor every stage of the isolation chain and adjust control parameters, such as filters and loop gains, as necessary.

11.7 Control scheme

The control scheme has three main purposes. One is to maintain alignment and positioning for all stages, against drifts such as caused by ambient temperature changes or tidal effects. These effects are extremely low frequency, with time-scales from tens of minutes to days. Therefore the control strategy consists of low gain integration feedback. The control system also has to maintain the test mass alignment to obtain a resonant cavity. The alignment of the test mass is controlled indirectly via pitch and yaw of the control mass stage. However, sensing is done with the control mass shadow sensors and directly from the test mass through the optical lever. The control is achieved by both proportional and integration control while aligning the cavity, and integration only when maintaining the cavity locked.

While the isolation design relies on passive isolation, some active damping is required for some ULF resonant modes of the pre-isolation, as well as low frequency torsional modes (yaw) of the entire chain. Additionally, the two normal modes of the test mass suspension (pitch and yaw) must be damped at least initially after any alignment or positioning offset. The extremely high Q-factor of the niobium suspension would otherwise result in several days of oscillations after any large perturbation. While controlling the alignment, the control system must also avoid driving the suspension resonant modes; this is done through carefully placed filters in the feedback gain.

At each stage the sensor signals are converted into orthogonal DOF by a sensing matrix, such that the control system consists of independent SISO systems, which simplifies the control strategy and requirements over a MIMO system (multiple-input, multiple-output). The various DOF of each stage relevant to the control system, as discussed in Section 11.6, are summarised in Table 11.2. Figure 11.18 illustrates the physical location for sensing and actuation of the feedback loops.

Particular care is needed for the control mass feedback which couples almost directly into the test mass, and also for automation of the transition between feedback from in-vacuum position sensors and the external optical lever. The yaw resonance is damped through the shadow sensor control loop when its amplitude is too large for the range of the optical lever, ~ 1 mrad. Although the oscillation couples weakly to the control mass, if there is sufficient signal-to-noise ratio to damp the test mass into a range where the optical lever becomes operable, then a much better level of control can be achieved ($\sim 10 \mu$rad). Low frequency resonances caused by the torsion of the suspension wires dominate the residual angular motion with either control method. The optical lever achieves better performance due to higher angular sensitivity, and decoupling from translational motion (X, Y, Z).

Table 11.2. *Stages and relevant degrees of freedom in the control scheme*

Stage	DOF
Inverse pendulum	X, Y, ϕ
LaCoste stage	Z
Roberts linkage	X, Y
Control mass	X, Y, Z, ϕ, θ
Optical lever	ϕ, θ

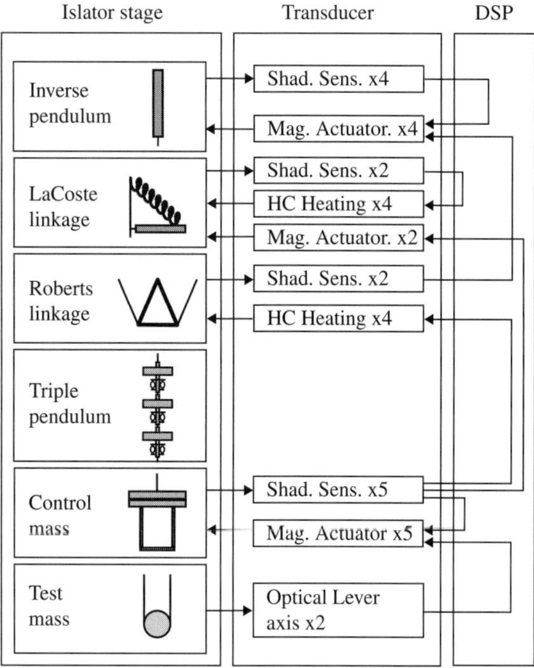

Figure 11.18 Block diagram of the isolation local control system. The signals from shadow sensors and a quadrant photodiode are used to feed back to several stages using magnetic actuators or high-current heating.

If the optical lever goes out of range, either due to a large offset or to a large amplitude in the suspension normal mode, the control system will damp the pitch and yaw modes with a strong velocity feedback gain. Integral and proportional feedback are used in the loop for cavity alignment; it also includes band-pass filters to damp the suspension modes. In particular, the phase of the control signal at the suspension resonances must be reversed, since the control mass coupled oscillation is out of phase with respect to the test mass oscillation. Once the optical lever is in range, the damping feedback automatically changes the source of its error signal to the optical lever signal (quadrant photodiode). This process is automated by the control system though a set of Boolean operations to determine the state of the system according to threshold values on the range of the optical lever and shadow sensors. The transition from one loop to the other is 'smooth' in the sense that the DC force resulting from integral and proportional set-points LS passed on from one loop to the other.

11.8 Conclusion

A vibration isolation design that adopts novel concepts including multiple pre-isolation stages, a LaCoste linkage, a Roberts linkage, self-damping pendulums, Euler springs and modular niobium ribbon suspension was implemented. A pair of such isolators was used to form a suspended optical cavity. It was shown that these systems have long-term stability, and are responsive to the controls required to operate long optical cavities (Barriga *et al.*, 2009). A digital local control system was implemented which provides feedback for position control, cavity alignment, and damping of normal modes. Three DOF are controlled by ohmic thermal tuning of the length of pendulum wires in the Roberts linkage and the spring constant of the LaCoste linkage. Large dynamic-range shadow sensors and actuators allow more than ± 3 mm three-dimensional position control of the test mass. Feedback forces are applied at the pre-isolation stages and at the control mass, but not directly on the test mass. Without using direct test mass control, it was possible to lock the cavity and maintain lock, with a residual motion of 3 nm per test mass above 1 Hz (Barriga *et al.*, 2009). Angular control of the test mass is aided by an optical lever, with an automatic smooth transition between local sensor feedback and optical lever feedback (Dumas *et al.*, 2009c).

Acknowledegments

J-C. Dumas gratefully acknowledges financial support from the West Australian Government Centres of Excellence scheme, the Australian Research Council and the University of Western Australia.

References

Abbott, R., *et al.* 2004. *J. Quantum Grav.*, **21**, S915–S921.
Barriga, P., *et al.* 2009. *Review of Scientific Instruments*, **80**, 114501.
Chin, E. J., *et al.* 2004. *Class. Quantum Grav.*, **21**, S959–S963.
Chin, E. J., *et al.* 2005. *Phys. Lett. A*, **336**, 97–105.
Daw, E. J., *et al.* 2004. *J. Quantum Grav.*, **21**, 2255–2273.
Dumas, J. C., *et al.* 2004. *Class. Quantum Grav.*, **21**, S965–S971.
Dumas, J. C., *et al.* 2009a. *Rev. Sci. Instrum*, **80**, 114502.
Dumas, J. C., *et al.* 2009b. *Phys. Lett. A.* **374**, 3705.
Dumas, J.-C., *et al.* 2009c. *Review of Scientific Instruments*, **80**, 114502.
Fritschel, P., *et al.* 2001. LIGO internal document E990303-03-D, available at http://admdbsrv.ligo.caltech.edu/dcc/. LIGO Internal Report.
Garoi, F., *et al.* 2003. *Rev. Sci. Instrum.*, **74**, 3487–3491.
Hardham, C., *et al.* 2004. Page 127 of: *Spring Topical Meeting on Control of Precision Systems*. Proc. ASPE, vol. 32.

Hardham, C., *et al.* 2006. In: *MECHATRONICS 2006*. 4th IFAC-Symposium on Mechatronic Systems, nos. P060028–00–Z, available at http://admdbsrv.ligo.caltech.edu/dcc/.

Hardham, C. 2005. *Quiet hydraulic actuators for LIGO*. Ph.D. thesis, Stanford University. LIGO document #P050031-00-R, available at http://admdbsrv.ligo.caltech.edu/dcc/.

Hollander, B., *et al.* 2007. LIGO internal document G070156-00-R, available at https://dcc.ligo.org/. Design report from High Precision Devices.

Hua, W., *et al.* 2004a. Page 194 of: *Gravitational Wave and Particle Astrophysics Detectors*. Proc. SPIE, vol. 5500.

Hua, W., *et al.* 2004b. Pages 109–114 of: *Spring Topical Meeting on Control of Precision Systems*. Proc. ASPE, vol. 32.

Hua, W. 2005. *Low Frequency Vibration Isolation and Alignment System for Advanced LIGO*. Ph.D. thesis, Stanford University.

Kawamura, S. and Chen, Y. 2004. *Phys. Rev. Lett.*, **93**, 211103.

Kissel, J. 2010. *Calibrating and Improving the Sensitivity of the LIGO Detectors*. Ph.D. thesis, Louisiana State University.

Kramar, J., *et al.* 1999. Pages 1017–1028 of: Singh, B. (ed.), *Metrology, Inspection, and Process Control for Microlithography XIII*, vol. 3677. SPIE.

LaCoste, L. J. B. 1934. *Physics*, **5**, 178–180.

Lantz, B. 2009. LIGO internal document T0900450-v3, available at https://dcc.ligo.org/. LIGO Internal Report.

Lantz, B., Mason, K., *et al.* 2008. LIGO internal document T080236-v2, available at https://dcc.ligo.org/. LIGO Internal Report.

Lantz, B., *et al.* 2009. *Bulletin of the Seismological Society of America*, **99**, 980–989.

Lantz, B., Abbott, B., *et al.* 2010. LIGO internal document G1000412-v1, available at https://dcc.ligo.org/. LIGO Internal Report.

Lee, B. H., Ju, L. and Blair, D. G. 2005. *Phys. Lett. A*, **339**, 217–223.

Lee, B. H., Ju, L. and Blair, D. G. 2006. *Phys. Lett. A*, **350**, 319 – 323.

Matichard, F., *et al.* 2009. LIGO internal document L0900222-v2, available at https://dcc.ligo.org/. LIGO Internal Report.

Mercadal, M., *et al.* 2001. Tech. rept. SSL #15-91-R. MIT Space Systems Laboratory.

Robertson, N. A., *et al.* 1982. *Journal of Physics E: Scientific Instruments*, **15**, 1101.

Robertson, N. A., *et al.* 2002. *Classical and Quantum Gravity*, **19**, 4043–4058.

Robertson, N. A., *et al.* 2004. Page 81 of: Hough, J. and Sanders, G. (eds), *Gravitational Wave and Particle Astrophysics Detectors*. Proc. SPIE, vol. 5500.

Sims, C. S. 1973. Pages 383–387 of: Parker, S. (ed.), *7th Asilomar Conference on Circuits, Systems, and Computers*. Western Periodicals: USA.

Smith, K., *et al.* 2004. LIGO internal document C040120-A-D, available at http://antares.ligo.caltech.edu/dcc. Design report from Alliance Spacesystems Inc.

Teague, E. C., How, J. P. and Parkinson, B. W. 1998. *Journal of Guidance Control and Dynamics*, **21**, 673–683.

Torrie, C. 2001. *Development of Suspensions for the GEO 600 Gravitational Wave Detector*. Ph.D. thesis, University of Glasgow.

Webb, S. C. 2002. Seismic Noise on Land and on the Sea Floor. In: Lee, W. H. K., *et al.* (eds), *International Handbook of Earthquake and Engineering Seismology*. Academic Press: UK.

Winterflood, J. 2001. *High performance vibration isolation for gravitational wave detection*. Ph.D. thesis, The University of Western Australia.

Winterflood, J., Barber, T. and Blair, D. G. 2002. *Class. Quantum Grav.*, **19**, 1639–1645.

12

Interferometer sensing and control

P. Barriga

Advanced gravitational wave detectors require sophisticated sensing and control systems in order to acquire and maintain synchronous operation of the various optical cavities. This chapter will firstly provide some mathematical background, before discussing the sensing and control of both the lengths and angular orientation of the various components of the optical configuration. We will also describe how the modulation frequencies are calculated before discussing further issues of control relevant to the signal recycling cavity and the readout scheme.

12.1 Introduction

The topology of the first generation of interferometers is based on that of a Michelson interferometer with Fabry–Perot cavities coupled as arm cavities and a power recycling cavity (PRC) to increase its sensitivity. As we have learned in previous chapters, the design of advanced gravitational wave (GW) detectors will be based on a configuration known as dual-recycling through the addition of a signal recycling (extraction) cavity (SRC). The improvement also includes input laser power a factor of ten greater than that of the first generation, stable recycling cavities, as well as very complex suspensions and seismic isolation systems.

The power recycling mirror (PRM) creates a composite cavity (PRC) with the common mode of the arm cavities, while the signal recycling mirror (SRM) creates another composite cavity (SRC) with the interferometer differential mode. The transmittance and reflectivity of the compound mirror formed by the SRM and the input test mass (ITM) is dependent on frequency. In signal recycling (SR), this cavity is tuned so that the GW signal will produce a lower transmittance (higher reflectivity) than that of the ITM alone (Meers, 1988). This controls the effective number of round trips over which the GW sidebands are summed, increasing their storage time, and determining the bandwidth of the interferometer. The position of the SRM controls which GW sideband frequencies will add constructively and which will add destructively, thus determining the tuning of the interferometer and its frequency response.

The bandwidth of an interferometric GW detector with Fabry–Perot arm cavities, without a SRM, is set by the bandwidth of the cavities. For a constant input power an increased storage time in the arm cavities will reduce their bandwidth, at the same time increasing the circulating power. Even with a reduction of the cavity bandwidth the shot-noise-limited sensitivity is still better than under a short storage time scheme at frequencies below the cavity pole. At frequencies above the cavity pole these two effects cancel out. It is, however, this increase in circulating power that imposes a limit to the storage time due to radiation pressure effects and increased difficulty for lock acquisition.

The case for advanced detectors is more complex due to the addition of the SRC. The circulating power in the arm cavities depends on the input power and the losses in these arm cavities. Assuming that the optical losses of the recycling cavities are smaller than the arm cavity losses (which is normally the case), we can then assume that the quantum noise depends fundamentally on the circulating power in the arm cavities. As a consequence the PRC gain can be adjusted independently of the arm cavities' finesse so as to obtain the same circulating power. Therefore a given quantum noise spectrum can be achieved with different arm finesses, by suitable selection of the SRM transmission and SRC detuning.

Resonant sideband extraction (RSE) was proposed by Mizuno *et al.* (1993) as a new optical configuration to overcome this limitation. In this mode of operation the extra cavity at the output of the interferometer is usually referred to as the signal extraction cavity (SEC). The purpose is to reduce the storage time for the GW signal, allowing for long storage times in the arm cavities without sacrificing the detector bandwidth (Heinzel *et al.*, 1996; Freise *et al.*, 2000). The tuning of the SEC in this case results in a bandwidth wider than that of an interferometer without a SRM. It is then possible to create a compound mirror with a transmissivity higher (lower reflectivity) than that of the ITM alone. This reduces the storage time of the signal frequencies of interest, resulting in an increased bandwidth for the interferometer. Since the interferometer bandwidth is not limited by the bandwidth of the arm cavities, Fabry–Perot cavities with high finesse, and narrow bandwidth can be used to maximise the stored energy. This also has the advantage of reducing the amount of light power which must be transmitted through the optical substrates of both ITMs and the beam splitter (BS) in order to obtain the same amount of energy stored in the arm cavities. As a consequence, the thermally induced distortion inside the substrates is effectively reduced.

In general the operating point of a dual-recyled interferometric GW detector is such that the carrier has to be resonant in both arm cavities, resonant in the PRC, and detuned by the right amount in the SRC. In RSE the SRM is positioned to make it anti-resonant with the carrier; therefore RSE can be considered a strongly detuned case of SR. Both modes of operation (SR and RSE) also have the possibility of operating in a detuned mode. Conventionally the special case of maximum response at zero signal frequency is termed *tuned* (or broadband); all other cases, with peak response at a finite frequency, are called *detuned*. This detuning is usually described by either the frequency of peak response or by the shift of the SRM away from the tuned point, often in terms of an optical (one-pass) phase shift. This positioning or detuning of the SRM within a wavelength of the carrier light allows for some narrowing of the detection bandwidth at the expense of loss of sensitivity outside the bandwidth. This could be valuable in searches for the continuous wave sources of gravitational radiation discussed in Chapter 2.

The spatial eigenmodes of both recycling cavities have to match the spatial eigenmode of the arm cavities. This ensures an efficient extraction of the signal or GW induced sidebands and a good mode matching between the spatial modes of the carrier and the spatial modes of phase modulation radio frequency (RF) sidebands used to control the longitudinal and angular degrees of freedom. It also improves the coupling of the carrier field into the arm cavities. The first generation of large scale interferometric GW detectors, LIGO and Virgo, used marginally stable power recycling cavities in an essentially near flat-flat configuration. These cavities do not confine the spatial modes of the RF sidebands if there are small mirror distortion such as thermal lensing, which can lead significant spatial mode mismatch between them and the carrier (Gretarsson *et al.*, 2007). A sophisticated thermal correction system was necessary in order to overcome these problems during the commissioning phase of the interferometers (Lawrence *et al.*, 2002).

In order to confine the spatial modes of the RF sidebands, stable PRC were first proposed by the University of Florida (Mueller, 2005; Arain, 2006) as part of their input optics work (Arain *et al.*, 2007). A group at the California Institute of Technology discovered that marginally stable signal recycling cavities will also reduce the amplitude of the GW sidebands by resonantly enhancing the scattering of light into higher order spatial modes (Pan, 2006). As a consequence, stable recycling cavities have been thoroughly studied and they are an integral part of the optical design of the next generation of interferometric GW detectors (Abbott *et al.*, 2008; Arain and Mueller, 2008; Barriga *et al.*, 2009b).

As we can see, the next generation of interferometric GW detectors are a collection of optical cavities coupled together. Each cavity is locked with such a precision that allows for the detection of gravitational waves. However, to obtain a working interferometer it is necessary to keep these cavities locked at all times. This requires a sophisticated sensing and control system, which has to bring the interferometer from an unlocked state to a configuration appropriate for collecting science data. This process can be divided into three steps. First is the lock acquisition, where all initially uncontrolled length degrees of freedom are globally controlled and brought to their operating point. Second is the transition from a locked interferometer with all degrees of freedom controlled to a configuration where science data can be collected. Third is to maintain the science mode and the data collection.

Different sensors are installed around the interferometer in order to obtain the input signals for the sensing and control systems. Tuned signals that are band-pass filtered before being synchronously demodulated are used for detection of the various degrees of freedom. The outputs of the length and sensing control system, and the alignment and sensing control system, are processed by the global control system. They are then distributed to each of the core optics components, through the suspension local control system, and to the pre-stabilised laser (PSL) for actuation on the laser frequency.

12.2 Mathematical background

Phase modulation is the most commonly used technique to generate the RF sidebands. A phase-modulated laser field can be described by a pair of sidebands separated from the central frequency component (the carrier) by a modulation depth of δ (radians) and an angular modulation frequency ω_m (radians/second):

$$E_{in} = E_0 \exp\{i(\omega_0 t + \delta \sin(\omega_m t))\}$$

$$E_{in} \approx E_0 \left\{ \exp(i\omega_0 t) + \frac{\delta}{2} \exp[i(\omega_0 + \omega_m)t] - \frac{\delta}{2} \exp[i(\omega_0 - \omega_m)t] \right\} \tag{12.1}$$

where E_0 is the amplitude and ω_0 is the angular frequency of the original laser field. The amplitude of each sideband is proportional to the modulation index δ provided that it is sufficiently small. Larger modulation indices would also create higher harmonics at $\omega_0 \pm N \times \omega_m$.

The reflected, transmitted, or internal pick-off field (a field extracted from within the power recycling cavity, such as by the pick-off shown later in Figure 12.3) in any interferometer can be described as a linear combination of these frequency components with amplitudes and phases modified by the frequency-dependent transfer function $T(\omega)$ of the interferometer. The transfer function is evaluated at three particular values: ω_0, $\omega_0 + \omega_m$ and $\omega_0 - \omega_m$.

The photocurrent that such a field generates in a fast photo-detector is proportional to:

$$|E_0|^2 + 2\delta|E_0|^2 \left\{ \Re[T_0(T_+^* - T_-^*)]\cos(\omega_m t) + \Im[T_0(T_+^* + T_-^*)]\sin(\omega_m t) \right\} \tag{12.2}$$

where the asterisk is used to denote complex conjugation. This signal will be demodulated with an ac coupled electronic mixer followed by a low-pass filter to yield the signal:

$$S = 2\Re[T_0(T_+^* - T_-^*)]\cos(\alpha) + 2\Im[T_0(T_+^* + T_-^*)]\sin(\alpha) \tag{12.3}$$

where 0, +, and − label the carrier, the upper and the lower sideband fields, respectively, and α is the tunable demodulation phase. The first term, proportional to $\cos\alpha$, is usually called the quadrature (Q) signal. It is especially sensitive to differences in the amplitudes of the sidebands. The second component, proportional to $\sin\alpha$, is the in-phase (I) signal. It is primarily sensitive to phase changes of the carrier relative to the sidebands. Both parts of the signal disappear only if the carrier or both of the sidebands disappear, or if the transfer function of the carrier is real and the transfer functions of the sidebands are complex conjugated to each other (modulus over all phase in all three transfer functions). If this is not the case, as for example in a detuned configuration, the intensity oscillates with the modulation frequency, and only one specific demodulation phase causes the signal to disappear. This is similar to a simple heterodyne interferometer in which a single additional frequency component is added to the carrier. In such cases the signal obtained after demodulating yields:

$$S = 2|T_1 E_1||T_2 E_2|\cos(\varphi - \psi) \tag{12.4}$$

where $E_{1(2)}$ and $T_{1(2)}$ are the electric field and the transfer function for the carrier (1) and the sideband (2) fields, φ is the demodulation phase and ψ corresponds to the phase difference between the two transfer functions $[\arg(T_1) - \arg(T_2)]$.

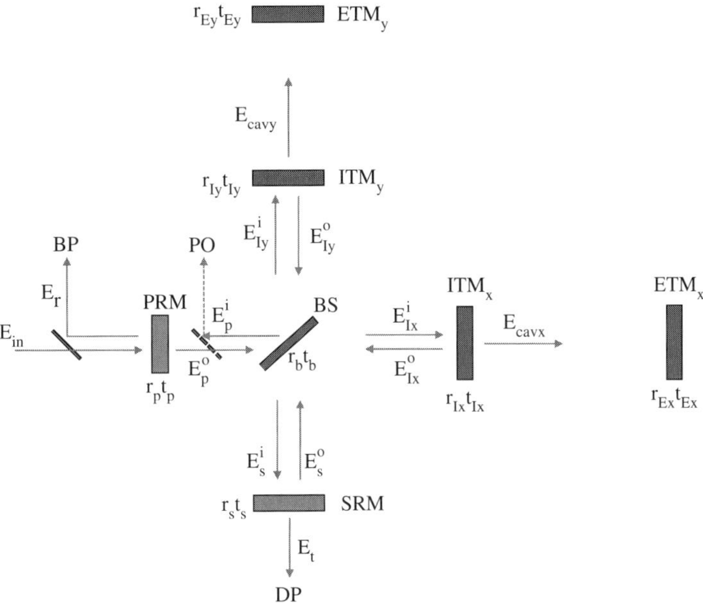

Figure 12.1 Fields at the different locations in the interferometer. E_r is the field that is reflected at the power recycling mirror outside of the interferometer. It will be detected at the bright port (BP). E_t is the field that is transmitted through the whole interferometer which will be detected at the dark port (DP). The subscript n of all the other fields E^m_n denotes the mirror at which the field is calculated. The superscript m denotes the direction of the field. The pick-off (PO) field is proportional to E^o_p.

Although the mixer output is non-zero in general, the appropriate choice of φ will set the mixer output to zero in the static case. A perturbation to a pathlength in the interferometer that introduces a differential phase modulation of E_1 and E_2 will then appear as a non-zero demodulated output that can be used as a signal to correct that length. The general application of this idea is to use one of the fields as a phase reference, or local oscillator, for the other. This implies that the local oscillator field will somehow not be affected by the perturbation that impresses the signal onto the other field. Achieving this goal is the primary aim of signal extraction design for the interferometer.

The transfer functions for the carrier and each sideband are required to enable the calculation of the error signals. These transfer functions can be calculated from the equations that express the propagation of the fields for each cavity in the interferometer, as presented in Figure 12.1. The full calculation of these transfer functions can be found in Strain *et al.* (2003). The final set of linearised results is given by the following set of equations at each of the sensing ports.

At the reflected port:

$$T_r(k_c) = T^0_r(k_c) + i(T_1\phi_+ + 2\Phi_+)t^0_r(k_c), \tag{12.5}$$

$$T_r^0(k_c) \equiv \frac{-T_p T_I + 2T_I \bar{a} + 4\bar{A}}{T_p T_I + 2T_I \bar{a} + 4\bar{A}}, \tag{12.6}$$

$$t_r^0(k_c) \equiv \frac{4T_p T_I}{(T_p T_I + 2T_I \bar{a} + 4\bar{A})^2}, \tag{12.7}$$

At the dark port:

$$T_t(k_c) \approx t_t^0(k_c)[i(T_I \phi_- + 2\Phi_-) + \Delta A + T_I \Delta a], \tag{12.8}$$

$$t_t^0(k_c) \equiv \frac{4t_s t_p T_I}{(T_s T_I + 2T_I \bar{a} + 4\bar{A})^2}, \tag{12.9}$$

At the pick-off port:

$$T_p(k_c) \approx T_p^0(k_c) - i(T_I \phi_+ + 2\Phi_+)t_p^0(k_c), \tag{12.10}$$

$$T_p^0(k_c) \equiv \frac{2t_p T_I}{T_p T_I + 2T_I \bar{a} + 4\bar{A}}, \tag{12.11}$$

$$t_p^0(k_c) \equiv \frac{4t_p T_I}{(T_p T_I + 2T_I \bar{a} + 4\bar{A})^2}. \tag{12.12}$$

For these calculations it was assumed that the carrier is at resonance in both recycling cavities and that the round-trip losses in the arm cavities are much smaller than the transmission of the input mirror, i.e. $\bar{A}/T_I$ and $\Delta A/T_I$ are much smaller than unity. For A, the losses in the cavities due to imperfections of both mirrors are defined as $\bar{A} = (A_x + A_y)/2$ and $\Delta A = (A_x - A_y)/2$. T_I corresponds to the intensity transmissivity of the input mirror assuming both ITMs have similar values. Here $a_{x,y}$ are the losses in each of the short Michelson interferometer arms caused by the bulk absorption in the ITMs and BS. They also define $\bar{a} = (a_x + a_y)/2$ and $\Delta a = (a_x - a_y)/2$. In the previous equations, k_c corresponds to the wavenumber for the carrier. The longitudinal degrees of freedom are defined by their phase changes (microscopic), which are proportional to the macroscopic lengths of the cavities given in Table 12.1.

From the equations (12.5), (12.8) and (12.10) we can clearly see that the common mode cavity detuning Φ_+ and the detuning of the PRC ϕ_+ can be detected at the pick-off and reflected ports but not at the dark port. In these common ports the transfer functions of the carrier are independent from the differential degrees of freedom Φ_- and ϕ_- and from the SRC degree of freedom ϕ_s. In contrast to this, the transmitted carrier field at the dark port depends only on the two differential degrees of freedom Φ_- and ϕ_-.

The GW signal should be detected with the highest possible sensitivity. Therefore the carrier transmitted to the dark port will be used to sense Φ_-. The carrier should also be used to detect Φ_+. The role of the additional frequency components injected into the system is to provide us with the necessary local oscillators at all detection ports and to generate additional control signals for the auxiliary degrees of freedom (ϕ_-, ϕ_+, ϕ_s).

Table 12.1. *Five relevant longitudinal degrees of freedom in an advanced interferometer*

Description	Symbol	Physical distance
Differential arm cavity	Φ_-	$2k(L_x - L_y)$
Common arm cavity	Φ_+	$2k(L_x + L_y)$
Differential Michelson	ϕ_-	$k(l_x - l_y)$
Power recycling cavity	ϕ_+	$k(2l_{pr} + l_x + l_y)$
Signal recycling cavity	ϕ_s	$k(2l_{sr} + l_x + l_y)$

Note: It is convenient to describe the two arm cavities by use of their average or common length and their length difference or differential length instead of the individual lengths. $k = 2\pi/\lambda$ is the wave number.

12.3 Length sensing and control

In principle two RF modulation frequencies are enough to control all five length degrees of freedom in an advanced dual-recycled interferometer. These two RF modulation frequencies (f_{m1} and f_{m2}) will generate two RF sidebands each at $\pm f_{m1}$ and $\pm f_{m2}$ from the carrier frequency (f_0). These RF sidebands allows us to obtain length signals using single demodulation at f_{m1} and f_{m2}, differential demodulation at $f_{m1} \pm f_{m2}$, and double demodulation, first at f_{m1} and then at f_{m2}.

As seen in Section 12.2, the RF signal readout from the photo detectors is split and demodulated at $0°$ (in-phase) and $90°$ (quadrature-phase), giving two signals per detector. Each of these signals is fed to an input matrix where each length degree of freedom can be calculated. These are the differential arm length (DARM or L_-), Michelson length (MICH or l_-), common arm length (CARM or L_+), power recycling cavity length (PRCL or l_+), and signal recycling cavity length (SRCL or l_{src}). Figure 12.2 shows their definitions in an advanced dual-recycled interferometer configuration. The signals are digitally filtered and converted through an output matrix into control signals for the core optics. This is done by feeding the control signals to the local control of the suspension system, where the signals are added to the appropriate degrees of freedom of the corresponding optic.

In general, as shown in Table 12.1, the length degrees of freedom that need to be controlled are defined as:

DARM differential arm length: $L_- = (L_x - L_y)$
CARM common arm length: $L_+ = (L_x + L_y)$
MICH michelson length: $l_- = (l_x - l_y)$
PRCL power recycling cavity length: $l_+ = l_{\mathrm{pr}} + (l_x + l_y)/2$
SRCL signal recycling cavity length: $l_{\mathrm{src}} = l_{sr} + (l_x + l_y)/2$

Before the interferometer longitudinal degrees of freedom can be controlled it is necessary to initially align the cavity axes to one another. At this stage, signals from the quadrant

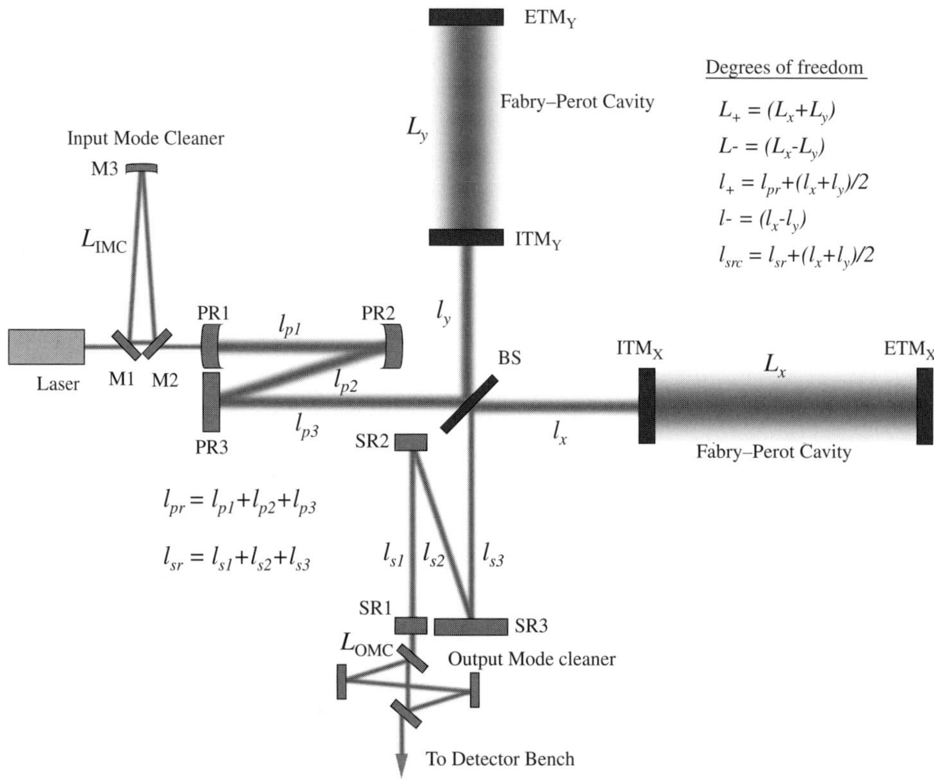

Figure 12.2 Dual-recycled advanced interferometric GW detector. The figure shows stable recycling cavities and length degrees of freedom, including input mode cleaner and output mode cleaner. In general the input mode cleaner is part of the input optics subsystem; while the output mode cleaner is part of the output optics. They are not part of the interferometer control system.

photodetectors (QPD) and the optical levers (and the camera images) will be used to centre all beams on all mirrors and achieve an initial alignment stage. This will reduce the angular and lateral mismatches between the various cavity eigenmodes to be smaller than their angular divergences and beam sizes. At this stage there is no need to operate the laser at full power, which could compromise lock acquisition by radiation pressure effects. The objective of the initial alignment is to attain a sufficiently high power build-up in the cavities of the interferometer. This allows for strong enough length sensing error signals for locking the longitudinal degrees of freedom (Evans *et al.*, 2002).

The next step is to lock the central part of the interferometer comprising the ITMs, BS, PRMs and SRMs. This corresponds to three longitudinal degrees of freedom: the MICH, the PRCL and the SRCL. With the central interferometer locked, it is possible to lock the arms through the CARM and the DARM signals, locking the whole interferometer (Abbott *et al.*, 2008).

Standard Pound–Drever–Hall (Drever *et al.*, 1983) signals generated by the beating between carrier and sidebands are strongly dependent on the behaviour of the carrier inside the arm cavities and on the interferometer losses. Control signals can be obtained from the beating of the first sideband (f_{m1}) with the second sideband (f_{m2}) by demodulating the signal twice. Since the double demodulation scheme does not depend on the carrier, the amplitude and polarity of the control signals obtained from the sidebands are not affected by the lock or unlock status of the arm cavities. However, since double demodulation signals do not work effectively far from the locking point they are not ideal for lock acquisition. Moreover, in the configuration for broadband detection, there is no detuning of the SRC. In such a configuration the two sidebands can no longer beat and there is no double demodulation signal at all. The TAMA group (Arai and the TAMA Collaboration, 2002) proposed and studied the use of signals demodulated at three times the modulation frequencies ($3f$) in order to obtain signals that are independent from the CARM offset necessary to keep the arm cavities out of lock. These signals are produced by the beating between $2f$ and f sidebands and between $3f$ sidebands with the carrier. The second contribution is typically smaller than the first, so that the $3f$ signal depends very little on the carrier behaviour. Signals detected in reflection and demodulated at $3f_{m1}$ and $3f_{m2}$ are in fact very good signals for PRCL, MICH and SRCL.

As can be deduced from equations (12.5), (12.8) and (12.10), the CARM degree of freedom is sensed at the reflected port while the DARM degree of freedom is sensed at the anti-symmetric port. The DARM degree of freedom is controlled by feedback to the end mirrors, while the CARM degree of freedom is expected to be controlled by feedback to the laser frequency. With such a locking scheme the interferometer is brought to the operating point by passing through stable states, which allow the activation of a frequency servo even during lock acquisition.

Alternative procedures for cavity locking are still under investigation. In general the suspension system will isolate the test masses from seismic noise and ground motion at frequencies above a few hertz. However, there will be very little or no attenuation at micro-seismic noise frequencies and the resonant frequencies of the suspension, so there is then the risk of not being able to acquire lock of the arm cavities. As presented before, it is necessary to lock the arm cavities independently from the central cavities. The current procedure is not trivial and therefore the use of an auxiliary dual-wavelength laser (532 nm and 1064 nm) has been proposed to smooth the process (de Vine *et al.*, 2009).

This scheme facilitates the locking of the central cavities since it allows for an independent method of locking these cavities. The introduction of a 532 nm heterodyne locking loop allows one to control the detuning of the arm cavity from the carrier resonance. Therefore the cavity can be held off resonance to allow the locking of the central degrees of freedom (PRCL, MICH and SRCL). Once these degrees of freedom are locked, the auxiliary system can bring the arm cavities into resonance with the carrier and hand over the feedback to the global sensing and control system.

The 1064 nm beam is used to ensure that the auxiliary laser is phase locked to the PSL of the same frequency (Mullavey *et al.*, 2010), while the 532 nm beam is used to lock the auxiliary laser to the arm cavity. Once each arm cavity is locked to its auxiliary laser, the 532 nm beams transmitted by the arm cavities (through the ITMs) are extracted at PR2 (after going through the BS and PR3) and routed to the PSL table. Two heterodyne measurements are performed: one between the X arm transmission and a doubled sample of the carrier beam; the other is between the X arm and Y arm transmissions. The outputs of these two measurements are used to provide error signals for the CARM and DARM degrees of freedom, respectively. The CARM is controlled through the input mode cleaner common mode board servo, while the DARM is controlled by feedback to the ETMs.

In order for this scheme to be implemented it is necessary to modify the mirror coatings adding a specified reflectivity at 532 nm creating dichroic cavities. In the current Advanced LIGO design the reflectivities are 99% for the ITMs and $\sim$95% for the ETMs, with a resulting finesse of $\sim$100. At the time of writing, plans for testing this method are underway (Fritschel and Coyne, 2011).

12.4 Angular sensing and control

The diagram in Figure 12.3 shows the interferometer sensing and control ports which, as mentioned before, will be equipped with QPD and wavefront sensors (WFS). Angular misalignment can be performed through a phase modulation–demodulation technique known as wavefront sensing. It is also possible to obtain DC outputs for fast beam stabilisation. In order to determine beam pointing (pitch and yaw), each quadrant of the WFS and QPD is processed individually. These signals are then passed through a matrix that converts them into signals for the alignment of the optics through its suspension. In addition, optical levers are used for initial angular control of the core optics, as well as imaging cameras to locate the beam position on the high reflectivity surfaces of all core optics. The length sensing and control (LSC) system monitors the cavity and the interferometer arm lengths and controls them using a feedback mechanism (Miyakawa *et al.*, 2006; Abbott *et al.*, 2008).

At the end of the lock acquisition sequence the longitudinal degrees of freedom become controlled by the LSC system and transition to the detection/science mode begins. Once the lock of all interferometer axial degrees of freedom is acquired, the angular degree of freedom control loops engage, providing the static optical cavities alignment and beam centring on the optics as well as the suppression of the optics angular motion up to several hertz.

In detection/science mode the well known wavefront sensing scheme is used. It measures the amplitude and phase of beat signals between the TEM_{00} and TEM_{10} modes generated by the misalignment of the interferometer optics. The signal depends on the relative Gouy phase between TEM_{00} and TEM_{10} modes, which is a function of the longitudinal position of the detector along the axis. For this reason, angular misalignments of different mirrors can be distinguished by placing detectors at different locations along the optical path.

Figure 12.3 shows the sensing and control signals available for the control system. These are: the reflected port (reflection from the PRC), the dark port or asymmetric port, the X arm pick-off port (POX) and the power recycling pick-off port (POP), which can be measured through the PR2 mirror. The dark port signal can further be split into a pick-off of the dark port beam and the reflected field coming from the OMC.

The angular sensing and control system also provides damping of the angular optical spring resonance produced by radiation pressure from the main laser. At full power the

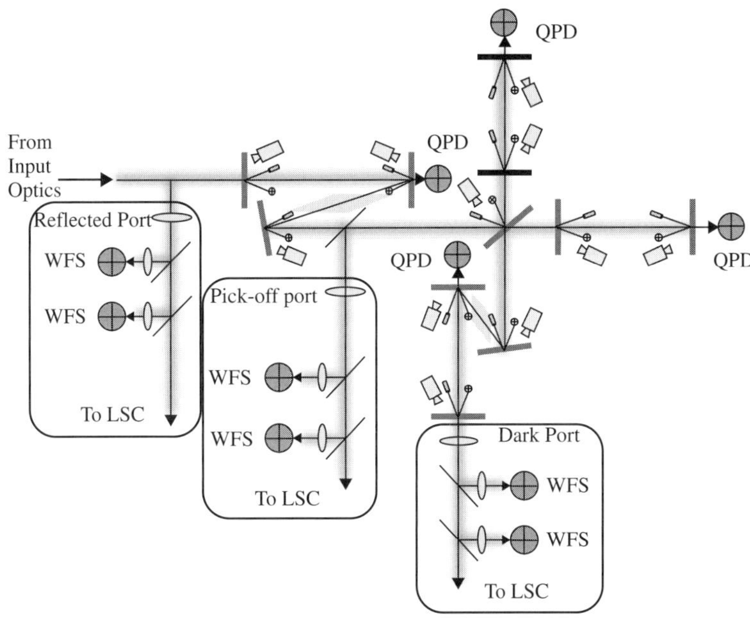

Figure 12.3 Distribution of the different sensing apparatus of an advanced interferometric GW detector. These include WFS and QPD. It also includes the aid of CCD cameras and optical levers, which are auxiliary lasers and QPD to monitor the optics alignment. (Figure courtesy of the LIGO Scientific Collaboration, Abbott *et al.* 2008.)

radiation pressure effects generated by a circulating power of the order of 700 kW in the arm cavities will induce a torque comparable to the restoring torque of the mirror suspension. As such we must think of the opto-mechanical response, instead of just the mechanical response of the system. The control strategy has to take these effects into account in order to maintain the alignment of the mirrors.

Analysis of a single cavity in resonance by Sidles and Sigg (2006) shows two coupled modes of the opto-mechanical system. In one case the torque induced by radiation pressure tends to make the mode stiffer, while in the other case it tends to make the mode less stiff (soft). In advanced detectors, the tilt of the beam axis in the stiff mode forces the radiation pressure to work to enhance the mechanical restoring force. On the other hand, in the soft mode, the shifting beam axis causes the radiation pressure to work against the restoring force from the suspension. If the power exceeds a critical value, then this mode becomes unstable. In this case active feedback is required to stabilise it. This can be easily extended to the arm cavities of the interferometer, where the error signals coming from the interferometer naturally distinguish between common and differential motions of the two arm cavities.

12.5 Local control system

In addition to high performance isolation within the detection bandwidth, it is critical for the operation of the interferometer that the isolator provides minimal residual motion at low frequencies. This facilitates cavity locking and minimises noise injection through actuation forces. Pendulum systems inherently have large Q-factors; it is therefore often necessary to damp the normal modes of the suspension system. A feedback loop injects control forces at various actuation points on the isolator. These forces are derived by applying appropriate filters to error signals from local transducers such as position sensors.

Typically, the local control system is responsible for several tasks, each requiring different bandwidths and different filters. For example, in the frequency band DC to $\sim$10 mHz the control system is responsible for drift correction, positioning, and alignment. In the frequency band up to $\sim$1 Hz the control system is needed for damping normal mode peaks. In this frequency band the micro-seismic noise has a large contribution (around 150 mHz), therefore a very high loop gain is required. For higher frequencies (hertz–kilohertz) the control system may be required to suppress high frequency noise in the form of active vibration isolation. Despite being conceptually simple, the operation of local control systems is complicated by resonant modes and mode interaction between isolation components and different degrees of freedom. As a result, the feedback scheme often requires complex filter designs to avoid noise injection at critical eigenmodes that would interfere with the noise budget of the test mass.

In the gravitational wave community there have been two broad approaches to the vibration isolation problem. The first, and most widely adopted, has been to create mostly passive vibration isolators based on mass–spring systems (Losurdo *et al.*, 1999; Plissi *et al.*, 2000; Marka *et al.*, 2002; Barriga *et al.*, 2009a). Primarily based on very low resonant frequencies, these are achieved with different damping solutions based on anti-spring designs and electromagnetic coupling, complemented with low bandwidth feedback control loops. The second approach, adopted by Advanced LIGO, is to implement a 'stiff' pre-isolation technique based on hydraulic systems distributed on each corner of the vacuum chamber, complemented by a two-stage six-degree-of-freedom active isolator in vacuum (Abbott *et al.*, 2004). These provide vertical and horizontal pre-isolation for the whole chamber, combining geophones and seismometers with high resonant frequencies in a high-bandwidth control loop. A four-stage pendulum test-mass suspension provides significant additional isolation in the GW band. In the first case, one allows the system itself to provide an inertial reference. In the second, the inertial reference is provided by the test mass of the seismometer. The Australia International Gravity Research Center (AIGRC) has extended the idea of the system itself being the inertial sensor, by using a pair of very low frequency stages that are designed specifically to allow relative sensing and feedback, to provide an additional means of active suppression of very low frequency seismic noise (Dumas *et al.*, 2009).

12.6 Modulation frequencies calculations

The resonance condition of the PRC will determine the modulation frequencies (f_m) for the RF sidebands. In this case the carrier sees an over-coupled arm reflectivity[1] (Siegman, 1986). Therefore if the carrier is resonant in the PRC the sidebands need an extra phase shift, thus the following relation applies to the modulation frequencies of the sidebands.

$$f_{\mathrm{m}} = \left(n_1 + \frac{1}{2}\right)\frac{c}{2L_{\mathrm{PRC}}}. \tag{12.13}$$

Here n_1 is an integer and L_{PRC} the average PRC length, which is defined by $l_{pr} + (l_1 + l_2)/2$, where l_1 and l_2 represent the distance from the BS to the ITM, l_1 on the perpendicular arm and l_2 on the inline arm as seen in Figure 12.2, and c is the speed of light in vacuum.

The selected modulation frequencies and the carrier frequency will all have to go through the IMC. Therefore the length of the IMC (L_{IMC}) has to be such that its FSR, defined as $FSR_{\mathrm{IMC}} = c/2L_{\mathrm{IMC}}$, allows the transmission of the carrier and both modulation frequency sidebands. Therefore the sideband frequencies need to additionally satisfy the following relation:

$$f_{\mathrm{m}} = n_2 \frac{c}{2L_{\mathrm{IMC}}}. \tag{12.14}$$

Here n_2 is an integer equal or bigger than 1. Therefore the two sideband modulation frequencies will be an integer multiple of the FSR_{IMC}.

Using equations (12.13) and (12.14) we can then calculate the maximum length of the IMC subject to the available space in the main lab and the constraints set by the vacuum envelope. A longer IMC will set a lower modulation frequency for the generation of the RF sidebands, with both frequencies going through the IMC and f_{m2} an integer multiple of f_{m1}. The maximum modulation frequency is determined by the bandwidth of the high efficiency photodiodes used in the readout. It has been shown that if the frequency of f_{m2} is too high, then the phase noise of the modulator would be increased and the efficiency of the differential length photodetector would be decreased (Miyakawa *et al.*, 2006).

By introducing a difference in the arm lengths in the recycling cavities, one can arrange for the output of the interferometer to be dark for the carrier but not dark for one of the sidebands (Meers and Strain, 1991). This asymmetry is required in order to have some signal in the SRC that allows for the control of that degree of freedom. This Michelson arm length difference, known as Schnupp asymmetry (Schnupp, 1998), is defined as $l_- \equiv |l_1 - l_2|$, which corresponds to the difference between BS - ITM$_X$ (inline) and BS - ITM$_Y$ (perpendicular) lengths. If we assume that the reflectivities of the mirrors in the PRC and SRC are the same, we can define this asymmetry as $l = c/(4f_{\mathrm{m2}})$. In reality the Schnupp asymmetry depends strongly on the parameters of the interferometer such as mirror reflectivities, beam splitter properties, detuning of the SRC, etc. Therefore the final value for the Schnupp asymmetry ends up being a compromise between the best error signal, for controlling the SRCL degree of freedom, and the power in the SRC. It also needs to take into account the different SRM detuning values considered for the different modes of operation based on the science cases (broadband or binary neutron star inspirals, for example).

The Schnupp asymmetry combined with a detuned SRC causes the control RF sidebands to be imbalanced in the interferometer. This leads to demodulation phase-dependent offsets in the error signals derived from these sidebands. It is frequency noise on RF sidebands

---

[1] A cavity in which the power transmittance of the coupling mirror dominates the total loss inside the cavity.

coupled with contrast defect (only due to imbalance) that makes the biggest contribution, since frequency noise on the carrier light is filtered out by the power recycled arm cavity. One way to avoid this problem is to use the carrier light as local oscillator. This implies a change from a heterodyne or RF readout scheme to a homodyne or DC readout scheme (see Section 12.7).

The lower modulation frequency (f_{m1}) needs to be resonant in the PRC and non-resonant in the arm cavities. This allows the control of the common mode signal and the PRC length. The higher modulation frequency (f_{m2}) needs to be resonant in the PRC and additionally in the SRC in order to be able to control the SRC length and the Michelson arms. There is, however, an ongoing investigation for a different selection criterion for the modulation frequencies that would allow for a lower high frequency to be selected without extending the Schnupp asymmetry (Somiya *et al.*, 2005).

Signal recycling cavity

The length for the SRC will depend on the operation scheme selected for the interferometer. For example for the narrow band detection of a compact binary inspiral one can select a peak frequency around 300 Hz (Cutler and Thorne, 2002). This has been extensively studied and it is not the purpose of this chapter to analyse in detail the theory behind the detuning of the SRC in order to obtain the desired resonance peak frequency. However, a thorough analysis can be found in Mizuno (1995) and Mason (2001) and in the references within this section. In this section we will study the case where no detuning ($\phi_{dt} = 0$) is required, which corresponds to a RSE configuration for the interferometer. Both modulation frequencies need to fulfill equations (12.13) and (12.14), but only the higher frequency (f_{m2}) will be use to determine the length of the SRC, as per the following relation (Mason and Willems, 2003):

$$L_{\mathrm{SRC}} + \Delta L_{\mathrm{SRC}} = \frac{c}{2\pi f_{m2}} \left(n_3 \pi + \phi_{\mathrm{dt}}\right). \tag{12.15}$$

Here $L_{\mathrm{SRC}} + \Delta L_{\mathrm{SRC}}$ corresponds to the length of the SRC plus its detuning, ϕ_{dt} corresponds to the signal recycling detuning in radians and n_3 an integer. In order to operate the interferometer in a broadband RSE configuration we need to select n_3 carefully. Since the modulation frequencies are multiples of each other, only one of the sideband frequencies (f_{m2}) needs to be resonant in the SRC. When operating in a detuned configuration all frequencies will be off-resonance.

The peak frequency response of the interferometer will depend on the SRC detuning. However, the peak frequency not only depends on the length of the SRC but also on the coupling of the cavities, which in turn depend on the transmission and losses of the mirrors that shape the interferometer. How these cavities are coupled (under-coupled, over-coupled or matched) will play an important role in the interferometer frequency response. At the same time the coupling combined with the length of the cavities, in particular the ratio between the lengths of the SRC and the arm cavities, will define the slope of the detuning.

The bandwidth of the frequency response and the highest possible frequency at which it will be possible to tune the interferometer will be defined by the transmission and losses of the mirrors. This is not a simple matter and has been studied in greater detail elsewhere (e.g. Mizuno, 1995; Mason, 2001).

12.7 Readout scheme

An incoming GW signal of frequency ω_g will induce sidebands on the carrier light (frequency of 2.8×10^{14} Hz). These sidebands will be located at $\pm\omega_g$ from the carrier light. To read such a high frequency signal, an optical oscillator is necessary in order to demodulate the output signal and read the gravitational wave signal.

The first generation of interferometric GW detectors favoured a heterodyne readout scheme, where RF sidebands are modulated on the carrier light before entering the interferometer. A large mirror offset (centimetre scale) to create a macroscopic asymmetry in the Michelson arms of the interferometer (also known as Schnupp asymmetry; see Schnupp, 1998) is required to allow leakage of the RF sidebands through the output port of the interferometer whilst maintaining a dark fringe. This signal can be used as the local oscillator. The next generation of interferometric GW detectors will, however, see the addition of a SRC. The operation of this cavity in a detuned configuration will introduce an imbalance on the control sidebands, reducing the sensitivity of the interferometer (Buonanno *et al.*, 2003). The sensitivity reduction will depend on the relative level of imbalance between the sidebands.

Since the carrier light enters the arm cavities of the interferometer they are effectively low-pass filtered, with the corner frequency set by the length of the arms. This is not the case for the sidebands which don't enter the arm cavities. The RF sidebands also experience some filtering, but the cavity pole of the power recycled Michelson interferometer is higher than the observation band. Therefore any noise on the modulation sidebands (independent of the source) will appear at the output of the interferometer, potentially injecting noise when using the sideband as local oscillator in a heterodyne readout scheme (Sigg, D. for the ISC Group, 1997).

A homodyne readout scheme requires a small fraction of carrier light to be used as the local oscillator. One approach is to obtain a signal from the incoming beam before it enters the interferometer, and feed this signal as the local oscillator at the output port. Another approach that also allows a small amount of light to appear at the output port is to deliberately introduce a very small offset (picometre scale) in the DARM degree of freedom. In this way the power of the signal at the output port of the interferometer will be directly proportional to the gravitational wave strain. This technique, known as DC readout, was proposed by Fritschel in 2000 (Fritschel and Sanders, 2000) and tested at the 40 m interferometer at the California Institute of Technology (Ward *et al.*, 2008). It also has the advantage of increasing the signal-to-shot-noise ratio by eliminating the vacuum fluctuations at twice the frequency of the modulation (Hild *et al.*, 2009). An analytical comparison between RF and DC readout schemes can be found in Somiya *et al.* (2006, 2007).

Acknowledgements

P. Barriga gratefully acknowledges financial support from the West Australian Government Centres of Excellence scheme, the Australian Research Council and the University of Western Australia.

References

Abbott, R., *et al.* 2004. *Class. Quantum Grav.*, **21**, S915–S921.

Abbott, R., *et al.* 2008. Technical Report T070247-01-I. LIGO.

Arai, K. and the TAMA Collaboration. 2002. *Class. Quantum Grav.*, **19**, 1843–1848.

Arain, M. A. 2006. LSC Meeting presentation G060155-00-Z. LIGO.

Arain, M. A., *et al.* 2007. Technical Report T060269-02-D. LIGO.

Arain, M. A. and Mueller, G. 2008. *Opt. Express*, **16**, 10018–10032.

Barriga, P., *et al.* 2009a. *Rev. Sci. Instrum.*, **80**, 114501.

Barriga, P., *et al.* 2009b. *Opt. Express*, **17**, 2149–2165.

Buonanno, A., Chen, T., and Mavalvala, N. 2003. *Phys. Rev. D*, **67**, 122005.

Cutler, C. and Thorne, K. S. 2002. In: Bishop, N. T. and Maharaj, S. D. (eds), *Proceedings of 16th International Conference on General Relativity and Gravitation (GR16)*. World Scientific, Singapore.

de Vine, G., *et al.* 2009. *Opt. Express*, **17**, 828–837.

Drever, R. W. P., *et al.* 1983. *Appl. Phys. B*, **31**, 97–105.

Dumas, J. C., *et al.* 2009. *Rev. Sci. Instrum.*, **80**, 114502.

Evans, M., *et al.* 2002. *Opt. Lett.*, **27**, 598–600.

Freise, A., *et al.* 2000. *Phys. Lett. A*, **277**, 135–142.

Fritschel, P. and Coyne, D. 2011. Presentation G1100505-v2. LIGO.

Fritschel, P. and Sanders, G. 2000. Technical Report T000062-00-D. LIGO.

Gretarsson, A. M., *et al.*, 2007. *J. Opt. Soc. Am. B*, **24**, 2821–2828.

Heinzel, G., *et al.* 1996. *Phys. Lett. A*, **217**, 305–314.

Hild, S., *et al.* 2009. *Class. Quantum Grav.*, **26**, 055012.

Lawrence, R., *et al.* 2002. *Class. Quantum Grav.*, **19**, 1803.

Losurdo, G., *et al.* 1999. *Rev. Sci. Inst.*, **70**, 2507–2515.

Marka, S., *et al.* 2002. *Class. Quantum Grav.*, **19**, 1605–1614.

Mason, J. E. 2001. *Signal extraction and optical design for an advanced gravitational wave interferometer*. Ph.D. thesis, California Institute of Technology.

Mason, J. E. and Willems, P. A. 2003. *Appl. Opt.*, **42**, 1269–1282.

Meers, B. J. 1988. *Phys. Rev. D*, **38**, 2317–2326.

Meers, B. J. and Strain, K. A. 1991. *Phys. Rev. A*, **44**, 4693–4703.

Miyakawa, O., *et al.* 2006. *J. Phys. Conf. Ser.*, **32**, 265–269.

Mizuno, J. 1995. *Comparison of optical configurations for laser-interferometric gravitational-wave detectors*. Ph.D. thesis, Max Planck Institut für Quantenoptik.

Mizuno, J., *et al.* 1993. *Phys. Lett. A*, **175**, 273–276.

Mueller, G. 2005. Technical Report G050423-00-Z. LIGO.

Mullavey, A. J., *et al.* 2010. *Opt. Express*, **18**, 5213–5220.

Pan, Y. 2006. *Topics of LIGO physics: Template banks for the inspiral of precessing, compact binaries, and design of the signal-reccyling cavity for Advanced LIGO*. Ph.D. thesis, California Institute of Technology, Pasadena, CA 91125, USA.

Plissi, M. V., *et al.* 2000. *Rev. Sci. Instrum.*, **71**, 2539–2545.

Schnupp, L. 1998. In: Talk at the *European Collaboration Meeting on Interferometric Detection of Gravitational Waves*.

Sidles, J. A. and Sigg, D. 2006. *Phys. Lett. A*, **354**, 167–172.

Siegman, A. E. 1986. *Lasers*. University Science Books: Sausalito, CA.

Sigg, D. for the ISC Group. 1997. Technical Report T970084-00-D. LIGO.

Somiya, K., *et al.* 2006. *Phys. Rev. D*, **73**, 122005.

Somiya, K., *et al.* 2007. *Phys. Rev. D*, **75**, 049905.

Somiya, K., *et al.* 2005. *Appl. Opt.*, **44**, 3179–3191.

Strain, K. A., *et al.* 2003. *Appl. Opt.*, **42**, 1244–1256.

Ward, R. L., *et al.* 2008. *Class. Quantum Grav.*, **25**, 114030.

13

Stabilising interferometers against high optical power effects

C. Zhao, L. Ju, S. Gras and D. G. Blair

High optical power is essential for improving the sensitivity of advanced detectors. Thermal lensing, Sidles–Sigg instability and parametric instability are three dominant effects that limit the optical power required to achieve the target sensitivity. This chapter summarises these three instabilities and how they can be controlled. In particular, we emphasise parametric instability and its control.

13.1 Introduction

The first generation laser interferometer gravitational wave detectors (LIGO, Chapter 6 and Virgo, Chapter 7) have achieved their target design sensitivity, and yet this sensitivity is estimated to be only sufficient to detect large rare gravitational wave signals. To achieve detection of signals from predicted sources, a ten-fold improvement in sensitivity is required. For LIGO, this would increase the sensitive range from ~ 14 Mpc today to about ~ 200 Mpc, leading to event rates of many per year (Cutler and Thorne, 2002). To achieve this improvement the Advanced LIGO (Fritschel, 1994) and other second generation detectors require a ~ 100-fold increase in circulating laser power, in addition to improvement in the interferometer configuration, test mass thermal noise, and vibration isolation. Such high power levels create three significant challenges. The first is the control of thermal lensing in the core optics substrates and dielectric coating layers. The second is the control of parametric instabilities in the form of radiation pressure mediated opto-mechanical oscillation. The third is the control of Sidles–Sigg instabilities, which are the angular instabilities of the suspended test mass that result from radiation pressure induced optical torque.

13.2 Thermal lensing and control

At extreme high optical power in the advanced detector, even very low levels of absorption in the core optics substrate and dielectric coatings induce significant thermal lensing (Winkler *et al.*, 1991). This thermal lensing causes wavefront distortion that would prevent the detector from achieving high sensitivity and could even lead to serious degradation of the operation of the interferometer (Strain *et al.*, 1994; Degallaix *et al.*, 2003; Lawrence, 2003). Various thermal compensation methods have been proposed and tested. Lawrence *et al.* (2004) demonstrated that a ring radiative heater, inserted near the test mass, could compensate axisymmetric thermal lensing effects by creating a negative thermal lens. A similar technique has been demonstrated on the GEO 600 GW interferometer (Lück *et al.*, 2004). Zhao *et al.* (2005) demonstrated compensation of thermal lensing using a compensation plate heated by a wire at its circumference inside a optical cavity. Fan *et al.* (2008) implemented a feedback control loop to control the thermal lensing on the same facility. A thermal compensation system consisting of a carbon dioxide (CO_2) laser with an annular heating pattern directly applied on the input test mass has been implemented on LIGO detectors (Smith *et al.*, 2004; Ballmer, 2006).

However, direct heating of the test mass by a CO_2 laser requires extremely high stability of the CO_2 laser power to avoid noise coupling in the advanced detectors, which is difficult to achieve (Willems, 2010). Instead, the Advanced LIGO thermal compensation system will combine a CO_2 laser-heated compensation plate and a radiant ring heater (Willems, 2009). The compensation plate will be suspended behind the arm cavity input mirror to compensate the thermal effect inside the recycling cavities. Radiant ring heaters will heat the rear surfaces of all arm cavity mirrors, allowing thermoelastic tuning of their radii of curvature.

13.3 Sidles–Sigg instability

In a suspended optical cavity the high circulating optical power can couple the torsional mode of the test mass suspension to the optical mode. This effect was first recognised by Solimeno *et al.* (1991). They analysed the phenomenon for a cavity consisting of one suspended mirror and one fixed mirror using a modal formalism. Some years later, Sidles and Sigg (2006) extended this analysis to a cavity with two suspended mirrors using a simplified geometric formalism, and they found that their results remained consistent with Solimeno *et al.* (1991) Sidles–Sigg (SS) theory predicts that optically induced, negative torsional stiffness could potentially be large enough to overcome the stiffness of the mirror suspensions, introducing an angular instability within the cavity of the interferometers. These instabilities depend upon the circulating power, cavity finesse and linewidth, and the detuning between the laser frequency and the cavity resonant frequency. SS theory indicates that choosing negative mirror g-factors (i.e. near concentric cavities) maximises the critical circulating power at which the cavity becomes unstable. Driggers *et al.* (2006) observed the optical torsion stiffness in a suspended three-mirror mode cleaner cavity and its dependency on the circulating power. Fan *et al.* (2009) studied the cavity g-factor dependence of optical torsional stiffness in a two-mirror cavity. Their results are in agreement with the SS geometric formalism model. Hirose *et al.* (2009) observed the radiation pressure effect on the angular sensing and control system of the LIGO interferometer.

13.4 Parametric instability

Ground-based gravitational wave detectors are designed to make extremely high precision measurements of the motion of test masses with perturbations limited by quantum measurement theory. To obtain high sensitivity, high laser power is required. If there is any mechanism through which a small amount of this power can be coupled directly into oscillatory vibration of the test masses, this can lead to their uncontrolled mechanical oscillation. One means by which this can occur is called *parametric instability*. The possibility of parametric instability (PI) in advanced laser interferometer gravitational wave detectors was first shown to be a potential problem by (Braginsky *et al.*, 2001, 2002). Later three-dimensional modeling showed that parametric instability was likely for current proposed interferometer configurations (Zhao *et al.*, 2005). If the high frequency acoustic modes in test masses have significant spatial overlap with high order optical cavity modes that satisfy a resonance condition, then PI could occur. It appears that PI is a potential threat to any interferometer which uses high optical power and low acoustic loss test masses. Instability causes test mass acoustic modes to ringup to a large amplitude in a time that could be in the range 50 ms to hundreds of seconds. The large amplitude destroys the interferometer fringe contrast. The phenomenon cannot be filtered or reduced through any post-processing of data.

Since the original predictions, parametric instability has been extensively modeled. Modeling has been undertaken by groups at the University of Western Australia (UWA) (Zhao *et al.*, 2005); (Ju *et al.*, 2006), Moscow State University (Gurkovsky *et al.*, 2007), the Japanese Large Scale Cryogenic Gravitational Wave Telescope (LCGT) project (Yamamoto *et al.*, 2007), and members of the Laser Interferometer Gravitational Observatory (LIGO) laboratory at the California Institute of Technology (Bantilan and Kells, 2006) and the Massachusetts Institute of Technology (Evans *et al.*, 2010).

Modeling requires detailed knowledge of optical modes and acoustic modes in test masses. Results are dependent on small changes in the test mass geometry (which change the acoustic mode spectrum) and on the mirror diameter and shape, which affect the optical mode spectrum. Small features such as small wedge angles in the test masses, and the placing of flats on the test mass sides have very strong effects because they break both the acoustic mode and optical mode degeneracy, acting in general to increase the number of potentially unstable modes. There is no significant disagreement between estimates of instability. However, the precise detail of instability is extremely sensitive to system parameters. For example, a change in mirror radius of curvature of one part in 10^4 is sufficient to modulate individual mode parametric gain by a large factor. Such changes are within the uncertainties of test mass material parameters, and 1–2 orders of magnitude larger than the mirror radius of curvature tolerance. Because of the high sensitivity of the resonant interactions to small parameter changes, only models which use the same finite element modeling (FEM) test mass meshing; the same acoustic losses of test masses including the contributions from coatings; the same model for optical cavity diffraction losses; and the same material parameters will give strictly identical results. Modeling results also contain significant uncertainties due to the limited precision of FEM (Strigin *et al.*, 2008). Thus modeling is unlikely to be able to predict the detailed instability spectrum of a real large-scale interferometer. However, modeling results are likely to present an accurate statistical picture. Best estimates currently predict an average of ~ 10 unstable modes per test mass in an Advanced LIGO type configuration (Gras *et al.*, 2010).

The most realistic modeling to date has been applied to a proposed Advanced LIGO configuration. Modeling takes into account test mass shape and acoustic losses due to mirror coatings. Results predict that between 0.2% and 1% of acoustic modes in the frequency range 10–100 kHz are likely to be unstable, depending on the precise instantaneous value of the effective mirror radius of curvature. The parametric gain for unstable modes varies between 1 and 10^3. Over a radius of curvature range of 30 meters (a range chosen to take into account thermally induced changes as well as manufacturing tolerances) current estimates indicate a positive parametric gain greater than unity for $\sim 20 - 40$ modes spread across four test masses. The number of modes and their gain fluctuates as the radius of curvature is thermally tuned. Over the 30 meters radius of curvature thermal tuning range more than 700 acoustic modes are parametrically excited in each test mass (at any single one radius of curvature the number of modes is only about 5–10 per test mass).

Parametric instability estimates were first obtained for interferometer arm cavities alone (Braginsky *et al.*, 2001). However, the real gravitational wave detector configuration is more complicated, with nested power recycling and signal recycling cavities (Meers, 1988). The effects of degenerate power and signal recycling cavities were then considered (Braginsky *et al.*, 2002; Gurkovsky *et al.*, 2007), including the realistic case of non-matched arm cavities. By selectively suppressing arm cavity modes the effect of using stable (non-degenerate) power recycling cavities can be estimated. To date all configurations analysed show instability.

The magnitude of parametric instability gain scales as the product of three Q-factors: those of the optical cavity main mode, the relevant cavity high order mode and the test mass acoustic mode. Unstable modes are generally in the frequency range 10 kHz to 100 kHz. Most methods suggested for controlling instability have focused on changing the value of the Q-factors of the acoustic modes.

In any room temperature interferometer in which thermal lensing is significant, parametric instability will be tuned by time-varying thermal lensing. Thus changes will occur over a thermal lensing time-scale, which is seconds to minutes in single crystal test masses (such as sapphire or silicon) and 1 hour for fused silica test masses. In 2007 the UWA group observed three-mode parametric interactions for the first time (Zhao *et al.*, 2008). The parametric gain was shown to be tunable through variation in the test mass radius of curvature, in agreement with predictions.

In a cryogenic interferometer, parametric instability is likely to be frozen into a particular configuration since the temperature coefficient for thermal lensing falls effectively to zero at cryogenic temperatures. However, the precise configuration is unlikely to be predictable in advance.

Instability control should be achievable without noise penalty because the instabilities are all narrow band and all occur outside the measurement band. However, for many reasons noise-free PI control is not simple. The following sections summarise the theory of parametric instability and the status of modeling results and explains why PI has not already been seen in LIGO and Virgo. We then discuss various approaches to PI control and show that simple methods have noise penalties, disadvantages and risks. Optical feedback control is

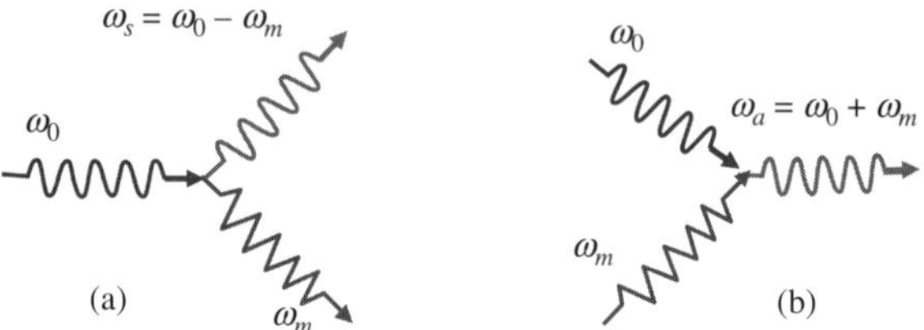

Figure 13.1 Parametric scattering of a photon of frequency ω_0 (a) into a lower frequency Stokes photon ω_s, and a phonon of frequency ω_m, and (b) into a higher frequency anti-Stokes photon ω_a, which requires the destruction of a phonon.

also discussed which, although complex, can be implemented as a fully automatic external suppression system.

13.5 Parametric instability theory and modeling

Summary of theory

Parametric interactions can be considered as classical ponderomotive interactions of optical and acoustic fields, or as simple scattering processes (Manley and Row, 1956), as illustrated, in Figure 13.1. In Figure 13.1(a) a photon of frequency 0 is scattered, creating a lower frequency (Stokes) photon of frequency s and a phonon of frequency m, which increases the occupation number of the acoustic mode. In Figure 13.1(b) a photon of frequency 0 is scattered from a phonon, creating a higher frequency (anti-Stokes) photon of frequency a which requires that the acoustic mode is a source of phonons, thus reducing its occupation number. The scattering could create entangled pairs of phonons and photons (Pirandola *et al.*, 2003).

Figure 13.2 illustrates three-mode parametric interactions in an optical cavity from a classical viewpoint. In this example energy stored in the form of a TEM_{00} mode is shown scattering into a TEM_{11} mode by a particular acoustic mode. The interactions can only occur strongly if two conditions are met simultaneously. Firstly, the optical cavity must support eigenmodes that have a frequency difference approximately equal to the acoustic frequency $|\omega_0 - \omega_1| \approx \omega_m$. Here ω_0 is the cavity fundamental mode frequency while ω_1 represents either a Stokes or an anti-Stokes high-order mode. Secondly, the optical and acoustic modes must have a suitable spatial overlap. The parametric interaction effect can be expressed by the parametric gain R. If $R > 1$, instability will occur.

The parametric gain in a simple cavity is given by Braginsky *et al.* (2002) as:

$$R = \pm \frac{4 P_0 Q_1 Q_m}{L M c \omega_m^2} \frac{\Lambda}{1 + (\Delta\omega/\delta_1)^2} \tag{13.1}$$

Here P_0 is the power stored in the TEM_{00} mode, which is the fundamental mode in the cavity; M is the mass of the acoustic resonator; L is the length of the cavity; ω_m is the acoustic mode frequency; $\Delta\omega = |\omega_0 - \omega_1| - \omega_m, \delta_1 = \omega_1/2Q_1$ is the half linewidth

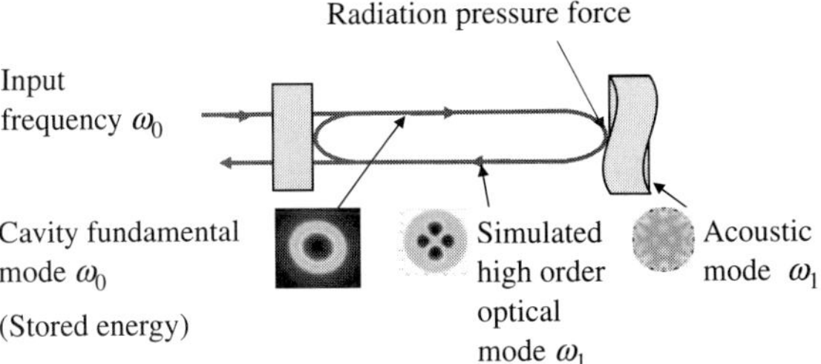

Figure 13.2 A cartoon of three-mode interactions in an optical cavity from a classical point of view, showing an acoustic mode scattering stored energy from the cavity fundamental mode into a high-order mode, while acting back on the test mass by radiation pressure.

(or damping rate) of the high-order optical mode; ω_1 is the frequency of the Stokes (anti-Stokes) high-order optical mode; and Q_1 and Q_m are the quality factors of the high-order optical mode and the acoustic mode, respectively. The factor Λ measures the spatial overlap between the electromagnetic field pattern and the acoustic displacement pattern, defined by Braginsky *et al.* (2002) as:

$$\Lambda = \frac{V(\int \psi_0(\vec{r}_\perp)\psi_1(\vec{r}_\perp)u_z d\vec{r}_\perp)}{\int |\psi_0|^2 d\vec{r}_\perp \int |\psi_1|^2 d\vec{r}_\perp \int |\vec{u}|^2 dV} \tag{13.2}$$

where ψ_0 and ψ_1 describe the optical field distribution over the mirror surface for the TEM$_{00}$ mode and higher-order modes, respectively, $\vec{u}$ is the spatial displacement vector for the mechanical mode, and u_z is the component of $\vec{u}$ normal to the mirror surface. The integrals $d\vec{r}_\perp$ and dV correspond to integration over the mirror surface and the mirror volume V, respectively.

It should be pointed out that there are often multiple mode interactions and the above equation should include a summation over all the possible modes (both Stokes and anti-Stokes modes)(Ju *et al.*, 2006). The positive and negative sign in R corresponds to Stokes and anti-Stokes modes, respectively.

From equation (13.1), it can be seen that parametric gain is a product of three parameters: P_0, Q_m and Q_1, corresponding to the arm cavity power, acoustic mode Q-factor, and high order optical mode Q-factor. Compared with the baseline parameters for Advanced LIGO, the initial LIGO arm cavity power ($P_0 \sim 20\,\text{kW}$) is ~ 40 times lower, the acoustic Q ($\sim 10^6$) is ~ 10 times lower and optical Q_1 (finesse of ~ 200) is ~ 2 times lower than those of Advanced LIGO ($P_0 \sim 800\,\text{kW}$, $Q_m \sim 10^7$, finesse of ~ 450). In addition, the test masses are smaller so that the acoustic mode density is lower at high frequency. The parametric gain in initial LIGO should be more than ~ 800 times lower than that in Advanced LIGO without signal recycling. With signal recycling, the PI gain of Advanced LIGO could be much higher if the high-order modes are at resonance inside the signal recycling cavity. It is clear that the risk of instability in initial LIGO is very low. Hence it is no surprise that initial LIGO has not observed PI.

It is important to point out that the derivation of equation (13.1) is based on a model that does not consider diffraction losses of the optical cavities. We found that in the case of a simple cavity, the addition of diffraction loss does not change the formulation and equation (13.1) is still valid. To incorporate diffraction losses in the analysis of parametric instabilities in interferometers with recycling cavities is mathematically difficult and has not yet been accomplished. However, for optical cavities with relatively low finesse, the diffraction losses of the low spatial order modes are found to be small compared with the coupling loss. Only very high-order modes have strong diffraction losses. The very high-order mode contribution to PI is usually small, thus diffraction losses can often be ignored. So the results presented here do not depend strongly on this issue.

Modeling approach

We have conducted many simulations of different interferometer configurations. In general we use finite element modeling to simulate several thousand acoustic modes of the test masses and their diffraction losses (Barriga *et al.*, 2007). We have used both modal analysis and fast Fourier transform (FFT) codes to model cavity optical modes. We have simulated interferometers with both signal recycling mirror (SRM) and power recycling mirror (PRM) configurations using Advanced LIGO parameters. This involves an elaborated form of equation (13.1) which takes into account both power and signal recycling (Strigin and Vyatchanin, 2007). We normally assume identical main cavities of 4 km length and a circulating power of 830 kW. We also assume that a power recycling cavity operates in a marginally stable scheme. For stable power recycling cavities one expects that the finesse of some high-order modes will increase, thereby increasing Q, and the associated parametric gain. Unfortunately we do not have the formalism available at present to analyse PI in interferometers with stable power recycling cavities.

Stokes and anti-Stokes modes are constructed using a set of transverse optical modes up to 11th order. In a typical analysis about 60 optical higher-order modes (HOM) and 5500 elastic modes of the test mass are taken into account. The HOM mode shapes and resonant frequency are obtained numerically using the eigenvalue method, whereas the elastic modes are calculated using finite element modeling. The test mass model takes into account most of the detailed structure. For our most precise model (see results below), test masses are modeled as 20 cm thick, 17 cm radius cylinders with flats on the circumference, including chamfers and a back face wedge of $0.5°$. In addition, optical modes frequency detuning due to the finite mirror geometry (Barriga *et al.*, 2007) is also taken into account. The Q-factor of the optical modes is determined by coupling losses, while the Q-factor of the elastic modes is based on substrate losses (Penn *et al.*, 2006) and coating losses (Martin *et al.*, 2008). Diffraction losses of the high-order modes are not significant for the low finesse design presented below.

Our analysis usually includes up to the 5th longitudinal mode number from the main cavity mode, which ensures that all possible interactions of optical modes with elastic mode are taken into account. However, in general only the first three mode numbers contribute significantly to the parametric gain. The PI analysis is carried out for different radii of curvature (RoC) of the end test mass (ETM) mirror. In the analysis presented below the ETM RoC is allowed to vary from 2.171 km to 2.201 km in steps of 0.1 m. The radius of curvature for the input test mass (ITM) is set as a constant of 1.971 km. For each ETM RoC data point, the resonance condition $\Delta\omega$ and the overlapping parameter Λ are calculated. Estimating PI for different RoC of the ETM enables us to simulate the thermal tuning of the interferometer and thus probe changes of the resonant conditions.

Modeling results

Our simulations of the Advanced LIGO detector with power and signal recycling cavities reveal that instability can occur over the whole RoC range considered. For this configuration there is a very strong dependence of the mirror radius of curvature on the parametric gain R of unstable modes. In the simulation, the transmissivity of the SRM, PRM, ITM, ETM are set to proposed Advanced LIGO values 20%, 2.5%, 1.4%, and 5 ppm, respectively; and both substrate and coating acoustic losses were taken into account for the Q-factor of the elastic modes. The simulation shows that at certain values of RoC, there are up to 20 unstable modes, while for other values only a few unstable modes are present. The number of unstable mode is on average 8 ± 3 for each test mass. However, the total number of unstable modes over the whole range is as high as 777. It also shows that the highest parametric gain can vary from ~ 2 up to ~ 1000. It should be noted that the instability condition for certain mechanical modes is very sensitive to change in the radius of curvature, in the range ~ 0.1 m. Therefore very small changes of the effective mirror curvature result in strong changes in the instability condition from one mode to another. This can be visualised by examining the movies located at http://www.gravity.physics.uwa.edu.au/docs/PI_Movies.htm. These movies show parametric gain for all modes during a sweep through the test mass radius of curvature. We recommend playing the sweep manually by dragging the tracker button, to allow a careful examination of the gain–frequency structure.

Statistically, the simulation shows that the majority of unstable modes have $R < 5$. However, there are a substantial number of modes with gain $R > 100$.

13.6 Possible approaches to PI control

Power reduction and thermal radius of curvature control

The simplest approach to PI control is to reduce the input power by a factor equal to the peak instability gain that might be encountered. To ensure stable operation over a reasonable range of mirror radii of curvature, the power would have to be reduced, say to 1% of nominal power, thereby increasing the detector shot noise by an order of magnitude.

A much better option would be to thermally stabilise the interferometer test masses at radii of curvature where the peak parametric gain is minimum. In the absence of any other damping scheme, a power level $\sim 10\%$ of the proposed peak power could be achievable.

Since power causes thermal radius of curvature tuning, the process of powering up sweeps the cavity through different parametric instability gain regimes on a test mass thermal time-scale (~ 1000 seconds for fused silica test masses). The ring up time constant of an unstable mode can be written as (Ju *et al.*, 2006):

$$\tau = \frac{2Q}{\omega_m (R-1)}.$$ (13.3)

For a test mass to ring up from thermal amplitude of $10^{-14}\,m$ to $10^{-9}\,m$ (the assumed breaking lock amplitude), the break lock time will be

$$t_B = \frac{23Q}{\omega_m (R-1)}.$$ (13.4)

For example, a 30 kHz mode with $Q = 10^7$ and $R = 20$ would cause the interferometer to break lock in about 1 min. While thermal actuation with CO_2 lasers or a ring heater may compensate for this, the process is complicated by the long time constants in fused silica test masses. Degallaix *et al.* (2007) derived the optimal time-dependent heating required to achieve the fastest possible actuation in sapphire and fused-silica substrates by simulations. It is not clear to us how to control simultaneously the instantaneous radius of curvature (which determines the parametric instability) and the thermal lensing environment (which defines the optical mode matching) with a single thermal actuation system. Because of the long thermal time constants, interferometer lock acquisition would need to use thermal lens-dependent power level control to navigate past instabilities without losing lock. Once operation is stabilised to a radius of curvature where the PI gain is low, the power level could be ramped up to just below the PI threshold. The ramping up of power would need to be slow enough that the radius of curvature stayed within a tight range. It could be possible to operate proposed interferometers at between 10% and 30% of the proposed power in the absence of other control schemes. After losing lock it might be a very slow process to regain stable operation because of the long thermal memory of the test masses.

PI control using ring dampers or resonant acoustic dampers

Ring dampers

The idea of ring dampers is to apply lossy strips on the circumference of test masses to suppress the Q-factors of the acoustic modes. This Q-factor reduction will not greatly increase the thermal noise of the test mass if the lossy parts are far from the laser spot at the test mass (Levin, 1998; Yamamoto *et al.*, 2002). By careful choice of the position of the lossy strips, it is possible to reduce the test mass mode Q-factor without greatly degrading the thermal noise performance of the interferometer. The lossy strips could be an optical coating or a layer of aluminum oxide (Al_2O_3), gold or copper applied by conventional ion-assisted deposition techniques. We have analysed the use of ring dampers; this work has been published elsewhere (Gras *et al.*, 2008). The ring damper method can effectively suppress many acoustic mode Q-factors by a factor of ~ 50. Most of the unstable modes have R values < 5. Thus by applying ring dampers to the test masses, most of the unstable modes can be suppressed. Unfortunately the effect of the ring damper is not uniform: some modes are only weakly damped and a few unstable modes are very difficult to suppress. The reduction of unstable modes is a function of RoC for a typical ring damper design. Here we assume a simple cavity interferometer without recycling cavities but with Advanced

LIGO parameters. The ring damper is a 2 cm wide, 20 μm thick strip with loss angle of 10^{-2}. The exact number of unstable modes at a particular RoC may differ from those when recycling cavities are considered, but the statistical results of the simulation are not altered. There is always a thermal noise penalty incurred by applying lossy ring dampers on the test masses. The model shown here contributes a Brownian thermal noise penalty of 5% (Gras *et al.*, 2008). If the noise penalty is allowed to increase to 20%, substantial instability-free windows appear in the RoC tuning range.

Resonant dampers

The use of resonant dampers, widely used in vibration isolation systems, was proposed by Evans *et al.* (2007). Preliminary analysis shows that in principle it is possible to attach a small lossy spring–mass resonator to a test mass to damp the resonant modes. If we choose the mass of the damper to be 1 g and the resonant frequency of the damper to be 20 kHz, it is easy to show that the Q-factor of the resonances of a 10 kg effective mass mirror, in the range of 30 kHz to 100 kHz, can be reduced from 10^7 to $< 5 \times 10^5$. This result assumes perfect coupling to the test mass modes.

To be usable in practice there are several factors that must be considered. For the damping to be effective, the damper should be placed sufficiently close to the antinodes of the resonant mode to be damped. Therefore, many dampers may be necessary to obtain good coupling to the large number of potentially unstable modes. Also from the thermal noise point of view, it is desirable that the dampers should be placed far away from the laser spot. This limits the effectiveness of the damping for some modes that have high amplitude near the laser spot. Detailed modeling is required to determine an optimum configuration of dampers, and their thermal noise contribution.

The preliminary experiment at LASTI (Gras, Evans and Fritschel, 2010) showed significant suppression of acoustic modes. However, the epoxy between the damper and the test mass contributed the most modes, and the extra thermal noise is also an issue.

It is clear in principle that strong damping can be achieved in the 10–100 kHz band where instabilities occur, with weak damping within the gravitational wave signal band. Thus this method is particularly worthy of further investigation.

Local control of PI by acoustic excitation sensing and feedback

Local feedback is a potentially simple solution to PI. The idea is to sense the acoustic excitation of a test mass, and provide a derivative feedback force using standard control theory. Motion sensing could use optical, radio frequency or electrostatic sensing, while feedback forces could be applied electrostatically. However, there are several issues with this method.

Firstly, the sensor must have sensitivity sufficient to maintain amplitudes small compared with the position bandwidth of the interferometer (position bandwidth is the linewidth expressed in displacement units). The position bandwidth of Advanced LIGO is ∼1 nm. The sensing noise must be small compared with this distance or else the feedback controller will excite all mechanical modes within its bandwidth (say 15–100 kHz) to an amplitude equal to the sensor noise floor. As long as the sensor noise floor is less than about 10^{-12} m/$\sqrt{\text{Hz}}$ this will not represent a severe degradation of fringe contrast. However, if the sensor noise was 10^{-10} m/$\sqrt{\text{Hz}}$ (typical of current local control sensors), this would represent a significant loss of fringe contrast. It is also worth pointing out that the requirement of test mass motion $< 10^{-20}$ m/$\sqrt{\text{Hz}}$ in the interferometer detection band (below a few kHz) means that the force noise from feedback actuation on the test mass should be sufficiently low to avoid injecting extra noise in the detection band. This imposes even tighter requirements on the sensor noise floor.

The second problem is the large number of test mass acoustic modes that are potentially excited, as shown in Section 13.5. Many of the acoustic modes have a complex mode structure. The overlap integral between each acoustic mode and an electrostatic sensor must be large to allow good signal coupling. It is often very difficult to excite test mass modes using electrostatic exciters because the exciter applies forces across a node so that the positive and negative displacements partially cancel, leading to small electromechanical coupling. In practice we have found it very difficult to achieve strong coupling to high-Q modes with high frequencies in our high-Q test masses in an 80 m cavity. To accommodate high coupling to a large set of modes the sensor and actuator would have to be broken up into a set of separately addressable elements. Each mode would require a different combination of actuator elements. Sensing suffers from the same problem, so that high signal-to-noise sensing will require sensing at different locations on the test mass.

The next problem is one of gain. The parametric gain we wish to suppress is typically 10, sometimes 100 and if we are unlucky it is more than 1000. Let us assume we choose a gain of 100 as the maximum we want to control. This means that the actuator must be able to excite the test mass mode at a rate 100 times faster than the ring-down time. For example, if the acoustic mode ring-down time is 10 seconds, the ring-up time will be 0.1 seconds. Such strong excitation is possible for small gap spacings and high excitation voltages, but in our experience such strong coupling is difficult to achieve. It would require a very high voltage, small gap spacing exciter, and it would have to be designed so that it did not supply residual in-band noise above the detector noise floor.

PI control using global optical sensing and electrostatic actuation

Parametric instability can will be very easily detected in the dark fringe signal from the interferometer, which is a much better probe of instability than any local sensor. The signal could be applied to all test masses simultaneously even though the unstable acoustic modes will be shared across four test masses. There should be no problem with driving test masses at frequencies where there is no instability, except in the unlucky circumstance that the frequency coincided with another test mass acoustic mode. If this occurred the system should also be able to automatically damp such a self-induced instability.

Using the global PI signal overcomes the SNR problem associated with local sensing, but there still remains the need to apply large enough forces to the test masses in the case of high gain instabilities. This is still limited by the problem of achieving a large overlap integral with the electrostatic driver, as discussed in Section 13.6. Miller (2010) showed in simulation that the quadrant electrostatic actuator used for the local positional and angular control of test masses can provide adequate overlap and forces to damp most test mass acoustic modes. To evaluate the active acoustic damping solution requires the experimental testing of suitable actuators to confirm adequate coupling to all of the predicted unstable test mass acoustic modes.

PI control using global optical sensing and direct radiation pressure

The force required to reduce the Q-factor of an acoustic mode is directly proportional to its amplitude. If a suppression system is required to act on large acoustic amplitudes (say 10^{-10} m), the force requirements are too large to use direct radiation pressure actuation. However, direct radiation pressure actuation is practical if instability is caught before the amplitude has grown too large. If we apply direct radiation pressure feedback control when the oscillation increases to 10 times the equilibrium thermal amplitude, the maximum force F_0 required to reduce the acoustic mode Q-factor is

$$F_0 = 10 \times \frac{\sqrt{2m\omega^2 k_{\mathrm{B}} T}}{Q_f}, \tag{13.5}$$

corresponding to a laser power of

$$P = \frac{cF_0}{2} = 5c\frac{\sqrt{2m\omega^2 k_{\mathrm{B}} T}}{Q_f}, \tag{13.6}$$

where k_{B} is Boltzmann's constant, c is the speed of light and Q_f is the desired Q-factor. Assuming an effective mass $m = 10\,\mathrm{kg}$, $\omega_m = 30\,\mathrm{kHz}$, $Q_f = 10^5$, $T = 300\,\mathrm{K}$, the required power is $P = 0.8\,\mathrm{W}$. Here we have used point mass approximation which assumes 100% overlap between the acoustic mode and the actuation force. For non-ideal overlap, corresponding to actuation far from an antinode of the acoustic mode, more power will be required.

In 2005 we used a laser walk-off delay line as a radiation pressure actuator (Feat *et al.*, 2005) and proposed to use direct radiation pressure as a means of PI control. Multiple reflections can reduce the power requirements or increase the force by ~ 30 times. In all cases the main problem is to obtain large overlap with all acoustic modes. The actuation phase depends on the location of the actuator and must be well defined for each relevant mode. If instability was not suppressed early enough such an actuator would be unable to achieve suppression. This threshold effect imposes strong requirements on the reliability and completeness of the suppression system.

Braginsky and Vyatchanin (2002) suggested using an external short optical 'tranquiliser' cavity to control instability. This method is also limited by the requirement of overlap between the actuator and the test mass acoustic modes. In Schediwy *et al.* (2008) the method was shown to be technically difficult but viable in principle. Again, in practice several such systems would be required for each test mass to obtain adequate overlap with the acoustic modes, greatly increasing the complexity of the interferometer system.

PI control by global optical sensing and optical feedback

The parametric interaction provides forces to the test mass via the high-order optical modes which are excited in the cavity. It should be possible to suppress the high-order mode in the cavity by introducing an antiphase high-order mode. Here we summarise the analysis of such a high-order mode interference system. It has the advantage that it could act very rapidly and suppress even very high gain instabilities. If it was practical and robust, it could eliminate the need for any other instability suppression scheme.

To model the system with optical feedback, Zhang *et al.* (2010) used a classical model of a cavity with fields including a high-order mode injection, $f_0(t)$, in addition to the main laser beam in the fundamental mode. The fundamental mode field (E with frequency ω_0) and the high-order mode field (f with frequency ω_1) contribute to create the radiation pressure force at the differential frequency of $\omega_0 - \omega_1$. This force acts back on the test mass, which vibrates at its internal mode frequency ω_m. The parametric instability comes from the interaction between the radiation pressure force and the mechanical mode vibration.

If no other external field is injected into the cavity apart from the fundamental mode, the high-order mode field f is produced by scattering of the fundamental mode into higher-order modes and their resonant build-up inside the cavity. The back-action force is determined by the product of f and E. As E is a constant depending on the input power and cavity parameters, the strength of the instability is solely determined by f and consequently by the parameters in equation (13.1) – the frequency difference between $\omega_0 - \omega_1$ and ω_m, the overlap between the high-order mode and the test mass internal mode, and the quality factors of the cavity and the test mass internal mode. If we inject another high-order mode optical field f_0 into the cavity, out of phase with the high order mode f created by scattering, it will destructively interfere with f to suppress the instability. By detecting the optical signal at the interferometer dark port it is possible to determine the amplitude, frequency and phase of f. One can then inject field f_0 with appropriate phase and amplitude to suppress the instability.

Detailed analysis and numerical results (Zhang *et al.*, 2010) show that in principle it should be possible to suppress PI in the next generation detector such as Advanced LIGO. This would only require external optical and electronic components at the corner station and would not require modification of the test masses or the local control systems. It has the advantage that it could act very rapidly and suppress even very high gain instabilities. The disadvantage of this method is that each high-order mode to be suppressed required a Mach–Zehnder arrangement and a phase mask for adding the high-order mode. This would increase the complexity if many modes were to be suppressed. However, it would be possible to use this technique combined with other methods for PI suppression. If the injection optics in advanced interferometers included the basic beam splitters required, particular problematic instabilities could be suppressed without internal modification of the interferometer. An experiment at the High Optical Power Facility (Ju *et al.*, 2004) at Gingin is planned to investigate this solution.

13.7 Conclusion

We have shown that unless active control measures are taken, advanced gravitational wave interferometers may suffer from parametric instabilities. These will render the devices inoperable at high power.

The stochastic nature of instability has been emphasised, with instability sometimes occurring suddenly, as the thermal conditions of test masses vary with time. The difficulty of making precise predictions has also been emphasised. The control of instability using simple active damping is shown to be difficult.

We have shown that passive ring dampers can significantly reduce parametric gain but do not eliminate instability. Simple modeling shows that a number of passive resonant dampers could be attached to test masses to reduce instability and probably control it, but this requires detailed modeling.

A less invasive technique based on optical feedback has been proposed. This technique allows instability control by selectively suppressing all large amplitude high-order modes. This technique enables all instabilities to be controlled but requires knowledge of the high-order mode, and a separate phase mask for each high-order mode that requires suppression. This is an attractive back-up solution that could be implemented during commissioning if required for particular high-order modes, without requiring mechanical changes to the core optics.

Acknowledgements

The authors wish to acknowledge financial support from the West Australian Government Centres of Excellence scheme, the Australian Research Council and the University of Western Australia.

References

Ballmer, S. 2006. *LIGO interferometer operation at design sensitivity with application to gravitational radiometry*. Ph.D. thesis, MIT.

Bantilan, H. S. and Kells, W. P. 2006. Tech. rept.

Barriga, P., *et al.* 2007. *J. Opt. Soc. Am. A*, **24**, 1731.

Braginsky, V. B. and Vyatchanin, S. P. 2002. *Phys. Lett. A*, **293**, 228.

Braginsky, V. B., *et al.* 2001. *Phys. Lett. A*, **287**, 331.

Braginsky, V. B., *et al.* 2002. *Phys. Lett. A*, **305**, 111.

Cutler, C. and Thorne, K. 2002. *Proceedings of the 16th International Conference on General Relativity and Gravitation*, edited by N.T Bishop and S. N. Maharaj. Singapore: World Scientific. p. 72.

Degallaix, J., *et al.* 2003. *Appl. Phys. B: Lasers Opt.*, **77**, 409.

Degallaix, J., *et al.* 2007. *J. Opt. Soc. Am. B*, **24**, 1336.

Driggers, J. , *et al.* 2006. Optomechanical alignment instability in LIGO mode cleaners, http://www.ligo.caltech.edu/docs/T/T060240-00.pdf

Evans, M., *et al.* 2007. Mechanical Mode Damping for Parametric Instability Control, http://www.ligo.caltech.edu/docs/G/G080541-00/

Evans, M., Barsotti, L. and Fritschel, P. 2010. *Phys. Lett. A*, **374**, 665.

Fan, Y., *et al.* 2008. *Rev. Sci. Instrum.*, **79**, 104501.

Fan, Y., *et al.* 2009. *Appl. Phys. Lett.*, **94**, 081105.

Feat, M., *et al.* 2005. *Rev. Sci. Instrum.*, **76**, 36107.

Fritschel, P. 1994. *Phys. Lett. A*, **194**, 124.

Gras, S., Blair, D. G. and Ju, L. 2008. *Phys. Lett. A*, **372**, 1348.

Gras, S., *et al.* 2010. *Classical and Quantum Gravity*, **27**, 205019.

Gras, S., Evans, M., and Fritschel, P. 2010. Passive Control of Parametric Instability, https://dcc.ligo.org/DocDB/0022/G1001023/002/

Gurkovsky, A. G., Strigin, S. E., and Vyatchanin, S. P. 2007. *Phys Lett. A*, **362**, 91.

Hirose, E., *et al.* 2009. *Appl. Opt.*, **49**, 3474.

Ju, L., *et al.* 2004. *Class. Quantum Grav.*, **21**, S887.

Ju, L., *et al.* 2006. *Phys. Lett. A*, **354**, 360.

Lawrence, R. 2003. Active Wavefront Correction in Laser Interferometric Gravitational Wave Detectors, Ph.D. thesis, MIT. LIGO-P030001-00-R

Lawrence, R., *et al.* 2004. *Opt. Lett.*, **29**, 2635.

Lück, H., *et al.* 2004. *Classical and Quantum Gravity*, **21**, S985.

Levin, Y. 1998. *Phys. Rev. D*, **57**, 659–663.

Manley, J. M. and Row, H. E. 1956. *Proc. IRE*, **44**, 904.

Martin, I., *et al.* 2008. *Class. Quantum Grav.*, **25**, 055005.

Meers, B. J. 1988. *Phys. Rev. D*, **38**, 2317.

Miller, J. 2010. *On non-Gaussian beams and optomechanical parametric instabilities in interferometric gravitational wave detectors.* PhD Thesis, University of Glasgow.

Penn, S. D., *et al.* 2006. *Phys. Lett. A*, **352**, 3.

Pirandola, S., *et al.* 2003. *Phys. Rev. A*, **68**, 062317.

Schediwy, S., *et al.* 2008. *Phys. Rev. A*, **77**, 0138131.

Sidles, J. A. and Sigg, D. 2006. *Phys. Lett. A*, **354**, 167.

Smith, M., Ottaway, D., and Willems, P. 2004. *LIGO-T040057-00* technical report, Heating Beam Pattern Optical Design CO2 Laser Thermal Compensation Bench.

Solimeno, S., *et al.* 1991. *Phys. Rev. A*, **43**, 6227.

Strain, K. A., *et al.* 1994. *Phys. Lett. A*, **194**, 124.

Strigin, S. E. and Vyatchanin, S. P. 2007. *Phys. Lett. A*, **365**, 10.

Strigin, S. E., *et al.* 2008. *Phys. Lett. A*, **372**, 5727.

Willems, P. 2010. TCS Actuator Noise Coupling, https://dcc.ligo.org/DocDB/0009/T060224/004/TCS Actuator Noise Couplings version 4.pdf

Willems, P. 2009. *Adaptive Optics: Methods, Analysis and Applications, OSA Technical Digest (CD).* Optical Society of America. Chap. Thermal Compensation in the LIGO Gravitational-Wave Interferometers, page aotha5.

Winkler, W., *et al.* 1991. *Phys. Rev. A*, **44**, 7022–7036.

Yamamoto, K., *et al.* 2002. *Phys. Lett. A*, **305**, 18.

Yamamoto, K., *et al.* 2007. Page 012015 of: *7th Edoardo Amaldi Conference on Gravitational Waves-Journal of Physics: Conference Series*, vol. 122.

Zhang, Z., *et al.* 2010. *Phys. Rev. A*, **81**, 013822.

Zhao, C., *et al.* 2005. *Phys. Rev. Lett.*, **96**, 231101.

Zhao, C., *et al.* 2005. *Phys. Rev. Lett.*, **94**, 121102.

Zhao, C., *et al.* 2008. *Phys. Rev. A*, **78**, 023807.

Part 4

Technology for third generation gravitational wave detectors

14

Cryogenic interferometers

J. Degallaix

This chapter discusses how mirrors at cryogenic temperature can be used to improve the sensitivity of advanced gravitational wave interferometers. We start by describing the most relevant physical parameters of sapphire substrates at low temperature. Then we discuss how lowering the temperature of the test masses can reduce thermal noise and suppress thermal aberration. We finish by describing plans for the Large Cryogenic Gravitational-Wave Telescope, an advanced cryogenic interferometer in Japan. Throughout, we will describe not only the advantages of cryogenic temperature for interferometers, but also the significant technical challenges that must be met.

14.1 Introduction

The strain sensitivity of advanced gravitational wave interferometric detectors is expected to be limited by quantum noise over most of the detection band. Unfortunately for room temperature interferometers, mirror thermal noise may be the dominant noise source in the hundreds of hertz region. This will result in degradation in the sensitivity and will prevent the successful use of squeezed light in this frequency band. One promising way to significantly decrease the magnitude of the thermal noise is to lower the temperature of the interferometer test masses. Lowering the sensor temperature has greatly extended the range of numerous astronomical detector, such as CCD camera and radio receivers. The technique can also be successfully applied to future gravitational wave detectors.

Cooling the detector mirrors will reduce the thermal noise and will also provide another essential benefit: the wavefront distortion induced by optical absorption will be greatly attenuated due to the properties of the mirror substrate at cryogenic temperature.

To understand the advantages and challenges of cryogenic interferometers, this chapter will be structured as follows. Firstly, the different physical properties of mirror substrates at low temperature will be described. Then the temperature dependence of the thermal noise magnitude and the level of thermal aberration will be explored. Finally, we will outline some specific problems and considerations for cryogenic interferometers.

14.2 Material properties at low temperature

Room temperature detectors extensively use fused silica as the substrate material in the main optics of the interferometer. Unfortunately, fused silica mirrors present many disadvantages at low temperatures (Rowan *et al.*, 2005). At cryogenic temperature the favoured material is sapphire, due to its high Q-factor and thermal conductivity.

In this section, particular attention will be paid to understanding some of the key parameters of sapphire at low temperature. For completeness the last part of this section will discuss some of the other candidate materials still under investigation.

Q-factor

The amplitude of the thermal noise, responsible for undesirable motion of the mirror surface, can be directly related to the acoustic loss of the substrate. Direct application of the fluctuation dissipation theorem (Callen and Greene, 1952) to derive the Brownian thermal noise of the test mass reveals that a low level of thermal noise is associated with low loss materials (Levin, 1998).

The inverse of the loss, called the Q-factor, is usually determined by measuring the ringdown decay of a resonant mechanical mode of the mirror (Nietzsche *et al.*, 2006). Such measurements can be performed in a cryostat with a controlled temperature, cooling the mirror down to a very low temperature. The Q-factor of sapphire as a function of temperature is shown in Figure 14.1. We see that lowering the substrate temperature from 300 K to 4 K, increases the Q-factor of sapphire by a factor greater than 45.

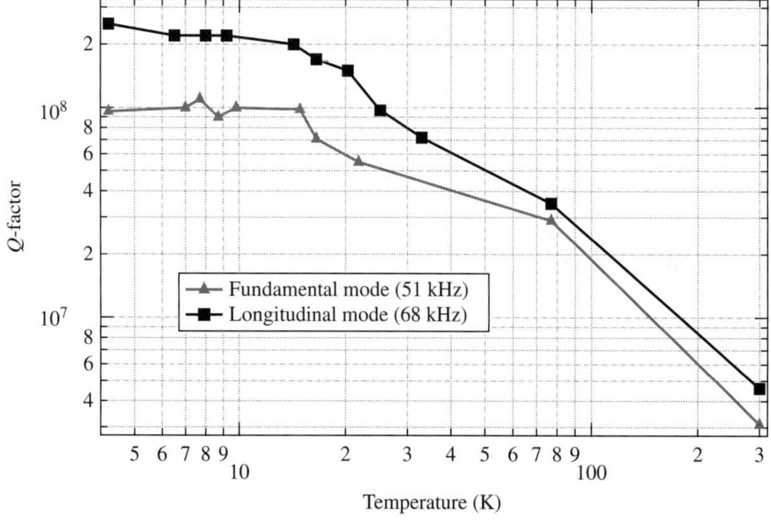

Figure 14.1 Evolution of the sapphire Q-factor as a function of temperature for two mechanical modes of the a sapphire test mass (Uchiyama *et al.*, 1999).

Coating loss angle

As is the case for substrates, internal dissipation inside the highly reflective coatings of the mirrors can produce excess displacement noise. The dielectric coating for the mirrors of gravitational wave laser interferometers is usually made of alternating layers of tantala (Ta_2O_5) and fused silica (SiO_2) – the tantala is responsible for much of the coating loss (Penn *et al.*, 2003).

Measurements of the coating loss angle as a function of temperature do not show any substantial loss reduction at low temperature (Yamamoto *et al.*, 2006; Nawrodt *et al.*, 2007b). However, loss measurements of a single layer of tantala show a dissipation peak at around 20 K, indicating potential extra coating loss around that temperature (Martin *et al.*, 2009). For this chapter, assuming multi-layer coatings, the coating loss is considered to be independent of temperature. The amplitude of the coating Brownian noise scales as the square root of the temperature. It is therefore expected that cryogenic interferometers will exhibit lower coating thermal noise than room temperature detectors.

Absorption

The problem of wavefront distortion, induced by optical power absorbed inside the test mass substrate, is a serious obstacle for advanced interferometers. Without any thermal compensation, system an interferometer's performance can be seriously limited by the presence of very high circulating light power.

Sapphire absorption has been measured at cryogenic temperatures and results do not differ significantly from those determined at room temperature (Tomaru *et al.*, 2001). Therefore, the optical absorption can be considered to be independent of temperature.

Specific heat capacity

The specific heat capacity of sapphire as a function of temperature is well described by the Debye model (Ekin, 2006). As illustrated in Figure 14.2, the specific heat capacity decreases rapidly with temperature, following a T^3 power law below room temperature.

Thermal conductivity

Substrate thermal conductivity is a key parameter in estimating the performance of an interferometer. High thermal conductivity is beneficial in reducing the aberration caused by optical absorption in the main optics. Unfortunately, it can also contribute towards large substrate thermoelastic noise.

The temperature dependence of the thermal conductivity can be derived by applying the Debye model to phonons within the crystal (Tritt, 2001). Inside the crystal, thermal energy is carried by propagating phonons and the thermal conductivity is inversely proportional to the resistance encountered by the moving phonons.

Two different regimes can be seen in the curve of the temperature dependence of thermal conductivity shown in Figure 14.3. Above 100 K, the phonon mean free path is dominated by phonon–phonon scattering processes and the thermal conductivity is inversely proportional

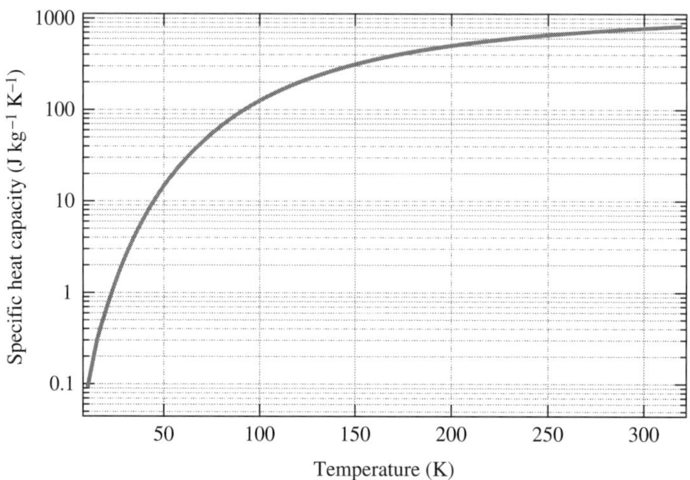

Figure 14.2 The temperature dependence of the specific heat of sapphire (Touloukian, 1970).

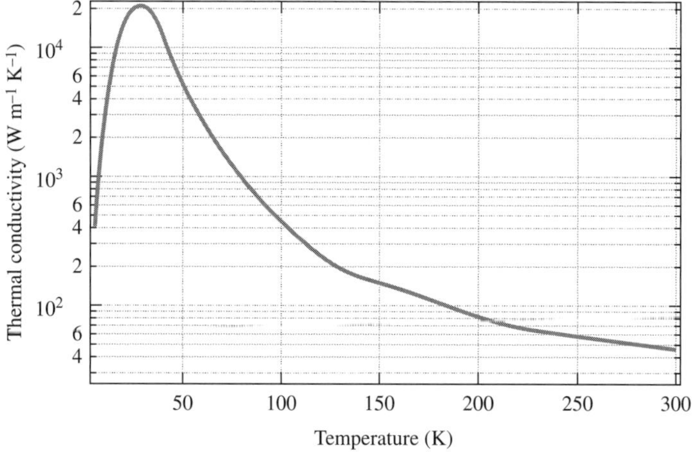

Figure 14.3 Thermal conductivity of sapphire as a function of temperature (Touloukian, 1970). By decreasing the temperature from 300 K to 20 K, the thermal conductivity has been increased by a factor of 400. For comparison, at 20 K, the thermal conductivity of sapphire is twice that of pure annealed copper (Powell *et al.*, 1959).

to the temperature. At low and cryogenic temperatures, the phonon mean free path is larger than the crystal dimension; the thermal conductivity therefore follows the temperature dependence of the heat capacity, decreasing as T^3. Between these two regimes, the thermal conductivity has a maximum at around 20 K, at which point the conductivity is 2000 times higher than that of room temperature.

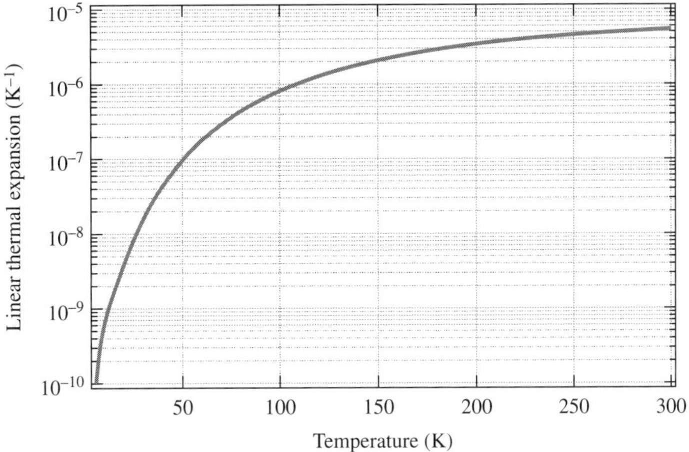

Figure 14.4 Coefficient of linear thermal expansion of sapphire as a function of temperature (White and Minges, 1997). The thermal expansion represents the averaged expansion over both parallel and perpendicular directions relative to the crystal axis.

Thermal expansion

The amplitude of test mass deformations due to temperature changes is directly proportional to the substrate thermal expansion. The temperature variations can originate from small thermodynamic local fluctuations (thermal noise) and also from the potentially large temperature gradient produced by optical absorption (thermal lensing).

With regard to specific heat, the coefficient of linear thermal expansion for sapphire decreases with temperature. The coefficient is always positive, and below 100 K it is proportional to T^3. This is shown in Figure 14.4.

Thermo-optic coefficient

The thermo-optic coefficient quantifies the change of refractive index with temperature. Typically, for low thermal expansion materials such as glass and sapphire, this coefficient is positive, signifying that the refractive index increases with temperature. The thermo-optic coefficient of sapphire decreases with temperature, as shown in Figure 14.5.

From room temperature to 20 K, the thermo-optic coefficient has been reduced by a factor 400. This low value was also confirmed by Tomaru (Tomaru *et al.*, 2001).

Other candidate materials

Sapphire is not the only material considered for cryogenic interferometers. Two other candidates have been investigated: calcium fluoride and silicon. Calcium fluoride (CaF_2) is a single crystal material with a high transmittance from ultraviolet to mid infrared wavelengths. It has been considered for cryogenic interferometers because it has a high quality factor at low temperature (Nawrodt *et al.*, 2007a) and also has good optical properties. Another promising material is silicon. Its widespread use in the semiconductor industry

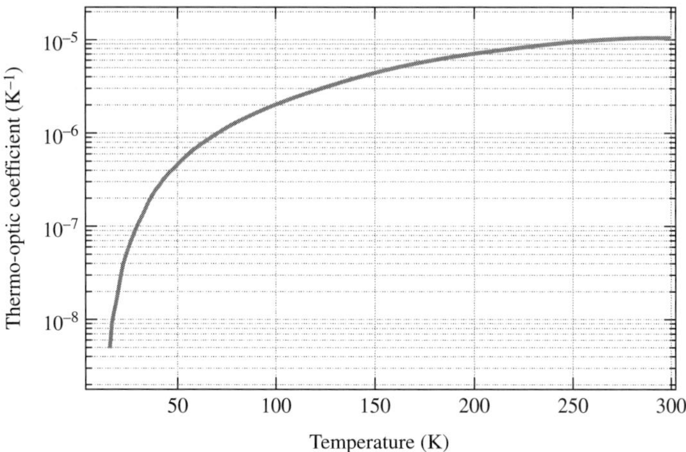

Figure 14.5 The thermo-optic coefficient of sapphire (Kaplan and Thomas, 2002).

means that pure silicon substrates are available in large size and ultraprecision polishing is possible. Like sapphire, silicon posseses excellent cryogenic properties (Nawrodt *et al.*, 2008; Degallaix *et al.*, 2006), but silicon has much lower optical absorption in its transparent region (Green and Keevers, 1995). One feature of silicon is a null coefficient of thermal expansion for temperatures of 20 K and 120 K. At such temperatures, thermal aberrations and potentially some thermal noise can be minimised. Silicon is, however, not transparent at 1064 nm, the laser wavelength of first generation of interferometers. The transparency region starts from 1300 nm. As the time of writing (2011), high power lasers that produce light with a wavelength of 1550 nm are being investigated to work with silicon test masses.

14.3 Reduction of mirror thermal noise

The main purpose of lowering the mirror temperature is to decrease the level of thermal noise. Assuming all other physical parameters are constant with temperature, the amplitude of the surface displacement of the mirror from thermal noise scales as a function of temperature, T, from $T^{\frac{1}{2}}$ up to $T^{\frac{3}{2}}$. However, as discussed in Section 14.2, the physical parameters of the mirror substrate also depend strongly on the temperature, and must therefore be taken into account to calculate accurately the thermal noise amplitudes.

To simulate the overall thermal noise in laser interferometers, an infinite size test mass is generally assumed for the noise formulae. This is a valid assumption since, in practice, to avoid unnecessary diffraction loss, the laser beam radius must be much smaller than the test mass radius. The common symbols and parameters used are defined in Table 14.1. Calculations generally consider four principal thermal noise sources:

1. *Substrate Brownian thermal noise.* This is also called structural damping and is related to the Q-factor (Q) of the mirror. Using the fluctuation dissipation theorem, the displacement power spectral density of the mirror front face due to bulk Brownian noise can be

Table 14.1. *Parameters used to calculate the amplitudes of the different types of thermal noises*

Parameter	Symbol	Units	Value
Boltzmann constant	k_b	$\mathrm{J\,K^{-1}}$	1.38×10^{-23}
Sapphire test mass			
Density	ρ	$\mathrm{kg\,m^{-3}}$	3970
Young modulus	E_{sub}	Pa	4.0×10^{11}
Poisson ratio	σ	-	0.29
Thermal expansion	α	$\mathrm{K^{-1}}$	see Figure 14.4
Thermal conductivity	κ	$\mathrm{W\,m^{-1}K^{-1}}$	see Figure 14.3
Specific heat	C	$\mathrm{J\,kg^{-1}K^{-1}}$	see Figure 14.2
Test mass coating			
Number of dual layer			10
Total thickness	d	m	6.2×10^{-6}
Young modulus	E_{coat}	Pa	1.1×10^{11}
Thermal expansion coefficient	$\bar{\alpha}$	$\mathrm{K^{-1}}$	3.2×10^{-6}
Thermorefractive coefficient	$\bar{\beta}$	$\mathrm{K^{-1}}$	7.0×10^{-6}
Interferometer			
Laser wavelength	λ	m	1064×10^{-9}
Laser beam radius	w_0	m	60×10^{-3}
Cavity arm length	L	m	4000

expressed as (Levin, 1998):

$$S_{\mathrm{B}}^{\mathrm{sub}}(f, T) = \frac{2k_b T}{\pi f} \frac{1 - \sigma^2}{w_0 E_{\mathrm{sub}}} \frac{1}{Q} . \tag{14.1}$$

The above formula is valid for homogeneous loss.

2. *Substrate thermoelastic noise.* This is due to statistical fluctuation of temperature inside the bulk of the mirror. The power spectral density of displacement for low and room temperature is given by Cerdonio *et al.* (2001) (including a correction from Black *et al.*, 2004):

$$S_{\mathrm{TE}}^{\mathrm{sub}}(f, T) = \frac{4k_b T^2}{\sqrt{\pi}} \frac{\alpha_{\mathrm{sub}}^2 (1 + \sigma)^2}{\kappa} w_0 J[\Omega] , \tag{14.2}$$

with $J[\Omega]$ given by:

$$J[\Omega] = \sqrt{\frac{2}{\pi^3}} \int_0^\infty du \int_{-\infty}^\infty dv \frac{u^3 e^{-u^2/2}}{(u^2 + v^2)[(u^2 + v^2)^2 + \Omega^2]} , \tag{14.3}$$

and $\Omega = 2\pi f / \omega_c$ with $\omega_c = 2\kappa / \rho C w_0^2$.

3. *Coating thermal noise.* This is related to the mechanical loss within the coating and is defined as (Harry *et al.*, 2002):

$$S_{\mathrm{B}}^{\mathrm{coat}}(f, T) = \frac{2k_b T}{\pi^2 f} \frac{d}{w_0 E_{\mathrm{sub}}} \left(\frac{E_{\mathrm{sub}}}{E_{\mathrm{coat}}} \phi_\| + \frac{E_{\mathrm{coat}}}{E_{\mathrm{sub}}} \phi_\perp \right) , \tag{14.4}$$

where $\phi_\parallel$ and $\phi_\perp$ are the coating mechanical loss angles in directions parallel and pedicular to the plane of the coating layers. Generally, $\phi_\parallel = \phi_\perp = 4 \times 10^{-4}$ as the losses are dominated by the tantala coating layers (Penn *et al.*, 2003).

4. *Coating thermo-optic noise.* This noise results from temperature variations being converted into displacement sensing noise by the coating thermal expansion and the thermorefractive effect of the coating layers (Evans *et al.*, 2008). The equivalent power spectral density of the displacement is given by:

$$S_{\mathrm{TO}}^{\mathrm{coat}}(f, T) = \frac{2k_{\mathrm{b}}T^2}{\pi^{\frac{3}{2}}} \frac{1}{w_0^2 \sqrt{\kappa \rho C f}} (\overline{\alpha} d - \overline{\beta} \lambda) . \tag{14.5}$$

Here $\overline{\alpha}$ and $\overline{\beta}$ are the effective thermal expansion and thermorefractive coefficient, respectively, as described in (Evans *et al.*, 2008). Equation (14.5) is a simplified formula used here to gain a basic understanding of the origins of the noise. In practice, and for results of thermal noise calculations shown later, a more exact formula is used (equation 4 in Evans *et al.*, 2008). We should note here that the level of the thermo-optic noise is not known at low temperature, since no formula has yet been derived for thermal diffusion lengths larger than the beam radius. Additionally, at the time of writing (2011), the physical properties of thin film tantala at cryogenic temperature have not yet been measured. However, this noise is not expected to be a limiting factor as it can be made as small as required with careful choice of material.

Thermal noise as a function of temperature

To quantify the benefits of reducing the test mass temperature, the test mass displacement due to thermal noise can be calculated as a function of temperature. This is shown in Figure 14.6 at a frequency of 100 Hz, around which thermal noise will limit the sensitivity of advanced room temperature interferometers.

At room temperature, the noise is dominated by the substrate Brownian thermal noise – a consequence of the relative low Q-factor used in our data: ($\sim 4 \times 10^7$, as shown in Figure 14.1). For comparison, a Q-factor of 2×10^8 has been measured at room temperature for a sapphire test mass (Gras *et al.*, 2004).

At 50 K, the thermal noise reaches a local maxima, dominated by the substrate thermoelastic noise. That is a direct consequence of sapphire's high thermal conductivity at low temperature. At around 20 K, the three principal sources of thermal noise – substrate Brownian noise, thermoelastic noise and coating Brownian noise – have the same order of magnitude. This suggests an optimised temperature operating point.

From room temperature to 20 K, the test mass displacement due to thermal noise is reduced by a factor 10. Such a gain is possible because the thermal noise of the sapphire test mass is not limited by the coating Brownian noise at 300 K. Since the coating loss angle is independent of the temperature, cooling the test mass from 300 K to 20 K lowers the amplitude of the coating Brownian noise by only a factor $\sqrt{(300/20)} \simeq 4$.

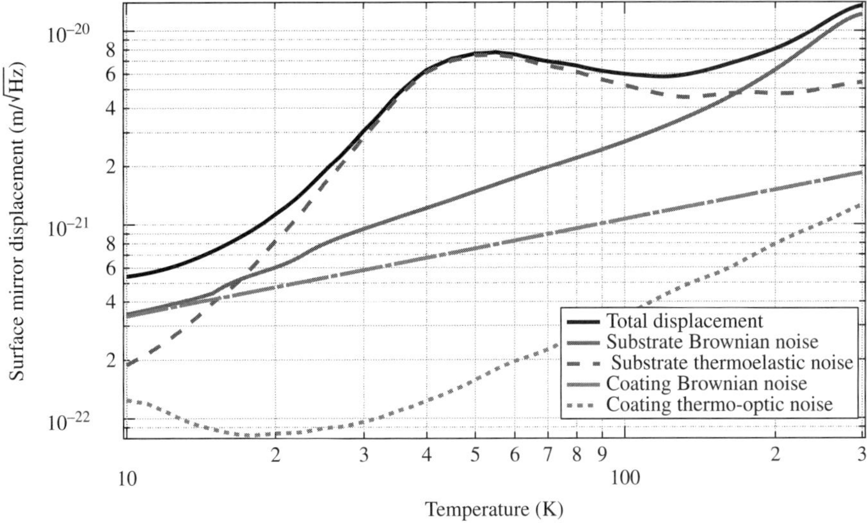

Figure 14.6 Amplitude spectrum of the test mass displacement due to thermal noise. The individual thermal noises have been calculated at a frequency of 100 Hz.

Thermal noise for cryogenic interferometers

The thermal noise budget for an advanced interferometer with a 4 km long arm cavity and a test mass cooled down to 20 K is shown in Figure 14.7. We see that at 100 Hz, the thermal noise is six times lower than the strain sensitivity of advanced interferometers.

We have not yet mentioned the thermal noise from the mirror suspension. The suspension system is composed of a multi-stage pendulum with the test masses at the bottom stage, suspended from thin fibres (Robertson *et al.*, 2002). Suspension thermal noise can be approximated by the sum of the pendulum thermal noise (Gretarsson *et al.*, 2000) and the thermal noise from the fibre violin modes (Saulson, 1990). For room temperature advanced interferometers, the suspension thermal noise may limit the sensitivity below 10 Hz, whereas above 40 Hz, the radiation pressure noise will be dominant (Whitcomb, 2008). As lowering the temperature will further reduce the suspension thermal noise, this noise is not considered to be a limiting factor for cryogenic interferometers (Arai *et al.*, 2009).

14.4 Elimination of thermal aberration

Another benefit from lowering the temperature of the mirror is the elimination of thermal aberration, also called 'thermal lensing'. The wavefront distortion induced by the optical absorption is a critical issue for room temperature interferometers, where specific actuators and control systems are required to limit the consequences of heat generated in the mirror. Fortunately, the substrate properties at low temperature, in particular very high thermal conductivity and low thermal expansion, can minimise the effect of the optical absorption if mirror temperatures are kept constant.

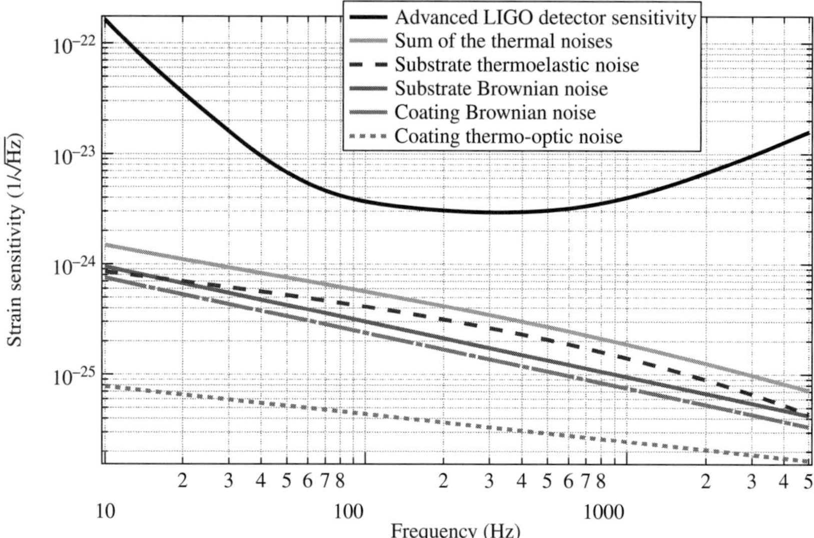

Figure 14.7 Thermal noise contribution for an advanced interferometer with a test masses cooled to 20 K. For reference, the design sensitivity of an advanced interferometer similar to Advanced LIGO is also shown

A simple model to understand thermal lensing

The principal results of thermal lensing at low temperature can be demonstrated with a simplified numerical model. For advanced interferometers, the total power absorbed, P_{abs}, inside the input test mass of the arm cavities can be expected to be around 1.4 W. This is composed of 1 W absorbed inside the substrate and 0.4 W absorbed by the coating (using the parameters quoted in Table 14.2). For comparison, the power absorbed by the substrate of a fused silica mirrors, as used for a room temperature interferometer, is around 10 times lower. Since the absorption is independent of temperature, the absorbed power is the same for both room temperature and low temperature interferometers.

The equivalent focal length in transmission of the thermal lens due to the absorption of a Gaussian beam, f_{TL}, is given by (Degallaix, 2007):

$$f_{\text{TL}} = \frac{\pi \kappa w_0^2}{P_{\text{abs}} \beta}. \tag{14.6}$$

Here, $P_{\text{abs}} = 2 \times l_{\text{m}} \times Abs_{\text{sub}} \times P_{\text{PRC}} + Abs_{\text{coat}} \times P_{\text{arm}}$, using the parameters defined in Table 14.2. At low temperature, the high thermal conductivity and low thermo-optic coefficient of sapphire substrates guarantee that thermal lensing is negligible. For example, assuming the same absorbed optical power, cooling the mirror from room temperature to 20 K can increase the focal length of the thermal lens by a factor 10^5.

A similar demonstration can also determine the thermal expansion of the test mass due to the power absorbed by the high reflective coating ($Abs_{\text{coat}} \times P_{\text{arm}}$). The change in sagitta,

Table 14.2. *Parameters used for thermal lensing simulations*

Parameter	Symbol	Unit	Value
Test mass dimensions			
Radius	r_{m}	m	0.15
Thickness	l_{m}	m	0.15
Test mass properties			
Substrate absorption	Abs_{sub}	ppm cm^{-1}	20
Coating absorption	Abs_{coat}	ppm	0.5
Thermal conductivity	κ	W m^{-1}K^{-1}	Figure 14.3
Thermal expansion	α	K^{-1}	Figure 14.4
Thermo-optic coefficient	β	K^{-1}	Figure 14.5
Interferometer			
Laser beam radius	w_0	m	60×10^{-3}
PRC circulating power	P_{PRC}	kW	1.6
Arm circulating power	P_{arm}	kW	800

Note: For comparison, the circulating power in LCGT (see Section 14.5) is around half of what is stated here.

δs, measured at the beam radius, induced by the absorption of a Gaussian beam, is given by:

$$\delta s = \frac{\alpha \times Abs_{\mathrm{coat}} \times P_{\mathrm{arm}}}{4\pi\kappa} . \tag{14.7}$$

At low temperature, the deformation is greatly reduced due to the high thermal conductivity of the substrate and the low thermal expansion. From 300 K to 20 K, the sagitta change is reduced by a factor 4×10^5.

For cryogenic interferometers, absorbed power is a major obstacle in keeping the temperature of the optics constant. At low temperature, radiative cooling is particularly inefficient; the heat generated inside the test mass can only be extracted along the suspension fibres. At room temperature, an increase in substrate temperature of 0.8 K is adequate for the test mass to radiate 1.4 W of heat. At 20 K, an increase of 70 K is necessary to radiate the same power. This increase in temperature can substantially increase the thermal noise.

With the simple models presented above, the main properties of thermal lensing at low temperature have been highlighted. For cryogenic interferometers, the temperature gradient inside the test mass is negligible, but special care must be taken to extract efficiently the heat generated inside the optics. This is the opposite to the situation at room temperature, where the thermal aberration due to the substrate gradient of temperature must be actively compensated, while the rise of temperature undergone by the test mass is not a concern.

Thermal lensing simulations

The ANSYS package (Moaveni, 1999) can be used to simulate the thermal gradient inside the optics and can also be used to calculate the deformation of the front surface. As reported in the previous section, the absorbed power is 1 W inside the substrate and 0.4 W by the

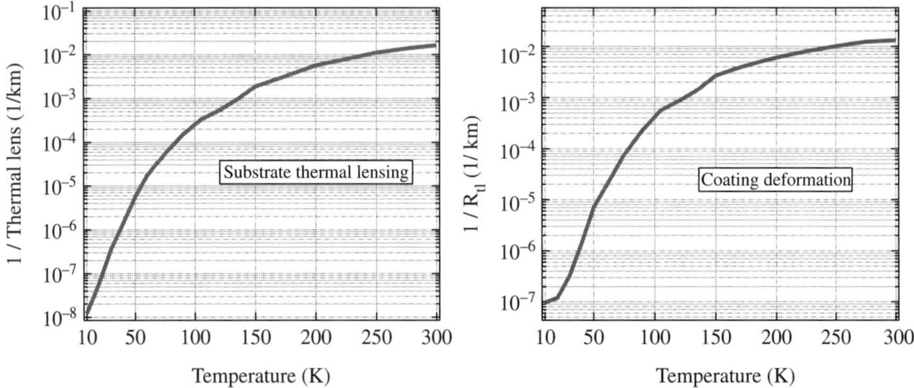

Figure 14.8 Substrate thermal lensing simulations (left) and coating deformations (right). The focal length of the thermal lens in the substrate is calculated in transmission for a total optical power absorbed of 1.4 W. The coating deformation is the equivalent radius of curvature of an initially flat test mass.

coating. At cryogenic temperatures, the heat generated inside the optics is extracted along the suspension fibre. Therefore, the average temperature of the test mass is kept at around 20 K.

To derive the equivalent focal length of the thermal lens, f_{TL}, the optical path difference induced by the absorption in the substrate in transmission is fitted by a quadratic polynomial. The same procedure is applied to calculate the equivalent radius of curvature, R_{TL}, of the high reflective coating due to the thermal expansion – it is assumed that the surface is flat prior to any heating power absorbtion.

The inverse of the focal length of the thermal lens is plotted in the left panel of Figure 14.8. As expected, the thermal lens vanishes at low temperature. Previous studies showed that below 70 K the shot-noise-limited sensitivity is not degraded by the effects of the optical absorption (Degallaix *et al.*, 2006).

Due to the thermal expansion, the change in the radius of curvature of the mirror from the designed curvature, R_{cold}, can be expressed as:

$$\frac{1}{R_{\mathrm{hot}}} = \frac{1}{R_{\mathrm{cold}}} - \frac{1}{R_{\mathrm{TL}}}. \tag{14.8}$$

Here, R_{hot} is the new radius of curvature of the mirror, including the thermal expansion due to the absorbed optical power. Typically, the sagitta change of the mirror due to thermal expansion is small in comparison with the initial cold sagitta (i.e. $R_{\mathrm{TL}} \gg R_{\mathrm{cold}}$). Thus, equation (14.8) can also be written as:

$$R_{\mathrm{hot}} = R_{\mathrm{cold}} \left(1 + \frac{R_{\mathrm{cold}}}{R_{\mathrm{TL}}} \right). \tag{14.9}$$

For advanced interferometers, the tolerance of the radius curvature of the arm cavities can be as low as 0.2%. Practically, for $R_{\mathrm{cold}} = 2\,\mathrm{km}$, R_{TL} must be larger than 1000 km for the thermal expansion of the reflective coating of the arm cavity test mass to be considered acceptable. The temperature dependence of R_{TL} is shown in the left panel of Figure 14.8. For temperatures below 125 K, the optical absorption in the optics is no longer an issue for the arm cavities.

14.5 LCGT

In this section, we focus on the most advanced cryogenic interferometer project, the Large Scale Cryogenic Gravitational Wave Telescope, known as LCGT. In particular, we will focus on specific issues related to low temperature interferometers.

LCGT is planned to be a dual recycled laser interferometer with Fabry–Perot arm cavities. It will use the technique of tuned resonant sidebands extraction to limit the circulating optical power in the power recycling cavity. Two particular elements will make LCGT unique. First, it will be located underground on the site of the Kamioka observatory, and second, the mirror will be cooled down to 20 K.

Cryogenic system

To house the mirror at a temperature of 20 K, the vacuum tank must also be served as a cryostat. The cryostat consists of two layers of radiation shield. The outer shield is actively cooled to a temperature of 100 K, whereas the inner shield is kept at 8 K. The long vacuum pipe used to house the circulating laser beam must also be partially cooled. To do this, 20 m of pipe before and after the test mass is cooled down to 100 K, serving as a cryogenic trap. Several days will be required to cool the whole system from room temperature to the final cryogenic temperature.

The cooling is achieved by the use of pulse-tube refrigerators similar to those currently used for the CLIO project (Uchiyama *et al.*, 2006). The cooling power required for one cryostat is 2 W for the inner shield, 80 W for the outer shield and 83 W for the cryogenic pipes (Uchiyama, 2005).

For cryogenic interferometers, not all the mirrors will have to be cooled – only the four mirror constituting the two arm cavities need to be at low temperature. The other optics, such as the power recycling mirror, the beam splitter and the signal recycling mirror, will remain at room temperature.

A cryogenic interferometer is not necessarily more expensive than an interferometer working a room temperature, since for LCGT, the price of the cryogenic system is equivalent to the price of the high power laser. In comparison to other subsystems it represents less than 5% of the total cost of the vacuum system (Uchiyama *et al.*, 2004).

Suspension design

A seismic suspension is used to isolate the mirror from the grounds seismic motion. The suspension system can be approximated by a system of cascaded pendula having low resonance frequency (around 1 Hz). Thus, at frequencies above 100 Hz, the mirror can be considered to be in free fall along the optical axis. For cryogenic interferometers the suspension has an additional essential role – it must be used as heat link to extract the heat generated inside the test mass by the absorption of the high power laser beam. Therefore in an ideal situation, the

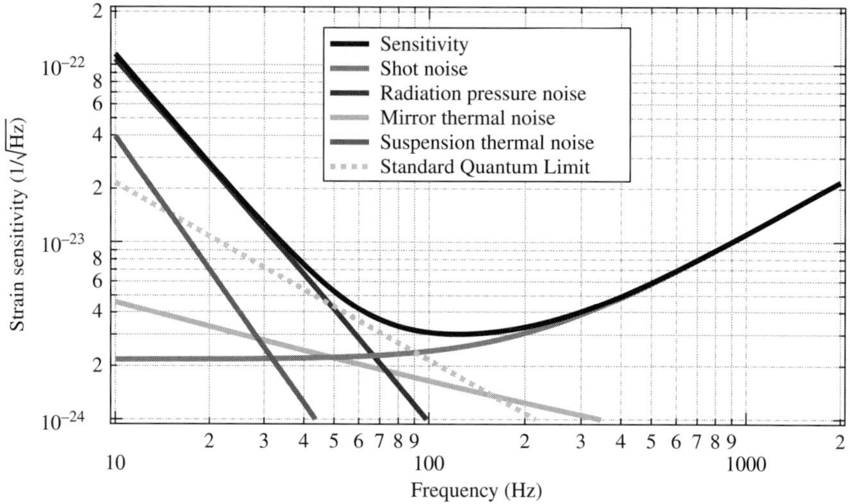

Figure 14.9 Noise budget for the LCGT cryogenic interferometer.

elements of a suspension will have very high thermal conductivity, whilst also possessing a very high Q-factor to avoid inducing large suspension thermal noise.

In LCGT, the mirror is suspended from the mass above it by sapphire optical fibres with very high thermal conductivity (Tomaru *et al.*, 2002) and high Q-factor (Uchiyama *et al.*, 2000). The heat generated in the mirror escapes along the fibre by conduction and reaches the upper mass, which is maintained at a temperature 5 K cooler than the mirror. The upper mass, which is also made of sapphire and serves as a mirror for a suspension point interferometer, is linked to the inner wall of the cryostat–itself cooled to 8 K by high thermal conductivity aluminum wires.

The upper mass is suspended from a suspension platform which is linked to the vibration isolation system. This system is similar in design to the current TAMA isolation system (Takahashi *et al.*, 2008). The vibration isolation system is composed of three anti-spring filters attached to an inverted pendulum. The whole system is kept at room temperature.

Sensitivity

The expected strain sensitivity of LCGT is presented in Figure 14.9. We note that due to the use of cryogenic test masses, the thermal noise has been reduced substantially and the sensitivity is now fully limited by quantum noise from 10 Hz.

14.6 Conclusion

We have seen that lowering the temperature of the test mass presents two main advantages: reduction of the thermal noise and elimination of thermal aberration. However, mirrors can only be kept at cryogenic temperature if the heat generated in the optics due to optical absorption can be properly conducted through the suspension system.

Although a number of challenges are still to be met, in particular, better knowledge of coating properties at low temperature and efficient techniques to dissipate the heat generated in the mirrors, cryogenic interferometers seem to represent the future for gravitational wave detectors. As we will see in Chapter 16, they are not only considered for advanced interferometers, but also for third generation interferometers.

Acknowledgements

J. Degallaix gratefully acknowledges support from the Max Planck Society (MPG) and the state of Lower Saxony in Germany.

References

Arai, K., *et al.* 2009. *Classical and Quantum Gravity*, **26**, 204020 (9pp).

Black, E. D., Villar, A., and Libbrecht, K. G. 2004. *Phys. Rev. Lett.*, **93**, 241101.

Callen, H. B. and Greene, R. F. 1952. *Phys. Rev.*, **86**, 702–710.

Cerdonio, M., *et al.* 2001. *Phys. Rev. D*, **63**, 082003.

Degallaix, J., *et al.* 2006. *Journal of Physics: Conference Series*, **32**, 404–412.

Degallaix, J. 2007. *Compensation of Strong Thermal Lensing in Advanced Interferometric Gravitational Waves Detectors*. Ph.D. thesis, University of Western Australia.

Ekin, J. 2006. *Experimental Techniques for Low Temperature Measurements*. Oxford: Oxford University Press.

Evans, M., *et al.* 2008. *Physical Review D*, **78**, 102003.

Gras, S., Ju, L. and Blair, D. G. 2004. *Classical and Quantum Gravity*, **21**, S1121–S1126.

Green, M. A. and Keevers, M. J. 1995. *Progress in Photovoltaics: Research and Applications*, **3**, 189–192.

Gretarsson, A. M., *et al.* 2000. *Physics Letters A*, **270**, 108–114.

Harry, G. M., *et al.* 2002. *Classical and Quantum Gravity*, **19**, 897–917.

Kaplan, S. G. and Thomas, M. E. 2002. *Proc. of SPIE*, **4822**, 41.

Levin, Y. 1998. *Phys. Rev. D*, **57**, 659–663.

Martin, I. W., *et al.* 2009. *Classical and Quantum Gravity*, **26**, 155012 (11pp).

Moaveni, S. 1999. *Finite Element Analysis: Theory and Application with Ansys*. Upper Saddle River, NJ, USA: Prentice Hall PTR.

Nawrodt, R., *et al.* 2007a. *Eur. Phys. J. Appl. Phys.*, **38**, 53–59.

Nawrodt, R., *et al.* 2007b. *New Journal of Physics*, **9**, 225.

Nawrodt, R., *et al.* 2008. *Journal of Physics: Conference Series*, **122**, 012008.

Nietzsche, S., *et al.* 2006. *Journal of Physics: Conference Series*, **32**, 445–450.

Penn, S. D., *et al.* 2003. *Classical and Quantum Gravity*, **20**, 2917–2928.

Powell, R. L., Roder, H. M., and Hall, W. J. 1959. *Phys. Rev.*, **115**, 314–323.

Robertson, N. A., *et al.* 2002. *Classical and Quantum Gravity*, **19**, 4043–4058.

Rowan, S., Hough, J., and Crooks, D. 2005. *Physics Letters A*, **347**, 25–32.

Saulson, P. R. 1990. *Phys. Rev. D*, **42**, 2437–2445.

Takahashi, R., *et al.* 2008. *Classical and Quantum Gravity*, **25**, 114036.

Tomaru, T., *et al.* 2001. *Physics Letters A*, **283**, 80–84.

Tomaru, T., *et al.* 2002. *Physics Letters A*, **301**, 215–219.

Touloukian, Y. S. 1970. *Thermophysical properties of matter*. IFI/Plenum.

Tritt, T. M. 2001. *Thermal Conductivity: Theory, Properties, and Applications (Physics of Solids and Liquids)*. Springer.

Uchiyama, T. 2005. Technical report of LCGT.

Uchiyama, T., *et al.* 1999. *Physics Letters A*, **261**, 5–11.
Uchiyama, T., *et al.* 2000. *Physics Letters A*, **273**, 310–315.
Uchiyama, T., *et al.* 2004. *Classical and Quantum Gravity*, **21**, S1161–S1172.
Uchiyama, T., *et al.* 2006. *Journal of Physics: Conference Series*, **32**, 259–264.
Whitcomb, S. 2008. *Classical and Quantum Gravity*, **25**, 114013 (15pp).
White, G. K. and Minges, M. L. 1997. *International Journal of Thermophysics*, **18**, 1572.
Yamamoto, K., *et al.* 2006. *Phys. Rev. D*, **74**, 022002.

15

Quantum theory of laser interferometer gravitational wave detectors

H. Miao and Y. Chen

Laser interferometers are quantum instruments. This chapter presents the quantum theory of laser interferometer gravitational wave detectors. We show the basics for analysing the quantum noise in the detector, and for deriving the associated standard quantum limit (SQL) for the sensitivity. By providing different perspectives on the origin of the SQL, we illustrate the motivations behind different approaches for surpassing the SQL.

15.1 Introduction

The most difficult challenge in building a laser interferometer gravitational wave (GW) detector is isolating the test masses from the rest of the world (e.g. random kicks from residual gas molecules, seismic activities, acoustic noises, thermal fluctuations, etc.) whilst keeping the device locked around the correct point of operation (e.g. pitch and yaw angles of the mirrors, locations of the beam spots, resonance condition of the cavities, and dark-port condition for the Michelson). Once all these issues have been solved, we arrive at the issue that we are going to analyse in this chapter: the fundamental noise that arises from quantum fluctuations in the system. A simple estimate (following the steps of Braginsky, 1968) will already lead us into the quantum world – as it will turn out, the superb sensitivity of gravitational wave detectors will be constrained by the *standard quantum limit* (SQL), which relates to the fundamental *heisenberg uncertainty principle*. Further improvements of detector sensitivity beyond this require us to manipulate the quantum coherence of light to our advantage. In this chapter, we will introduce how to analyse GW detectors quantum mechanically, and will describe several advanced configurations to surpass the SQL.

The outline of this chapter is as follows: in Section 15.2, we will make an order-of-magnitude estimate on the quantum noise in a typical GW detector and gain a qualitative understanding of the SQL; in Section 15.3, we will introduce the basic concepts to study the quantum dynamics of a detector, with a detailed analysis in Section 15.4. Section 15.5 will present a rigorous derivation of the SQL from a more general context of linear continuous quantum measurements. This can enhance the understanding of the results in the previous sections, and also illuminates different approaches for surpassing the SQL: (1) building correlations among quantum noises (Section 15.6); (2) modifying the dynamics of the test mass (Section 15.7). In Section 15.8, we will present an alternative point of view on the SQL which will introduce the idea of a speed meter for surpassing the SQL, with two possible configurations discussed in Section 15.9. Finally, we present our conclusions in Section 15.10.

15.2 An order-of-magnitude estimate

Here we first make an order-of-magnitude estimate of the quantum limit for the sensitivity. We assume that test masses have a reduced mass of m, and it is under the measurement of a laser with optical power I_0 and an angular frequency ω_0. Within a measurement duration τ, the number of photon is $N_\gamma = I_0\tau/(\hbar\omega_0)$. For a coherent light source (an ideal laser), the number of photon follows a Poisson distribution, and its root-mean-square fluctuation is $\sqrt{N_\gamma}$ (Scully and Zubairy, 1997). The corresponding fractional error in the phase measurement, also called the *shot noise*, would be $\delta\phi_{\rm sh} = 1/\sqrt{N_\gamma}$. For detecting GW with a period comparable to τ, the displacement noise spectrum of the shot noise is given by:

$$S_{\rm sh}^x \approx \frac{\delta\phi_{\rm sh}^2}{k^2}\tau = \frac{\hbar c^2}{I_0\omega_0}, \tag{15.1}$$

where $k \equiv \omega_0/c$ is the wave number.

Meanwhile, the photon number fluctuation also induces a random radiation pressure force on the test mass, which is the *radiation pressure noise* (also called the back-action noise). Its magnitude is $\delta F_{\rm rp} = \sqrt{N_\gamma}\,\hbar k/\tau$, which is equal to the number fluctuation multiplied by the momentum kick of a single photon $\hbar k$ over a duration τ. Since the response function of a free mass in the frequency domain is $-1/m\Omega^2$, the corresponding noise spectrum is given by:

$$S_{\rm rp}^x \approx \frac{\delta F_{\rm rp}^2}{m^2\Omega^4}\tau = \frac{I_0\omega_0}{c^2}\frac{\hbar}{m^2\Omega^4}. \tag{15.2}$$

The total noise spectrum is a sum of $S_{\rm sh}^x$ and $S_{\rm rp}^x$, namely,

$$S_{\rm tot}^x = S_{\rm sh}^x + S_{\rm rp}^x = \frac{\hbar c^2}{I_0\omega_0} + \frac{I_0\omega_0}{c^2}\frac{\hbar}{m^2\Omega^4} \geq \frac{2\hbar}{m\Omega^2}, \tag{15.3}$$

as illustrated in Figure 15.1. The corresponding lower bound that does not depend on the optical power is $S_{\rm SQL}^x \equiv 2\hbar/(m\Omega^2)$. In terms of GW strain h, this is:

$$S_{\rm SQL}^h = \frac{1}{L^2}S_{\rm SQL}^x = \frac{2\hbar}{m\Omega^2 L^2}, \tag{15.4}$$

where L is the arm length of the interferometer. This introduces us to the SQL (Braginsky, 1968; Caves *et al.*, 1980; Braginsky and Khalili, 1992), which arises as a trade-off between the shot noise and radiation pressure noise. In the rest of this chapter, we will develop the tools necessary to analyse the quantum noise of interferometers from first principles, in order to derive the SQL more rigorously. This will allow us to design GW detectors that surpass it.

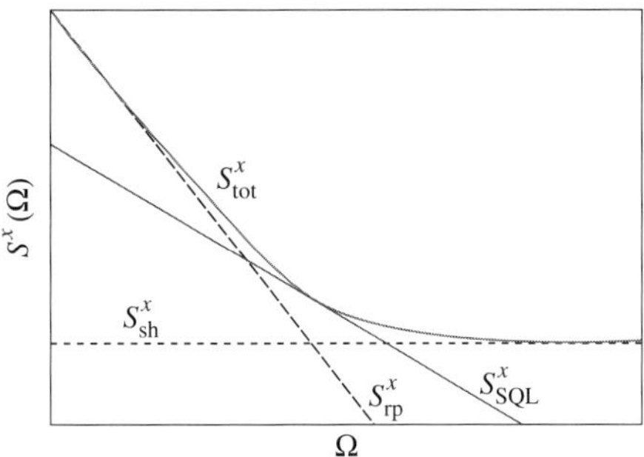

Figure 15.1 A schematic plot of the displacement noise spectrum for a typical interferometer. Increasing or decreasing the optical power, the power-independent lower bound of the total spectrum will trace over the SQL.

15.3 Basics for analysing quantum noise

To rigorously analyse the quantum noise in a detector, we need to study its quantum dynamics; the basics of this analysis are introduced in this section.

Quantisation of the optical field and the dynamics

For the optical field, the quantum operator of its quantised electric field is

$$\hat{E} = u(x, y, z) \int_0^{+\infty} \frac{\mathrm{d}\omega}{2\pi} \sqrt{\frac{2\pi \hbar \omega}{Ac}} \left[\hat{a}_\omega e^{ikz - i\omega t} + \hat{a}_\omega^\dagger e^{+i\omega t - ikz} \right]. \tag{15.5}$$

Here $\hat{a}_\omega^\dagger$ and $\hat{a}_\omega$ are the creation and annihilation operators, which satisfy $[\hat{a}_\omega, \hat{a}_{\omega'}^\dagger] = 2\pi \delta(\omega - \omega')$; A is the cross-sectional area of the optical beam; and $u(x, y, z)$ is the spatial mode, satisfying $(1/A) \int \mathrm{d}x \mathrm{d}y |u(x, y, z)|^2 = 1$.

For ground-based GW detectors, the GW signal that we are interested in is in the audio frequency range from $10\,\mathrm{Hz}$ to $10^4\,\mathrm{Hz}$. It creates sidebands on top of the carrier frequency of the laser ω_0 ($3 \times 10^{14}\,\mathrm{Hz}$). Therefore, it is convenient to introduce operators at these sideband frequencies for analysing the quantum noise. The upper and lower sideband operators are $\hat{a}_+ \equiv \hat{a}_{\omega_0 + \Omega}$ and $\hat{a}_- \equiv \hat{a}_{\omega_0 - \Omega}$, from which we can define the amplitude quadrature $\hat{a}_1$ and phase quadrature $\hat{a}_2$ as

$$\hat{a}_1 = (\hat{a}_+ + \hat{a}_-^\dagger)/\sqrt{2}, \quad \hat{a}_2 = (\hat{a}_+ - \hat{a}_-^\dagger)/(i\sqrt{2}). \tag{15.6}$$

They coherently create one photon and annihilate one photon in the upper and lower sidebands, for this reason this is also known as a two-photon formalism (Caves and Schumaker 1985). The electric field can then be rewritten as:

$$\hat{E}(x, y; z, t) = u(x, y, z) \sqrt{\frac{4\pi \hbar \omega_0}{Ac}} \left[\hat{a}_1(z, t) \cos \omega_0 t + \hat{a}_2(z, t) \sin \omega_0 t \right].$$

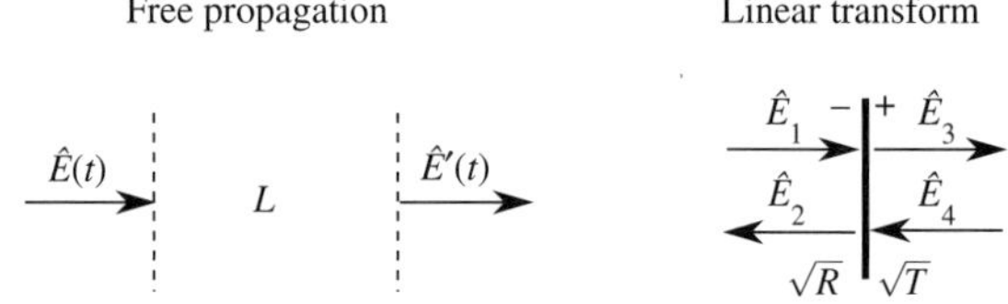

Figure 15.2 Two basic dynamical processes of the optical field in analysing the quantum noise of an interferometer.

with ω is approximated as ω_0 and the time-domain quadratures defined as:

$$\hat{a}_{1,2}(z,t) \equiv \int_0^{+\infty} \frac{\mathrm{d}\Omega}{2\pi} \left(\hat{a}_{1,2} \mathrm{e}^{-i\Omega t + ikz} + \hat{a}_{1,2}^{\dagger} \mathrm{e}^{i\Omega t - ikz} \right). \tag{15.7}$$

These correspond to amplitude and phase modulations in the classical limit.[1]

After having introduced the quantisation, we can further look at the dynamics of the optical field. The equations of motion that we will encounter turn out to be very simple, and only two are relevant, as shown in Figure 15.2:

1. *A free propagation.* Given a free propagation distance of L, the new field $\hat{E}'(t)$ is

$$\hat{E}'(x, y, z + L, t) = \hat{E}(x, y, z, t - \tau), \tag{15.8}$$

   with $\tau \equiv L/c$;

2. *Continuity condition on the mirror surface. The equations are:*

$$\hat{E}_2(t) = \sqrt{T} \hat{E}_4(t) - \sqrt{R} \hat{E}_1(t), \tag{15.9}$$

$$\hat{E}_3(t) = \sqrt{R} \hat{E}_4(t) + \sqrt{T} \hat{E}_1(t), \tag{15.10}$$

with transmissivity T and reflectivity R and a sign of convention as indicated in Figure 15.2. These equations relates the optical field before and after the mirror. Due to linearity of the system, these equations are both identical to the classical equations of motion.

In later discussions, different quantities of the optical field are always compared at the same location, and they all share the same spatial mode. In addition, the propagation phase shift can be absorbed into the time delay. Therefore, we will ignore the factors $u(x, y, z)\sqrt{\frac{4\pi \hbar \omega_0}{\mathcal{A}c}}$ and $\mathrm{e}^{\pm ikz}$ hereafter.

Quantum states of the optical field

To determine the expectation value and the quantum fluctuation of the Heisenberg operators (related to the quantum noise), e.g. $\langle \psi | \hat{O} | \psi \rangle$, not only should we specify the evolution of

---

[1] To see such correspondence, suppose the electric field has a large steady state amplitude A:

$$\hat{E}(z,t) = [A + \hat{a}_1(z,t)] \cos \omega_0 t + \hat{a}_2(z,t) \sin \omega_0 t \approx A \left[1 + \frac{\hat{a}_1(z,t)}{A} \right] \cos \left[\omega_0 t - \frac{\hat{a}_2(z,t)}{A} \right].$$

$\hat{O}$, but we also need to specify the quantum state $|\psi\rangle$. Of particular interest to us are the vacuum, coherent, and squeezed states.

Vacuum state

The vacuum state $|0\rangle$ is, by definition, the state with no excitation and for every frequency, $\hat{a}_{\Omega}|0\rangle = 0$. The associated fluctuation is given by:

$$\langle 0|\hat{a}_i(\Omega)\hat{a}_j^{\dagger}(\Omega')|0\rangle_{\text{sym}} = \pi\,\delta_{ij}\delta(\Omega - \Omega'), \quad (i, j = 1, 2). \tag{15.11}$$

Equivalently, the double-sided spectral densities [2] for $\hat{a}_{1,2}$ are given by:

$$\tilde{S}_{a_1}(\Omega) = S_{a_2}(\Omega) = 1, \quad S_{a_1 a_2}(\Omega) = 0. \tag{15.12}$$

Coherent state

The coherent state is defined as (cf. Blow *et al.*, 1990)

$$|\alpha\rangle \equiv \hat{D}[\alpha]|0\rangle \equiv \exp[\int \tfrac{d\Omega}{2\pi}(\alpha_{\Omega}\hat{a}_{\Omega}^{\dagger} - \alpha_{\Omega}^*\hat{a}_{\Omega})]|0\rangle, \tag{15.13}$$

which satisfies $\hat{a}_{\Omega'}|\alpha\rangle = \alpha(\Omega')|\alpha\rangle$. The operator $\hat{D}$ is unitary and $\hat{D}^{\dagger}\hat{D} = \hat{I}$. We can use it to make a unitary transformation for studying the problem

$$|\psi\rangle \rightarrow \hat{D}^{\dagger}|\psi\rangle, \quad \hat{O} \rightarrow \hat{D}^{\dagger}\hat{O}\hat{D}, \tag{15.14}$$

which leaves the physics invariant. This means that the coherent state can be replaced by the vacuum state, as long as we perform corresponding transformations of $\hat{O}$ into $\hat{D}^{\dagger}\hat{O}\hat{D}$. For the annihilation and creation operators, we have $\hat{D}^{\dagger}(\alpha)\hat{a}_{\Omega}\hat{D}(\alpha) = \hat{a}_{\Omega} + \alpha_{\Omega}$ and $\hat{D}^{\dagger}(\alpha)\hat{a}_{\Omega}^{\dagger}\hat{D}(\alpha) = \hat{a}_{\Omega}^{\dagger} + \alpha_{\Omega}^*$ with the original operators plus some complex constants.

An ideal single-mode laser with a central frequency ω_0 can be modeled as a coherent state and $\alpha_{\Omega} = \pi\,\bar{a}\,\delta(\Omega - \omega_0)$ with $\bar{a} = \sqrt{2I_0/(\hbar\omega_0)}$ and I_0 the optical power. Under transformation $\hat{D}$, the electric field reads (cf. equation 15.7 and also Appendix B of Kimble *et al.* 2001):

$$\hat{E}(t) = [\bar{a} + \hat{a}_1(t)]\cos\omega_0 t + \hat{a}_2(t)\sin\omega_0 t, \tag{15.15}$$

which is simply the sum of classical amplitude and quantum quadrature fields. This is what we intuitively expect for the optical field from a single-mode laser, namely 'quantum fluctuations' superimposed onto a 'classical carrier'. In Figure 15.3, we show $E(t)$ and the associated fluctuations in the amplitude and phase quadratures schematically. As we will see later, these fluctuations are attributable to the quantum noise, and cause the associated SQL.

---

[2] For any pair of operators $\hat{O}_1$ and $\hat{O}_2$, the double-sided spectral density is defined through:

$$\frac{1}{2\pi}\langle 0|\hat{O}_1(\Omega')\hat{O}_2^{\dagger}(\Omega)|0\rangle_{\text{sym}} \equiv \frac{1}{2\pi}\langle 0|\hat{O}_1(\Omega')\hat{O}_2^{\dagger}(\Omega) + \hat{O}_2^{\dagger}(\Omega)\hat{O}_1(\Omega')|0\rangle \equiv \frac{1}{2}S_{O_1 O_2}(\Omega)\delta(\Omega - \Omega').$$

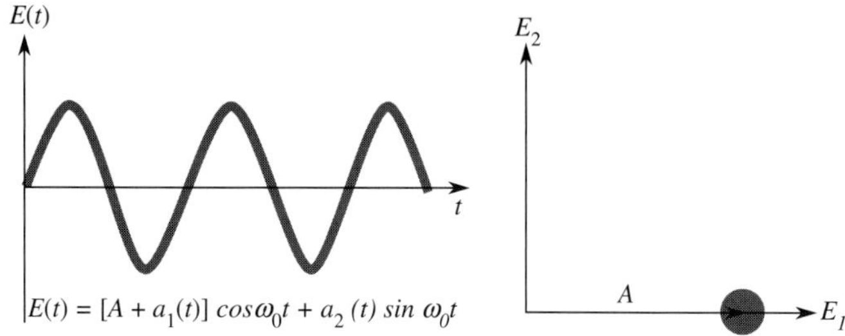

Figure 15.3 A schematic plot of the electric field and the fluctuations of amplitude and phase quadrature (shaded area). The left panel shows the time evolution of E, and the right panel shows E in the space expanded by the amplitude and phase quadratures (E_1, E_2).

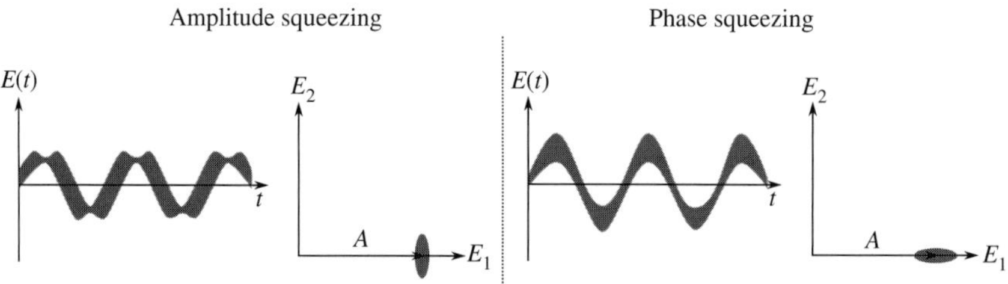

Figure 15.4 The fluctuations of the amplitude and phase quadratures (shaded area) of the squeezed state. The left two panels show the case of amplitude squeezing; the right two panels show phase squeezing.

Squeezed state

A more complicated state would be the squeezed state (cf. Blow *et al.*, 1990):

$$|[\chi]\rangle \equiv \exp[\int_0^{+\infty} \tfrac{d\Omega}{2\pi}\left(\chi_\Omega\, \hat{a}_+^\dagger \hat{a}_-^\dagger - \chi_\Omega^*\, \hat{a}_+\hat{a}_-\right)]|0\rangle \equiv \hat{S}[\chi]|0\rangle . \tag{15.16}$$

Similar to the coherent state case, we can also understand a squeezed state better by making a unitary transformation of the basis through $\hat{S}$. By redefining $\chi_\Omega \equiv \xi_\Omega\, e^{-2i\phi_\Omega}$ $(\xi_\Omega, \phi_\Omega \in \Re)$, for quadratures, this leads to:

$$\hat{S}^\dagger \hat{a}_1 \hat{S} = \hat{a}_1(\cosh\xi + \sinh\xi \cos 2\phi) - \hat{a}_2 \sinh\xi \sin 2\phi , \tag{15.17}$$

$$\hat{S}^\dagger \hat{a}_2 \hat{S} = \hat{a}_2(\cosh\xi - \sinh\xi \cos 2\phi) - \hat{a}_1 \sinh\xi \sin 2\phi . \tag{15.18}$$

Let us look at two special cases:

(1) $\phi = \pi/2$. We have:

$$\hat{S}^\dagger \hat{a}_1 \hat{S} = e^{-\xi}\hat{a}_1, \quad \hat{S}^\dagger \hat{a}_2 \hat{S} = e^{\xi}\hat{a}_2, \tag{15.19}$$

in which the amplitude quadrature fluctuation is squeezed by $e^{-\xi}$ while the phase quadrature is magnified by e^{ξ}.

(2) $\phi = \pi/2$. The situation will just be the opposite. Both cases are shown schematically in Figure 15.4.

Dynamic of the test mass

Similarly, due to linear dynamics, the quantum equations of motion for the test masses (relative motion) are formally identical to their classical counterparts:

$$\dot{\hat{x}}(t) = \hat{p}(t)/m, \quad \dot{\hat{p}}(t) = \hat{I}(t)/c + mL\ddot{h}(t). \tag{15.20}$$

Here $\hat{x}$ and $\hat{p}$ are the position and momentum operators, which satisfy $[\hat{x}, \hat{p}] = i\hbar$; $\hat{I}(t)/c$ is the radiation pressure, which is a linear function of the optical quadrature fluctuations; and $mL\ddot{h}(t)$ is the GW tidal force. Since the detection frequency ($\sim 100\,\mathrm{Hz}$) is much larger than the pendulum frequency ($\sim 1\,\mathrm{Hz}$) of the test masses in a typical detector, they can be treated as free masses.

Homodyne detection

In this section, we will consider how to detect the phase shift of the output optical field which contains the GW signal. To make a phase sensitive measurement, we need to measure quadratures of the optical field instead of the power. This can be achieved by a *homodyne detection* in which the output signal light is mixed with a *local oscillator* and produces a photon flux that depends linearly on the phase (the GW strain). Specifically, for a local oscillator $L(t) = L_1 \cos\omega_0 t + L_2 \sin\omega_0 t$ [3] and output $\hat{b}(t) = \hat{b}_1(t)\cos\omega_0 t + \hat{b}_2(t)\sin\omega_0 t$, the photocurrent is $i(t) \propto \langle\psi||L(t) + \hat{b}(t)|^2|\psi\rangle = \langle\psi|2L_1\hat{b}_1(t) + 2L_2\hat{b}_2(t)|\psi\rangle + \cdots$ [4] The rest of the terms represented by '$\cdots$' contain either frequency contents that are strictly DC and around $2\omega_0$, or terms quadratic in $\hat{b}$. In such a way, we can measure a given quadrature $\hat{b}_\zeta(t) = \hat{b}_1(t)\cos\zeta + \hat{b}_2(t)\sin\zeta$ by choosing the correct local oscillator, such that $\tan\zeta = L_2/L_1$.

In order to realise the above ideal superposition, there are two possible schemes: introducing the local oscillator from the injected laser (external scheme, as shown in the left panel of Figure 15.5); or intentionally offsetting the two arms at the very beginning, with a very small phase mismatch, which is the so-called DC readout scheme (shown in the right panel of Figure 15.5).

15.4 Quantum noise in a GW detector

After having introduced the basic concepts, we are now ready to analyse the quantum noise of a typical GW detector: a Michelson interferometer with Fabry–Perot arm cavities and a power recycling mirror (PRM).

Input–output relation of a simple Michelson interferometer

To illustrate the procedure for analysing the quantum noise, we first consider a simple Michelson interferometer as shown schematically in Figure 15.6. Later in this section we

---

[3] Here, the local oscillator is approximated as a classical field, and we have ignored its quantum fluctuation, which makes a negligible contribution to the photocurrent.

[4] Depending on the quantum state of the light, the photocurrent would be different (cf. Section 4.4.2 of Scully and Zubairy (1997)).

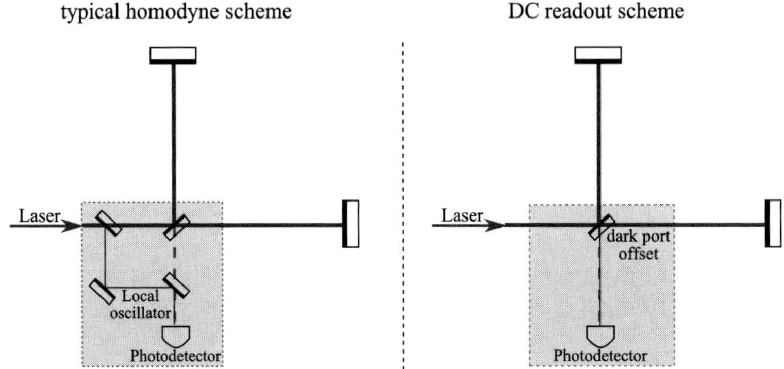

Figure 15.5 A schematic plot of homodyne readout schemes.

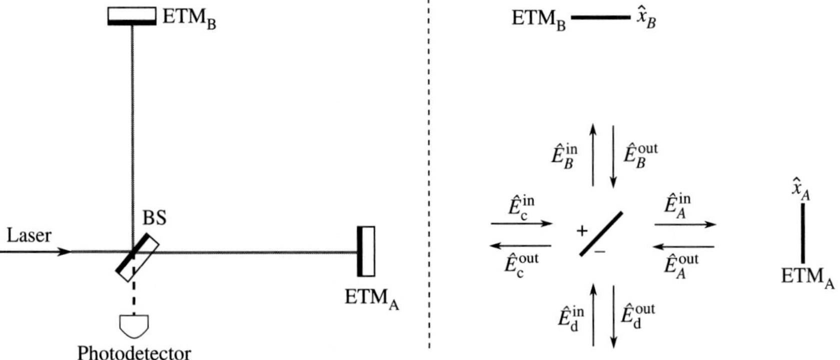

Figure 15.6 A schematic plot of a simple Michelson interferometer (left) and its mathematical model with propagating optical fields (right). ETM is short for end test mass.

will add the power recycling mirror and input test masses. Ideally, the interferometer is set up to have identical arms, so that at the zero working point of the interferometer (lock to dark fringe), the optical field entering from each port will only return to that port.

The laser-pumped input optical field into the common port is given by:

$$\hat{E}_c^{\text{in}}(t) = [\sqrt{2I_0/(\hbar\omega_0)} + \hat{c}_1(t)]\cos\omega_0 t + \hat{c}_2(t)\sin\omega_0 t. \tag{15.21}$$

Given no laser pumping, the input field into the differential port is simply

$$\hat{E}_d^{\text{in}}(t) = \hat{a}_1(t)\cos\omega_0 t + \hat{a}_2(t)\sin\omega_0 t. \tag{15.22}$$

The fields after the half–half beam splitter and propagating towards ETM$_A$ and ETM$_B$ are $\hat{E}_{A,B}^{\text{in}}(t) = [\hat{E}_c^{\text{in}}(t) \mp \hat{E}_d^{\text{in}}(t)]/\sqrt{2}$. The fields returning from the ETM are $\hat{E}_{A,B}^{\text{out}}(t) = \hat{E}_{A,B}^{\text{in}}(t - 2\tau - 2\hat{x}_{A,B}/c)$, where $\tau \equiv L/c$ is the propagation time through the arm. The output at the dark port is

$$\hat{E}_d^{\text{out}}(t) = [\hat{E}_B^{\text{out}}(t) - \hat{E}_A^{\text{out}}(t)]/\sqrt{2} \equiv \hat{b}_1(t)\cos\omega_0 t + \hat{b}_2(t)\sin\omega_0 t. \tag{15.23}$$

To the leading order in $\hat{x}_{A,B}$, the input–output relation reads

$$\hat{b}_1(t) = \hat{a}_1(t - 2\tau), \tag{15.24}$$

$$\hat{b}_2(t) = \hat{a}_2(t - 2\tau) - \sqrt{2I_0/(\hbar\omega_0)}\,(\omega_0/c)\,\hat{x}_d(t - \tau). \tag{15.25}$$

where we have assumed, $\omega_0 L/c = n\pi$, with n an integer and have defined the differential mode motion $\hat{x}_d(t) \equiv \hat{x}_B(t) - \hat{x}_A(t)$.

Radiation pressure forces acting on the two test masses have both common and differential components, which are proportional to $\hat{c}_1$ and $\hat{a}_1$, respectively. If test masses have nearly the same mass, m, then $\hat{c}_1$ ($\hat{a}_1$) will only induce common-mode (differential-mode) motion. Mathematically, we have, up to leading order in fluctuations/modulations,

$$\hat{F}_{A,B}(t) = 2(I_0/c)\{1 + \sqrt{\hbar\omega_0/2I_0}[\hat{c}_1(t - \tau) \mp \hat{a}_1(t - \tau)]\}. \tag{15.26}$$

For the differential mode, the radiation pressure force is $\hat{F}_B(t) - \hat{F}_A(t) = \sqrt{8\hbar\omega_0 I_0/c^2}\,\hat{a}_1(t - \tau)$. The motion of the differential mode $\hat{x}_d$ under both the radiation pressure force and the GW tidal force $F^h_{A,B} = \mp mL\ddot{h}(t)/2$ is given by

$$m\ddot{\hat{x}}_d(t) = \sqrt{8\hbar\omega_0 I_0/c^2}\,\hat{a}_1(t - \tau) + mL\ddot{h}(t). \tag{15.27}$$

We can solve equations (15.24), (15.25) and (15.27) by transforming them into the frequency domain. Thus we obtain:

$$\begin{bmatrix} \hat{b}_1(\Omega) \\ \hat{b}_2(\Omega) \end{bmatrix} = e^{2i\Omega\tau} \begin{bmatrix} 1 & 0 \\ -\kappa & 1 \end{bmatrix} \begin{bmatrix} \hat{a}_1(\Omega) \\ \hat{a}_2(\Omega) \end{bmatrix} + \begin{bmatrix} 0 \\ e^{i\Omega\tau}\sqrt{2\kappa} \end{bmatrix} \frac{h(\Omega)}{h_{\mathrm{SQL}}} \tag{15.28}$$

with $\kappa = 4I_0\omega_0/(mc^2\Omega^2)$ and $h_{\mathrm{SQL}} = \sqrt{4\hbar/(m\Omega^2 L^2)}$.

The output phase quadrature $\hat{b}_2(\Omega)$ contains both signal $h(\Omega)$ and noises which include the shot noise ($\hat{a}_2$ term) and the radiation pressure noise ($\hat{a}_1$ term). The signal-referred noise spectrum is simply (cf. equation (15.12)):

$$S^h(\Omega) = \left[\frac{1}{\kappa} + \kappa\right]\frac{h^2_{\mathrm{SQL}}}{2} \geq h^2_{\mathrm{SQL}} \tag{15.29}$$

with the lower bound given by the SQL.

Interferometer with power recycling and arm cavities

In order to decrease the shot noise, we need to increase the optical power. It would be difficult to achieve a high optical power by solely increasing the input power. Instead, we can add a power recycling mirror (PRM) as first proposed by Drever (1983) (see Figure 15.7). The output optical field gets coherently reflected back into the interferometer and amplify the circulation power to $I'_0 \equiv 4I_0/T_{\mathrm{PRM}}$ with T_{PRM} the power transmissivity of the PRM. Additional improvement comes from the arm cavities formed by ITMs and ETMs. These cavities further increase the optical power circulating in the arms, and also coherently amplify the GW signal by increasing the effective arm length.

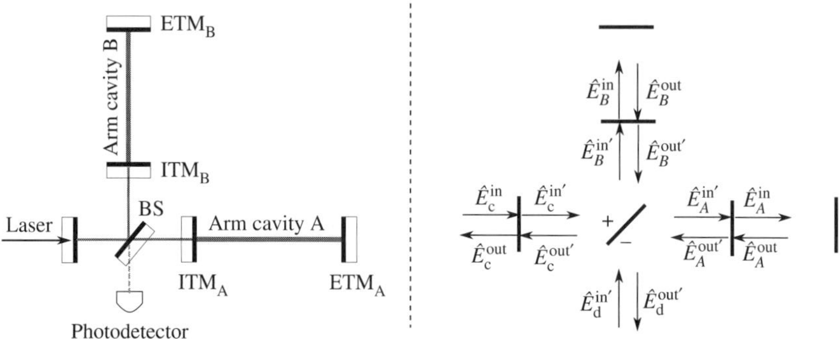

Figure 15.7 A schematic plot of a Michelson interferometer with a power recycling mirror (PRM) and additional input test masses (ITMs) to form arm cavities (left). The corresponding propagating fields are indicated on the right.

The corresponding sensitivity can be obtained from the new input–output relation at the differential port by using equation (15.24) and the following connection:

$$\hat{E}_d^{\mathrm{in}} = \sqrt{R_I}\,\hat{E}_d^{\mathrm{out}} + \sqrt{T_I}\,\hat{E}_d^{\mathrm{in}'}, \tag{15.30}$$

$$\hat{E}_d^{\mathrm{out}'} = \sqrt{T_I}\,\hat{E}_d^{\mathrm{out}} - \sqrt{R_I}\,\hat{E}_d^{\mathrm{in}'}. \tag{15.31}$$

The general expression is rather cumbersome. We will focus on the case in which the transmissivity T_I is small (a high-finesse cavity). In addition, since the GW sideband frequency is around 100 Hz, $\Omega\tau$ is much smaller than unity even for a cavity length of 4 km. Therefore, we can make a Taylor expansion of the new input–output relation as a series of the dimensionless quantities T_I and $\Omega\tau$. Up to the leading order, this leads to:

$$\begin{bmatrix} \hat{b}_1'(\Omega) \\ \hat{b}_2'(\Omega) \end{bmatrix} = e^{2i\phi}\begin{bmatrix} 1 & 0 \\ -\mathcal{K} & 1 \end{bmatrix}\begin{bmatrix} \hat{a}_1'(\Omega) \\ \hat{a}_2'(\Omega) \end{bmatrix} + e^{-i\phi}\begin{bmatrix} 0 \\ \sqrt{2\mathcal{K}} \end{bmatrix}\frac{h(\Omega)}{h_{\mathrm{SQL}}}, \tag{15.32}$$

where $\phi \equiv \arctan(\Omega/\gamma)$, $\mathcal{K} \equiv 2\gamma\,\iota_c/[\Omega^2(\Omega^2 + \gamma^2)]$ and $h_{\mathrm{SQL}} \equiv \sqrt{8\hbar/(m\Omega^2 L^2)}$ with the cavity bandwidth $\gamma \equiv T_I c/(4L)$, parameter $\iota_c \equiv 8\omega_0 I_c/(mLc)$, and intra-cavity power $I_c \equiv 8I_0/(T_{\mathrm{PRM}}T_I)$. Compared with previous definition for h_{SQL}, the factor of two difference comes from the fact that the reduced mass of each arm is $m/2$. By detecting the phase quadrature, the signal-referred noise spectrum reads

$$S^h(\Omega) = \left[\frac{1}{\mathcal{K}} + \mathcal{K}\right]\frac{h_{\mathrm{SQL}}^2}{2}. \tag{15.33}$$

To illustrate this sensitivity curve, we choose the following specifications for different parameters (close to those of the Advanced LIGO): mass of individual test mass $m = 40$ kg, intra-cavity optical power $I_c = 800$ kW, arm cavity length $L = 4$ km, optical angular frequency $\omega_0 = 1.9 \times 10^{15}\,\mathrm{s}^{-1}$ (wavelength equal to 1 μm), arm cavity bandwidth $\gamma/(2\pi) = 100$ Hz. The corresponding GW strain-referred sensitivity is shown in Figure 15.8 with $f \equiv \Omega/2\pi$, and it touches the SQL round 100 Hz. As we can see, there is a degradation of sensitivity at high frequencies, and this is due to the low-pass filtering of the cavity.

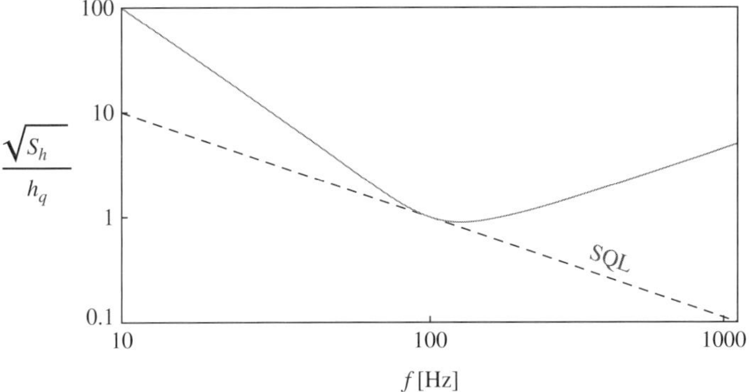

Figure 15.8 The GW strain-referred sensitivity of an advanced GW detector with power recycling mirror and arm cavities given the specifications detailed in the main text. Here we have normalised the spectrum by h_q, which is defined as h_{SQL} at 100 Hz.

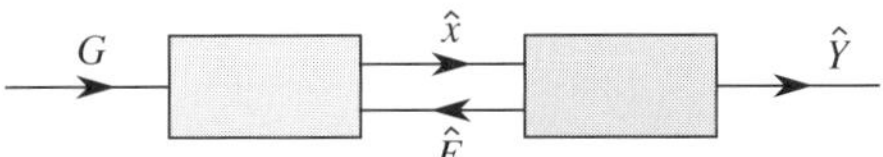

Figure 15.9 A schematic model of a linear continuous measurement process.

Basically, the cavity cannot respond efficiently to the test mass motion when the motion frequency becomes higher than the cavity bandwidth (here 100 Hz).

15.5 Derivation of the SQL: A general argument

Through the previous discussions, we have learnt that there are two types of noise, namely shot noise and radiation pressure noise, which together give rise to the SQL. In reality, the SQL exists in general linear continuous measurements, and it is directly related to the fundamental Heisenberg uncertainty principle (Braginsky and Khalili, 1992).

The model of a typical measurement process is shown schematically in Figure 15.9. The signal – a classical force G (e.g. the GW) – is driving the probe (e.g. the test mass), which is in turn coupled to an external detector (e.g. the optical field). The detector reads out the probe motion by monitoring its displacement $\hat{x}$, and at the same time it exerts a back-action force $\hat{F}$ onto the conjugate momentum of the probe. The signal force G can then be extracted from the detector output $\hat{Y}$, which contains both the signal and the fundamental measurement noise $\hat{Z}$ (i.e. the shot noise). The displacement-referred output $\hat{Y}$, i.e. the measurement result, can be written as:

$$\hat{Y}(t) = \hat{Z}(t) + \int_{-\infty}^{t} dt' R_{xx}(t - t')[\hat{F}(t') + G(t')], \tag{15.34}$$

with $R_{xx}(t)$ the response function of the probe to the external force. The corresponding displacement-referred noise spectrum is then given by

$$S^{x}(\Omega) = S_{ZZ}(\Omega) + 2\Re[R_{xx}(\Omega) S_{ZF}(\Omega)] + |R_{xx}(\Omega)|^{2} S_{FF}(\Omega). \tag{15.35}$$

According to the theory of linear continuous measurements (Braginsky and Khalili, 1992), the shot noise and the back-action force satisfy

$$[\hat{Z}(t), \hat{F}(t')] = i\hbar \delta(t - t') . \tag{15.36}$$

This commutator dictates the Heisenberg uncertainty relation among the spectral densities $S_{ZZ}(\Omega)$, $S_{FF}(\Omega)$ and $S_{ZF}(\Omega)$:

$$S_{ZZ}(\Omega)S_{FF}(\Omega) - |S_{ZF}(\Omega)|^2 \geq \hbar^2 . \tag{15.37}$$

When there is no correlation between the shot noise and the back-action noise $S_{ZF} = 0$, the displacement noise spectrum is bounded by the SQL:

$$S^x(\Omega) \geq 2|R_{xx}(\Omega)|\sqrt{S_{ZZ}(\Omega)S_{FF}(\Omega)} \geq 2\hbar|R_{xx}(\Omega)| \equiv S_{\text{SQL}}^x . \tag{15.38}$$

In the case of GW detection, the GW tidal force on the test masses is $G(t) = mL\ddot{h}(t)$. The corresponding GW signal-referred SQL is given by:

$$S_{\text{SQL}}^h = \frac{2\hbar}{m^2\Omega^4 L^2 |R_{xx}(\Omega)|} . \tag{15.39}$$

For a free mass, $R_{xx}(\Omega) = -1/(m\Omega^2)$, it recovers the free-mass SQL.

From the above derivation, we immediately realise that there are two possible approaches to surpass the free-mass SQL:

1. *Creating correlations between the shot noise and the back-action noise* such that $S_{ZF} \neq 0$, The inequality in equation (15.38) is not satisfied and the noise spectrum will not be bounded by the SQL;
2. *Modifying the dynamics of the probe*. For an oscillator with a resonant frequency ω_m and a decay rate of γ_m, $R_{xx}(\Omega) = 1/[-m(\Omega^2 - \omega_m^2 + i\gamma_m\Omega)]$ and around resonance frequency

$$\frac{S_{\text{SQL}}^h|_{\text{oscillator}}}{S_{\text{SQL}}^h|_{\text{free mass}}} = \frac{\sqrt{(\Omega^2 - \omega_m^2)^2 + \gamma_m^2\Omega^2}}{\Omega^2}\bigg|_{\Omega=\omega_m} = \frac{\gamma_m}{\omega_m} . \tag{15.40}$$

Therefore, an oscillator can surpass the free-mass SQL by a significant amount of the mechanical quality factor ω_m/γ_m around its resonance frequency.

15.6 Beating the SQL by building correlations

Since the shot noise and the radiation pressure noise are related to the fluctuations in the amplitude and phase quadratures, there are three approaches to create the correlation between them: (1) signal recycling; (2) squeezed input; (3) variational readout. These three approaches are the main topics of this section.

Signal recycling

The correlation is naturally built up by adding a signal recycling mirror as shown in Figure 15.10. The output fields are reflected back into the interferometer with additional

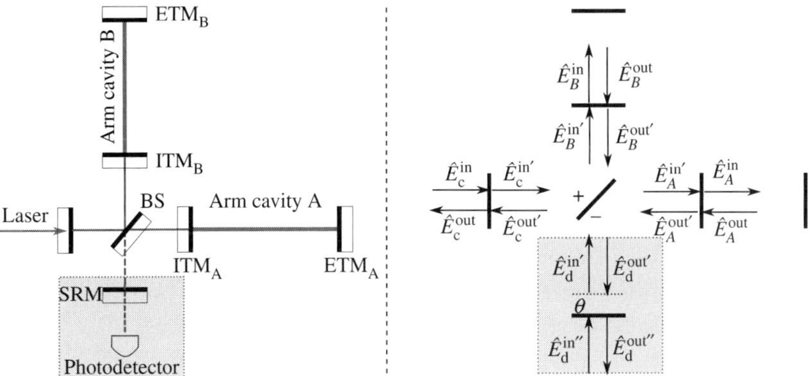

Figure 15.10 A Michelson interferometer with power recycling mirror (PRM) and signal recycling mirror (SRM). The corresponding propagating fields are indicated in the right part, with θ the detuned phase.

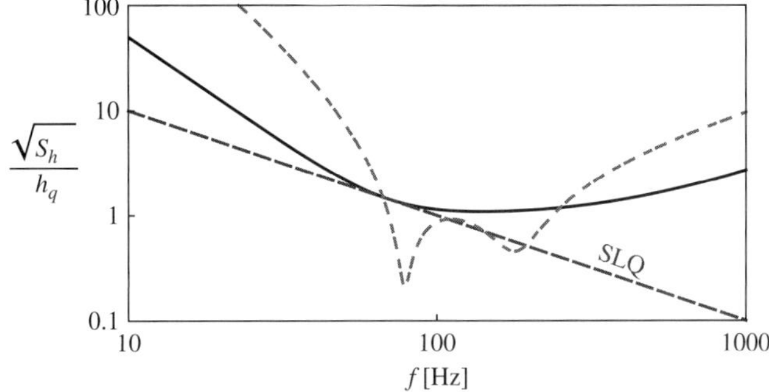

Figure 15.11 The GW strain-referred sensitivity of a signal recycling GW detector given a phase quadrature detection. The solid curve corresponds to the resonant sideband extraction scheme with $\theta = \pi/2$ and $\sqrt{R_S} = 0.6$. The dashed curve shows the case with $\theta = 1.1$ and $\sqrt{R_S} = 0.9$.

phase due to the cavity detune and we have

$$\hat{E}_{\rm d}^{\rm in'}(t) = \sqrt{R_S}\,\hat{E}_{\rm d}^{\rm out'}(t - 2\theta/\omega_0) + \sqrt{T_S}\,\hat{E}_{\rm d}^{\rm in''}(t - \theta/\omega_0), \tag{15.41}$$

$$\hat{E}_{\rm d}^{\rm out''}(t) = \sqrt{T_S}\,\hat{E}_{\rm d}^{\rm out'}(t - \theta/\omega_0) - \sqrt{R_S}\,\hat{E}_{\rm d}^{\rm in''}(t). \tag{15.42}$$

Here R_S and T_S are the reflectivity and transmissivity of the signal recycling mirror, respectively. This rotates the amplitude and phase quadratures, therefore, the shot noise and radiation pressure noise naturally gain correlations (Buonanno and Chen, 2003)). In Figure 15.11, we show the resulting sensitivity of a signal-recycled interferometer. If the detuned phase is neither zero nor equal to $\pi/2$, we can surpass the SQL significantly.

Squeezed input

As pointed out in the pioneering work of Kimble *et al.* (2001), a frequency-dependent squeezed input can be used to surpass the SQL; in which case the shot noise and radiation-pressure noise are naturally correlated as a result of the correlation between the amplitude and phase quadratures of the input squeezed light. Given a squeezed input, the amplitude and phase quadrature will be transformed according to equations (15.17) and (15.18). If the squeezing angle ϕ has a frequency dependence of $\phi(\Omega) = -\text{arccot}\,\mathcal{K}(\Omega)$, the amplitude and phase fluctuations are squeezed at low and high frequencies, respectively, and the noise spectrum will be reduced by an overall squeezing factor of e^{2q}, namely

$$S_{\text{sqz}}^h = e^{-2q}\left[\mathcal{K} + \frac{1}{\mathcal{K}}\right]\frac{h_{\text{SQL}}^2}{2}, \tag{15.43}$$

which surpasses the SQL around the most sensitive frequencies. A frequency-dependent squeezing can be realised by filtering frequency-independent squeezing through detuned Fabry–Perot cavities with prescribed detuning (Kimble *et al.*, 2001).

Variational readout: back-action evasion

Another approach to build correlations is to measure a certain combination of the amplitude $\hat{b}_1$ and phase quadrature $\hat{b}_2$ (Kimbe *et al.*, 2001), namely (cf. equation (15.32) for expressions of $\hat{b}_{1,2}$):

$$\hat{b}_\zeta = \hat{b}_1 \cos\zeta + \hat{b}_2 \sin\zeta$$

$$= (\hat{a}_1 \cos\zeta + \hat{a}_2 \sin\zeta) - \hat{a}_1 \mathcal{K}\sin\zeta + \sqrt{2\mathcal{K}}\,\frac{h}{h_{\text{SQL}}}\sin\zeta. \tag{15.44}$$

The shot noise $(\hat{a}_1 \cos\zeta + \hat{a}_2 \sin\zeta)$ and the radiation pressure noise $\hat{a}_1 \mathcal{K}\sin\zeta$ have non-zero correlation. If the detection angle ζ has the following frequency dependence: $\zeta(\Omega) = \text{arccot}\,\mathcal{K}(\Omega)$, we can completely evade the effect of back action, and obtain a sensitivity limited only by shot noise (Figure 5.12), namely

$$S_{\text{var}}^h = \frac{h_{\text{SQL}}^2}{2\mathcal{K}}. \tag{15.45}$$

The required frequency dependency can be achieved in a similar way to that of frequency-dependent squeezing.

One significant issue with this scheme is its sensitivity to the optical loss. As shown by Kimble *et al.* (2001), around the SQL touching frequency, the SQL beating ratio of a variational readout scheme with loss is given by $\mu \equiv \sqrt{S_{\text{SQL}}^h/S_{\text{var}}^h} \approx \sqrt[4]{\epsilon}$, only if the interferometer can manage a factor of $1/\sqrt{\epsilon}$ times stronger optical power. Given a typical loss of 0.01, this produces a factor of 0.3 and a power 10 times greater, which is rather challenging. Therefore, a low-loss optical setup is essential for implementation.

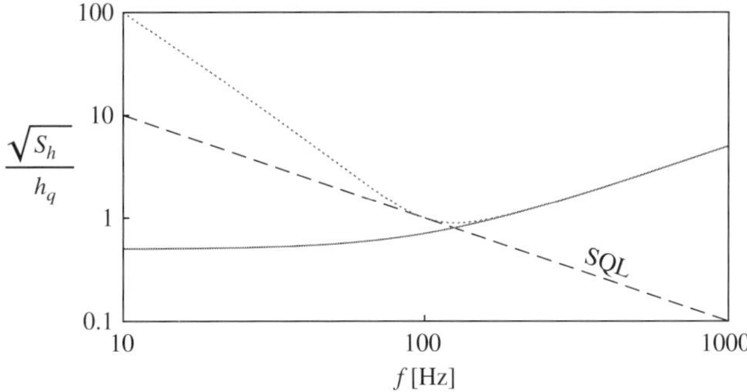

Figure 15.12 The GW strain-referred sensitivity of a variational-readout scheme (solid curve). The low frequency back-action noise is completed evaded, thus achieving a sensitivity limited only by shot noise.

15.7 Optical spring: Modification of test mass dynamics

Apart from building correlations, as shown by equation (15.40), the free mass SQL can also be surpassed by modifying the dynamics of the test mass. This can be achieved with the optical-spring effect in a detuned signal recycling interferometer. A similar idea but with a different configuration was first proposed by Braginsky *et al.* (1997), and is the so-called 'optical bar' GW detector.

Qualitative understanding of optical-spring effect

The optical-spring effect can be understood qualitatively by looking at a single detuned cavity. The displacement of the end mirror (test mass) x will change the intracavity power I_c, which in turn changes the radiation pressure force. In the adiabatic limit, for a frequency detuning of Δ, the intra-cavity power as a function of x is given by:

$$I_c(x) = \frac{\gamma^2 I_c^{\max}}{\gamma^2 + [\Delta + (\omega_0 x/L)]^2},\tag{15.46}$$

which is shown in the left part of Figure 15.13. Since the radiation pressure force is equal to $F(x) = I_c(x)/c$, such a position-dependent force will introduce a rigidity which is the opposite of the derivative of the force, $-\mathrm{d}F(x)/\mathrm{d}x$, around the equilibrium point $x = 0$. Depending on the sign of the detuning, it will create either negative or positive rigidity. For the case of a detuned signal recycling configuration for Advanced LIGO, a strong optical-spring effect can shift the pendulum frequency from 1 Hz up to the detection band. Around the new resonant frequency, we can surpass the free-mass SQL. This actually accounts for the low-frequency dip in Figure 15.11 (see Buonanno and Chen, 2002, for a detailed discussion of the mechanical resonance due to this optical-spring effect).

Due to a delayed response of the intra-cavity power to the test mass motion, the optical spring also has a friction component. For $\Delta < 0$, such a delayed response gives a positive damping. For $\Delta > 0$, the damping is negative, which will destabilise the system. The

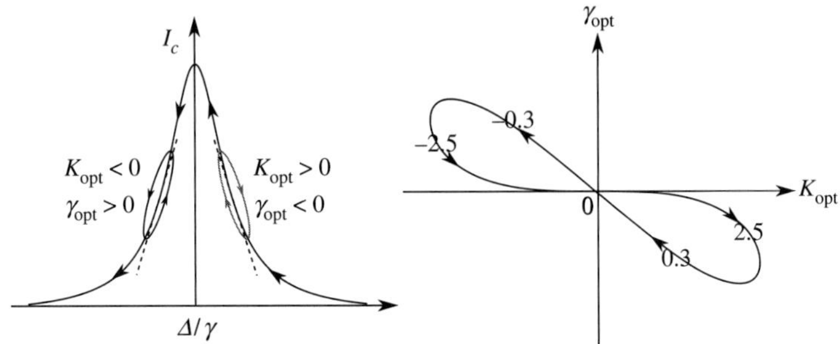

Figure 15.13 The optical power as a function of cavity detuning Δ (left) and a double optical spring (right), of which the total rigidity is sum of individual values $K_{\rm tot}(\Omega) = K_1(\Omega) + K_2(\Omega)$.

expression for an optical spring in the frequency domain is (Buonanno and Chen, 2003):

$$K(\Omega) = -\frac{2I_c\omega_0}{Lc}\frac{\Delta}{(\Omega - \Delta + i\gamma)(\Omega + \Delta + i\gamma)}. \tag{15.47}$$

For frequencies $\Omega < \Delta, \gamma$, we can perform a Taylor expansion and obtain

$$K(\Omega) = \frac{2I_c\omega_0}{Lc}\left[\frac{\Delta}{\gamma^2 + \Delta^2} + \frac{2i\gamma\Delta}{(\gamma^2 + \Delta^2)^2}\Omega\right] \equiv K_{\rm opt} - i\gamma_{\rm opt}\Omega. \tag{15.48}$$

We have introduced the rigidity $K_{\rm opt}$ (real part of K) and the damping $\gamma_{\rm opt}$ (imaginary part). The positive (negative) rigidity is always accompanied by a negative (positive) damping. In either case, the system is unstable. To stabilise the system, one can use a feedback control as described by Buonanno and Chen (2001). An interesting alternative is to use a double optical spring by pumping the cavity with two lasers at different frequencies (Corbitt *et al.*, 2007; Rehbein *et al.*, 2008). One laser with a small detuning provides a large positive damping while another with a large detune, but with a high power, provides a strong restoring force. The resulting system is self-stabilised with both positive rigidity and positive damping, as shown in Figure 15.13.

15.8 Continuous state demolition: Another viewpoint on the SQL

In the previous section [cf. Sec. 15.5], we derived the SQL by focusing on the quantum nature of the detection (the optical field). In this section, we will derive the SQL from another perspective – continuous state demolition, which is the early argument of Braginsky (1968). This will guide us to find new approaches to surpass the SQL.

In the Heisenberg picture, suppose we attempt to measure the position of a free mass successively at discrete times separated by τ. If we measure $\hat{x}$ at time t_1, *then immediately after* t_1, the uncertainty in the test mass displacement mass is comparable to the individual measurement error ϵ, or

$$\Delta x(t_1 + 0) = \epsilon. \tag{15.49}$$

The value of ϵ decreases indefinitely as individual measurement sensitivity increases. Applying free mass quantum mechanics for the duration of $t \in (t_1, t_2)$, we have [$\hat{x}(t_1 +$

$0), \hat{x}(t_2 - 0)] = [i\hbar(t_2 - t_1)]/m \equiv i\hbar\tau/m$. The decreasing ϵ will lead to an increasing variance in $\Delta x(t_2 - 0)$ immediately before the second measurement simply due to Heisenberg uncertainty principle which dictates that $\Delta x(t_1 + 0) \cdot \Delta x(t_2 - 0) \geq \hbar\tau/(2m)$ and

$$\Delta x(t_2 - 0) > \hbar\tau/(2m\epsilon). \tag{15.50}$$

If the successive measurements are done without coordination, i.e. if the *meters* that collapse the mirror's states at t_1 and t_2 are not correlated, the demolition will cause an additional noise. The noise of each measurement is

$$\Delta x \geq \max\left(\epsilon, \frac{\hbar\tau}{2m\epsilon}\right) \geq \sqrt{\frac{\hbar\tau}{2m}}. \tag{15.51}$$

This provides us with the scale of the SQL. In fact, if for any pulse with duration τ, which can vary at all scales, our measurement error is always $\Delta x \sim \sqrt{\hbar\tau/(2m)}$, then the *noise spectral density* of the device is characterised by $S^x(\Omega) \sim \frac{\hbar}{2m\Omega^2}$. From this point of view, the SQL can be traced back to the fact that the test mass positions at different times do not commute with themselves.

15.9 Speed meters

This alternative viewpoint on the SQL naturally brings up the idea of a speed meter which measures the speed (momentum) instead of the position of the test mass. Since momentum of a free mass is a conserved dynamics quantity and its Heisenberg operators at different times commute with each other, one can measure it continuously without imposing additional noise, thus allowing us to surpass the SQL. Here we will discuss two realisations, both found as prototypes in the early papers of Khalili and Levin (1996) and Braginsky *et al.* (2000), but later gradually deformed into the shape of km-scale laser interferometers (Purdue and Chen, 2002; Purdue, 2002; Chen, 2003).

Realisation I: Coupled cavities

A possible Michelson variant is shown in Figure 15.14. An additional sloshing cavity is added after the interferometer output. It has an input mirror with transmissivity T_s and a totally reflecting end mirror. There is another extraction mirror (with a transmissivity of T_0) between the interferometer output and the sloshing cavity, through which we read out the signal. This configuration emerges from the two-resonator model of Braginsky and Khalili, where it was pointed out that if two resonators are coupled, then a sloshing of signal light between the two cavity cancel with each other, leaving only sensitivity to the change in mirror position, i.e. the speed. The characteristic sloshing frequency is given by $\omega_s = \sqrt{T_s}c/(2L)$. The explicit expression for the response function of the output to the test mass motion can be found by analysing the input–output relation, giving:

$$\mathcal{T}(\Omega) = \frac{\omega_0}{L}\frac{i\Omega}{\Omega^2 - \omega_s^2 + i\gamma\Omega}. \tag{15.52}$$

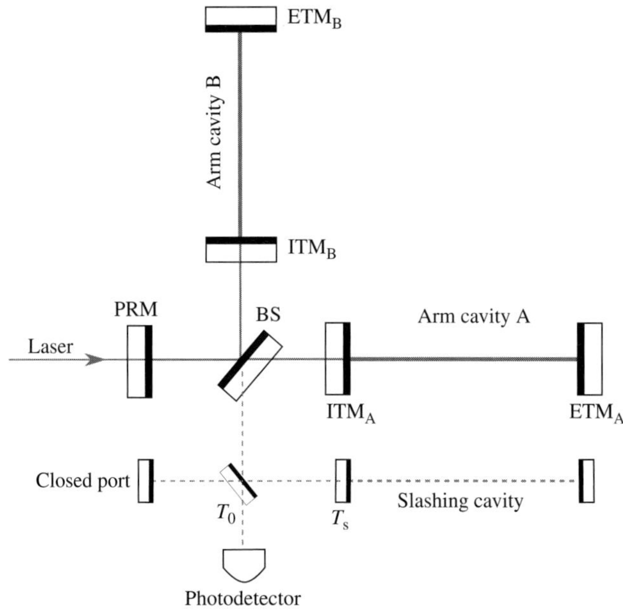

Figure 15.14 A schematic plot of a speed meter configuration with coupled cavities. A sloshing cavity is added beyond the dark port.

For frequencies $\Omega < \omega_s$, the response is proportional to $-i\Omega x(\Omega)$, which is the test mass speed in the time domain.

The GW strain sensitivity of such a configuration can be derived by using the input–output formalism we have introduced. Given a frequency-independent squeezing (phase squeezing factor e^{-2q}), it reads (refer to Purdue and Chen, 2002, for more details)

$$S^h = \left[\frac{e^{-2q}}{2\mathcal{K}_{\mathrm{sm}}} + \frac{e^{2q}(\cot\zeta - \mathcal{K}_{\mathrm{sm}})^2}{2\mathcal{K}_{\mathrm{sm}}} \right] h_{\mathrm{SQL}}^2, \tag{15.53}$$

where ζ is the readout quadrature angle, and

$$\mathcal{K}_{\mathrm{sm}} = \frac{16\omega_0 \gamma I_c}{mcL[(\Omega^2 - \omega_s^2)^2 + \gamma^2\Omega^2]}. \tag{15.54}$$

The first term in S^h is the shot noise term and the second term is the radiation pressure noise. At low frequencies, $\mathcal{K}_{\mathrm{sm}}$ is almost a constant. By choosing $\cot\zeta = \mathcal{K}_{\mathrm{sm}}(0)$, the low-frequency radiation pressure noise (back-action noise) can be completely evaded. We show the resulting GW strain sensitivity in Figure 15.15.

There is, however, a subtle issue: the original argument by Khalili and Levin (1996) and Braginsky *et al.* (2000) stated that momentum can be measured without additional noise – yet in speed meters, we still need back-action evasion: is this still consistent? The answer is yes, because when speed is coupled to an external observable, it ceases to be

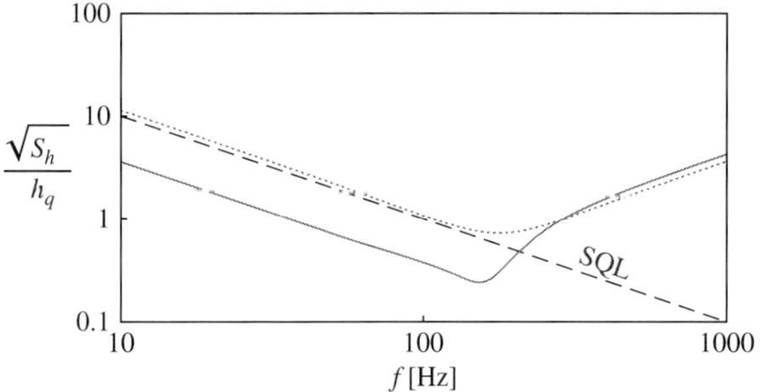

Figure 15.15 The GW strain-referred sensitivity of the speed meter scheme shown in Figure 15.14. We have chosen $\gamma/2\pi = 210\,\text{Hz}$, $\omega_s/2\pi = 180\,\text{Hz}$ and $I_c = 800\,\text{kW}$. The dotted curve is the case without a phase-squeezed input and the solid curve with $e^{-2q} = 0.1$ (a 10 dB squeezing).

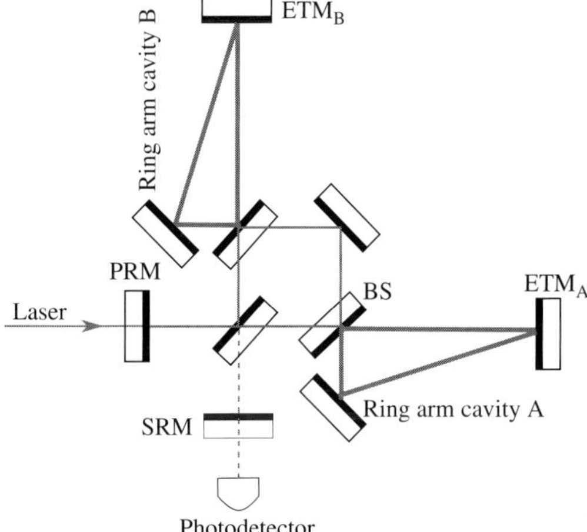

Figure 15.16 A schematic plot of a Sagnac-type speed meter configuration.

proportional to the conserved, canonical momentum, but has an additional constant contribution from the external observable (back action), which we need to subtract out from the output.

Realisation II: Zero-area Sagnac

Another possible speed-meter configuration is a zero-area Sagnac interferometer, which is shown schematically in Figure 15.16. This has a different optical topology from Michelson, where the light travels through two opposite loops in the interferometer. To understand its response to the test mass motion, we can look at a single trip. The light propagating towards arm A first picks up a phase shift proportional to the displacement of ETM$_A$, and it accumulates another phase shift due to motion of ETM$_B$ but at $t - \tau$ (τ is the delay time). A similar situation holds for the light propagating towards arm B but with the roles of ETM$_A$

and ETM_B swapped. When they recombine at the beam splitter, the total phase shift is simply:

$$\phi_{\text{tot}}(t) \propto \hat{x}_A(t) + \hat{x}_B(t-\tau) - \hat{x}_B(t) - \hat{x}_A(t-\tau) \approx [\dot{\hat{x}}_A(t) - \dot{\hat{x}}_B(t)]\tau. \qquad (15.55)$$

It naturally has no response to a static change in arm length, but only to the differential speed of two test masses. Therefore, it is a natural speed meter. This has been recognised by the GW community, but connection with the 'QND'(quantum non-demolition) speed meter has never been made. As shown in (Chen, 2003), the sensitivity of such a configuration is identical to that of the previous configuration with a sloshing cavity.

15.10 Conclusions

We have given a detailed introduction on how to analyse the quantum noise in advanced GW detectors by using the input–output formalism. In addition, we have discussed the origin of the standard quantum limit for GW sensitivity from both the detection and the test mass dynamics. This has led us to different approaches for surpassing the SQL: (1) building correlations between the shot noise and back-action noise; (2) modifying the dynamics of the test mass, e.g. through the optical-spring effect; (3) measuring the conserved dynamical quantity of the test mass – the momentum in the free mass case. For each of these approaches, we have also discussed in detail the most feasible configurations to achieve them. This not only serves as a review of advanced configurations for future GW detectors, but it additionally provides valid examples to help clarify many subtle issues in continuous quantum measurement.

Acknowledgements

Y. Chen's research has been supported by the NSF, and also by the David and Barbara Groce startup fund at Caltech. H. Miao is supported by the Australian Research Council.

References

Blow, K. J., *et al.* 1990. *Phys. Rev. A*, **42**, 4102.
Braginsky, V. B. 1968. *JETP*, **26**, 831.
Braginsky, V. B. and Khalili, F. Y. 1992. *Quantum Measurement*. Cambridge: Cambridge University Press.
Braginsky, V. B., Gorodetsky, M. L. and Khalili, F. Y. 1997. *Phys. Lett. A*, **232**, 340.
Braginsky, V. B., *et al.* 2000. *Phys. Rev. D*, **61**, 044002.
Buonanno, A. and Chen, Y. 2001. *Phys. Rev. D*, **64**, 042006.
Buonanno, A. and Chen, Y. 2002. *Phys. Rev. D*, **65**, 042001.
Buonanno, A. and Chen, Y. 2003. *Phys. Rev. D*, **67**, 062002.
Caves, C. M. and Schumaker, B. L. 1985. *Phys. Rev. A*, **31**, 3068.
Caves, C. M., *et al.* 1980. *Rev. Mod. Phys.*, **52**, 341.
Chen, Y. 2003. *Phys. Rev. D*, **67**, 122004.
Corbitt, T., *et al.* 2007. *Phys. Rev. Lett.*, **98**, 150802.
Drever, R. W. P. 1983. *Gravitational Radiation*. Amsterdam: North-Holland.
Khalili, F. Y. and Levin, Y. 1996. *Phys. Rev. D*, **54**, 004735.

Kimble, H. J., *et al.* 2001. *Phys. Rev. D*, **65**, 022002.
Purdue, P. 2002. *Phys. Rev. D*, **66**, 022001.
Purdue, P. and Chen, Y. 2002. *Phys. Rev. D*, **66**, 122004.
Rehbein, H., *et al.* 2008. *Phys. Rev. D*, **78**, 062003.
Scully, M. O. and Zubairy, M. S. 1997. *Quantum Optics*. Cambridge: Cambridge University Press.

16

ET: A third generation observatory

M. Punturo and H. Lück

Plans for a third generation interferometric gravitational wave (GW) detector are epitomised by the Einstein Telescope proposal. We start by describing the motivation for building third generation instruments, followed by a description of the different science objectives that can be achieved by such an observatory. In the next section we discuss the technological challenges that must be met to achieve third generation sensitivities. The final section outlines a possible timeline for the development of this detector and various detector configurations that are being considered.

16.1 Introduction to the third generation of GW observatories

As described in the previous chapters and based on the current models of GW sources, the next generation of advanced interferometric GW detectors (the 'second' generation of GW interferometers, such as 'Advanced LIGO' and 'Advanced Virgo') promise the detection of GW in the first year of operation close to the target sensitivity. For example, at the nominal sensitivity of these apparatuses, it is expected that a few tens of coalescing neutron stars will be detected each year. But, apart from extremely rare events, the expected signal-to-noise ratio (SNR) of these events, in the advanced detectors, is too low for precise astronomical studies of the GW sources and for complementing optical and X-ray observations in the study of fundamental systems and processes in the Universe.

These evaluations and the need for observational precision in GW astronomy have led the GW community to start a long investigative process into the future evolution of advanced detectors to a new ('third') generation of apparatuses (Punturo *et al.*, 2009), with a considerably improved sensitivity. To realise this third generation of GW observatories, with a target of roughly a factor of ten improvement in sensitivity, over a wide frequency range, with respect to the advanced detectors, several limitations of the technologies prepared for the second generation interferometers must be overcome and new solutions must be developed to reduce the fundamental and technical noises that will limit the next generation machines. In effect, the jump from the second to the third generation of GW interferometers is expected to be larger than the step made from the initial to the advanced detectors. The evolution from the first generation to the second generation is essentially a matter of updating, the technologies adopted in the initial machines, for example through an increase of the injected laser power; an improved seismic noise filtering (in Advanced LIGO); a reduction of the low frequency thermal noise by both improving the suspension design and suspension fiber material and adopting the monolithic fused silica design pioneered in GEO 600; and a reduction in the mirror thermal noise by selecting a better substrate material and coating (done in the intermediate step of the enhanced detectors, test mass coating for example e-LIGO and Virgo+). However, the transition from second to third generation will require a complete redesign of the observatory, starting from the infrastructure and the site hosting the interferometers.

The next sections of this chapter describe a subset of the scientific goals of a third generation GW detector and provide a short overview of the technological challenges introduced by this new generation of machines, currently under evaluation within the framework of the Einstein Telescope (ET) design study (ET design study, 2009). It is worth noting here that, within the ET scientific community, it has been suggested that the word 'observatory' should be used rather than 'detector' when referring to the third generation of interferometers, to underline the fact that the main component (cost-wise) of these apparatuses will be the infrastructure, which will last for decades and will be made compatible with subsequent upgrades of the hosted detectors. In addition, the main scientific target will be the *observation* of the GW sources, rather than their *detection*, which is expected to be achieved with the advanced detectors.

16.2 Scientific potential of a third generation GW observatory

The sensitivity expected of a third generation GW observatory is a trade-off between the scientific targets, the technological achievements and the financial costs of the project. The Einstein Telescope design study (ET design study, 2009) identified an targets an improvement in sensitivity by a factor of 10, with respect to the advanced detectors, and the possibility to access the 1–10 Hz frequency range, still inaccessible for the second generation. This resulted in a possible target sensitivity (named 'ET–B'), shown in Figure 16.1.

Such a sensitivity will open a new window for understanding the physics of extreme phenomena in the Universe. Some of the possible scientific targets, discussed in more detail in the ET science vision document (Amaro-Seoane *et al.*, 2009), are listed below.

- *Astrophysics:* Measure in great detail the physical parameters of compact stars, i.e. neutron stars (NS) and black holes (BH), in a binary system (Broeck, 2006; Broeck and Sengupta, 2007), constrain the equation of state of NS and solve the enigma of gamma ray bursts (GRB) (Nakar, 2007; Amaro-Seoane *et al.*, 2009).
- *General relativity:* Test general relativity by comparing observations of massive binary star systems with numerical relativity (NR) predictions and constrain alternative theories of gravity (such as the Brans–Dicke theory) through the observation of NS–BH coalescences (Amaro-Seoane *et al.*, 2009).

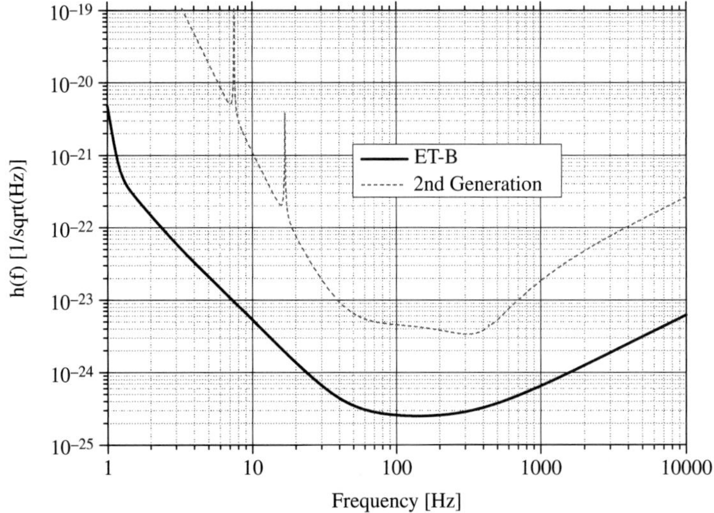

Figure 16.1 Possible sensitivity (solid curve) of an underground, long suspension, cryogenic, signal and power recycled single third generation gravitational wave observatory (see Table 1 in Hild *et al.*, 2008) compared with a typical sensitivity curve of an advanced detector (dashed curve). Note that the evaluation of the possible noise level of a third generation GW observatory is an ongoing activity within the ET design study. For this reason the curves are updated regularly and labeled with letters, here ET–B. This curve corresponds to a single wide-band detector; the suspension thermal noise contribution is missing.

- *Cosmology:* Measure cosmological parameters from standard gravitational sirens (Schutz, 1986; Sathyaprakash *et al.*, 2009) and probe the primordial Universe through the measurement of the GW stochastic background (Amaro-Seoane *et al.*, 2009).
- *Astroparticle physics:* Measure or constrain the neutrino (Arnaud *et al.*, 2002) and graviton masses through the detection of the GW emitted in a supernova.

Binary systems

Binary systems of neutron stars (BNS) or black holes (BH–BH) are the main targets of the advanced GW detectors. The expected rate of BNS coalescence events within the sight distance of $\lesssim 400$ Mpc is of the order of few tens per year, which means that detection should be guaranteed, although the low signal-to-noise ratio (SNR $\gtrsim 8$) will obstruct a detailed measurement of the physical properties of the GW signal sources. The enhanced sensitivity of the third generation GW observatories will permit detailed measurements of the characteristics of the GW source, which will have an impact on cosmology. BNS are considered to be standard sirens (Schutz, 1986; Sathyaprakash *et al.*, 2009); in fact, by measuring the GW signal with a network of (at least) three detectors, it should be possible to evaluate the chirp mass M (because the chirp rate depends on M) and the amplitude, allowing a direct evaluation of the BNS luminosity distance D_L. However, GW cannot determine the redshift z of a source (and an ambiguity between intrinsic and observed total

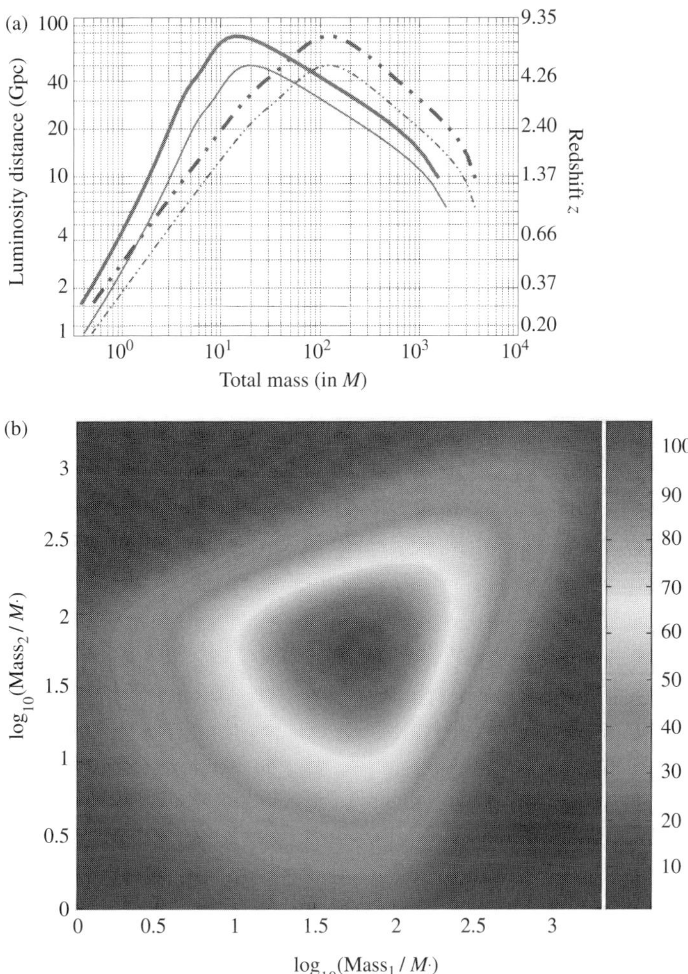

Figure 16.2 (a) The detection distance of ET-B (at an SNR = 10) for inspiral signals as a function of the *intrinsic* (solid lines) and *observed* (dashed lines) total mass. The thicker lines correspond to a mass ration $\nu = 0.25$ whereas the thinner lines correspond to $\nu = 0.10$. (b) SNR for binaries at a distance of 3 Gpc as a function of the component masses (Amaro-Seoane *et al.*, 2009; Punturo *et al.*, 2009).

BNS mass remains) and hence an electromagnetic counterpart is needed. Figure 16.2(a) shows the BNS detection range for the ET–B sensitivity, revealing a sight distance of about $z \sim 2$ and an impressive SNR at large distances. (Figure 16.2(b) shows that at 3 Gpc a SNR of several tens is possible.) With this performance, about 10^6 BNS coalescences are expected per year in the ET detection range. Suppose that in about 1 in a 1000 of these events the redshift can be measured, identifying them with coinciding gamma-ray bursts (GRB). In the adopted cosmological model the luminosity distance D_L and the corresponding redshift

Table 16.1. *Errors in the determination of the cosmological parameters through the detection of 5192 realisations of a catalogue containing 1000 BNS merger events of known redshift*

Free parameters	$\sigma_{\Omega_\Lambda}/\Omega_\Lambda$		$\sigma_{\Omega_M}/\Omega_M$		$\sigma_w/|w|$	
3	4.2%	3.5%	18%	14%	18%	15%
2	$\Omega_\Lambda = 0.73$		9.4%	8.1%	7.6%	6.6%
1	$\Omega_\Lambda = 0.73$		$\Omega_M = 0.27$		1.4%	1.1%

Source: (Sathyaprakash *et al.*, 2009)
Note: The fractional $1 - \sigma$ width of the distributions $\sigma_{\Omega_\Lambda}/\Omega_\Lambda$, $\sigma_{\Omega_M}/\Omega_M$ and $\sigma_w/|w|$ are shown, accounting for the weak lensing errors in the left column and considering it corrected in the right column. Here Ω_M and Ω_Λ are the (dimensionless) energy densities of the dark matter and dark energy, respectively; w is the dark energy equation of state parameter ($w = 1$ corresponds to a cosmological constant).

z are related by:

$$D_L = \frac{c(1+z)}{H_0} \int_0^z \frac{dz'}{[\Omega_M(1+z')^3 + \Omega_\Lambda(1+z')^{3(1+w)}]^{1/2}}, \tag{16.1}$$

Where H_o is the Hubble constant.

Introducing these two measured quantities in equation (16.1) allows the evaluation of some of the fundamental cosmological parameters. These include the total mass density Ω_M, the dark energy density Ω_Λ and the dark energy equation of state parameter w. The expected errors in the evaluation of these cosmological parameters, computed in Sathyaprakash *et al.* (2009), are listed in Table 16.1. These errors are comparable to those from other cosmological measurement attempts, such as the Joint Dark Energy Mission.

The detection distance and the SNR shown in Figure 16.2 are computed using templates produced in the so-called restricted post-Newtonian (PN) approximation, where only the lowest order (corresponding to the second harmonic of the orbital frequency) of the PN expansion is retained. Although the potential of the sub-dominant higher harmonics are not fully evaluated, it is clear (Broeck, 2006; Arun *et al.*, 2007; Broeck and Sengupta, 2007) that they could facilitate the observation of heavier binary systems (by enriching the high frequency part of the GW spectrum) as well as greatly improve the accuracy in the measurement of the GW source parameters (relaxing the requirement on the minimal number of detectors in the network needed to resolve a GW source). All this will allow the determination of the mass function of neutron stars and black holes, the maximum mass of a neutron star and its equation of state and provide a comprehensive history of the formation and evolution of compact binaries. As described in Gair *et al.* (2009a,b), such studies will give answers to important astrophysical questions such as the history of star formation, the birth of intermediate mass black holes and their growth, etc.

When the two massive stars (NS or better BH) in a binary system are close to the final coalescence, because of the ultra strong gravitational fields, the PN approximation is no longer valid and a fully relativistic treatment is needed. Thanks to the advances in analytical relativity (Buonanno and Damour, 1999, 2000; Damour *et al.*, 2002, 2003; Damour and Nagar, 2007a,b; Buonanno *et al.*, 2007) and in numerical relativity (Brügmann *et al.*, 2004; Pretorius, 2005; Baker *et al.*, 2007; Boyle *et al.*, 2007; Ajith *et al.*, 2008; Hannam and Hawke, 2009), it is possible to predict the characteristics of the signal emitted in the process of merger. Comparison of the predicted signal amplitudes and fluxes with the observation carried out by ET will enable new tests of general relativity. It should be possible to check whether the spacetime geometry of the merged object is that of a black hole or some other exotic object, whether black holes are enclosed in a horizon, and so on.

Isolated neutron stars

As explained in the previous chapters, neutron stars (NS) are one possible end state of massive stars that undergo gravitational collapse followed by a supernova of Type II, Type Ib or Type Ic. The composition of a NS is still under debate and the measurement of the GW emitted by such as body could provide important information about its internal structure. In particular, being the main mechanism for the emission of GWs by a NS due to the quadrupolar moment generated by its ellipticity ϵ, from its measurement it is possible to deduce important information on the equation of state of the ultra-dense matter composing the NS. According to current NS models, a solid core NS could sustain $\epsilon \lesssim 10^{-3}$, whereas the crust of the NS could support up to $\epsilon \simeq 10^{-6} - 10^{-7}$. As a result of its improved performance, a third generation GW observatory could be sensitive to the GW emitted by an isolated NS having a relatively small ellipticity, as shown in Figure 16.3, and allows a large fraction of our galaxy to be covered in a blind search (see Figure 16.4).

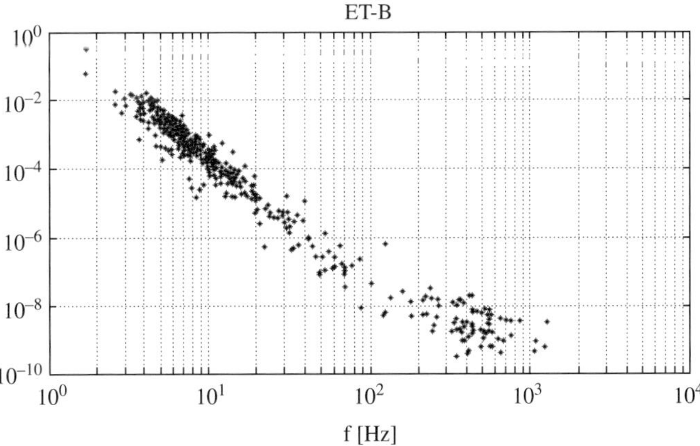

Figure 16.3 Smallest ellipticity of known pulsars detectable in ET (ET-B sensitivity) (Amaro-Seoane *et al.*, 2009).

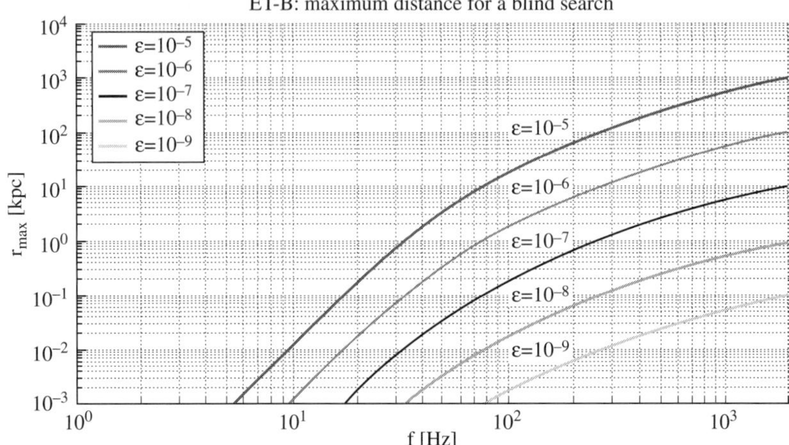

Figure 16.4 The detection range for a blind search of NS in ET (ET-B sensitivity) for different NS ellipticity (Amaro-Seoane *et al.*, 2009).

Supernova explosion

In a supernova explosion caused by the collapse of the core of a NS the energy radiated in GWs is expected to be a very small fraction of the total energy emitted (still under debate, but less than 10^{-7}–10^{-8} solar masses, according to the review by Ott (2009). This limits the detection potential of the initial and advanced detectors to the Milky Way and the closest galaxies, with an expected detection rate of about one event every two decades in the Local Group (Bergh and Tammann, 1991). ET could detect supernova explosions up to few Mpc, with an expected supernova explasion rate of one event every 1–2 years (Ando *et al.*, 2005) (if we consider a detection distance of 5–6 Mpc). If, as expected, in the time scale of the implementation of the third generation of GW observatorics, megaton-class neutrino detectors are operational and have a detection range comparable to ET (as indicated in Ando *et al.*, 2005), it will be possible to provide coincident observation with non-negligible detection rate, permitting important evaluations on the supernova explosion mechanisms and on the neutrino mass value.

16.3 Third generation sensitivity: How to suppress the noises limiting the advanced GW detectors

In Figure 16.1 a possible target sensitivity (so-called ET-B) of a third generation GW observatory is compared with the expected sensitivity of an advanced detector. To bridge the gap between the two curves, several noise sources must be suppressed. Obviously a conceptually easy way to improve the sensitivity is to realise a new interferometer with longer arms; indeed, the ET-B curve is computed assuming an arm length of 10 km. This is the first trivial indicator of the need of a new infrastructure for the third generation. But, to act on the various types of noise limiting the nominal sensitivity of the advanced detectors, in different frequency ranges, new and specific technological solutions must be found. At very low frequencies (below 3–4 Hz) the seismic and related gravity gradient noises must be suppressed. In the range 4–50 Hz the thermal noise of the optics suspension system and the quantum noise, related to the radiation pressure exerted on the suspended mirror by the photons in the main Fabry–Perot cavities, must be reduced. Above 40 Hz, the thermal noise (dominated by the mechanical dissipation of the high reflectivity dielectric coating) of the suspended mirror and the shot noise component of the quantum noise play a dominant role and must be reduced simultaneously to achieve the target sensitivity. The following subsections are devoted to the description of the limiting noises and to the possible technological solutions in a third generation GW observatory.

Seismic and gravity gradient noise

At low frequency, ground-based interferometric GW detectors are limited by the natural and anthropogenic vibration of the ground where the apparatus is built. Seismic excitation acts on the suspended test masses both indirectly, through the suspension chain, shaking each stage according to its transfer function, and directly, coupling the mass vibration in the soil layers, perturbed by the seismic waves, with the test mass displacement, via the mutual attraction force expressed by Newton's universal law of gravitation (so-called gravity gradient noise or Newtonian noise).

Advanced detectors will implement seismic filtering systems to reduce the test mass shaking due to the ground vibration. Advanced LIGO will implement a so–called *active* filtering system (Abbott *et al.*, 2002) where, in a chain of three subsystems, the displacements and accelerations caused by the seismic noise are read through position and acceleration sensors and are actively and hierarchically suppressed through hydraulic and electromagnetic actuators. Advanced Virgo will adopt, instead, the *passive* filtering philosophy already successfully implemented in the initial Virgo, based on a chain of harmonic oscillators filtering the seismic vibration horizontally (inverted and 'normal' pendulums) and vertically (blades). The so-called Virgo Super-Attenuator pushes the residual seismic noise below the thermal noise of a first generation detector like Virgo starting from about 4 Hz (Braccini *et al.*, 2005), and the compliance of this apparatus with Advanced Virgo has been demonstrated (Braccini, 2009).

In the ET design there are additional requirements that make the achievement of the sensitivity target very challenging. Over and above the request to reduce the noise level, roughly by a factor of ten, in the whole detection frequency band, the requirement to access the frequency region between 1 Hz and 10 Hz, excluded in the advanced detectors, is too difficult to be obtained just with an improved seismic filtering system; alternative solutions must be found. It is well known that underground sites are seismically quieter (i.e. see Beker *et al.*, 2009), and the possibility of realising an underground GW detector has been analysed and selected by the LCGT collaboration in Japan. The comparison between the seismic noise in the TAMA site (Tokyo) and in the LISM site (Sato *et al.*, 2004) (Kamioka mine, the prime candidate for an LCGT site) shows that by going underground there is a reduction in the low-frequency region by a factor of 100 in terms of acceleration and by two to three orders of magnitude in displacement spectral amplitude. A corresponding and even larger noise reduction has been reached in the output of the LISM interferometer, due to the fact that in going underground several other 'technical' noises, induced by external disturbances like wind, scattered light or temperature fluctuations, are suppressed by the quietness of the site. Hence, to achieve the ET-B target sensitivity an underground site is mandatory; for these conditions, it has been shown (Braccini, 2009) that the Virgo Super-Attenuator satisfies the ET requirements above 3–4 Hz, whereas to access the 1–3 Hz frequency range major technical upgrades of the suspension system have to be realised. This is the second, more important, indicator that a new infrastructure is a priority issue on the path towards the realisation of the third generation of GW observatories.

The gravity gradient noise effect has been modelled for an interferometer on the Earth's surface (Saulson, 1984; Beccaria *et al.*, 1998; Hughes and Thorne, 1998) and the predicted noise level is negligible for the initial and advanced detectors. In a third generation GW detector, the more stringent requirements at low frequencies accentuate the role of this noise source. It has been shown (Cella, 2009) that by going deep underground ($\gtrsim 100$ m) it is possible to effectively suppress the gravity gradient noise (at least above 2 Hz). Further reductions require a major effort to complement the seismic attenuation with the subtraction of the residual gravity gradient noise through signals extracted from a network of sensors located around the detector (Cella, 2009).

Thermal noise

As noted above the so-called thermal noise, generated by all the processes modulating the optical path of the light in the interferometer, coupling it to the Brownian fluctuation or to the stochastic fluctuation of the temperature field in the optical components, dominates the sensitivity of the initial and advanced detectors in the low and medium frequency range (4–200 Hz). To model and understand the thermal noise in the interferometers (in thermal equilibrium), two fundamental instruments are used: the equipartition theorem, which relates the temperature of a system to its average energies, and the fluctuation–dissipation (FD) theorem (Callen and Welton, 1951), which relates the power spectrum of the fluctuations of a system in thermal equilibrium to the dissipation processes, described by its mechanical impedance.

In the initial and advanced interferometers, the unique tool used to reduce the thermal noise effect on the detector has been the reduction of the dissipation processes through the selection of the best material available [i.e. C85 steel in Virgo (Cagnoli *et al.*, 1999) or fused silica in GEO 600 (Goßler *et al.*, 2004) and the advanced detectors (Amico *et al.*, 2002; Cagnoli *et al.*, 2006)] or through minimisation of the clamping losses [i.e. by optimising the clamping design (Cagnoli *et al.*, 1996) in initial detectors, or by introducing the monolithic design in GEO 600 and advanced detectors].

In the third generation of GW observatories a further reduction of the thermal noise limitation must be found. In both the ET design and the LCGT project, the additional methodology is based on the reduction of the interferometer operational temperature down to a few degrees kelvin. According to the equipartition theorem, temperature is directly proportional to the energy stored in each degree of freedom of the suspended system, allowing the fluctuation amplitude to be reduced by lowering the suspension temperature. Furthermore, at low temperature, some materials show a suppression of the dissipation mechanisms.

It is very important to select the right technology to cool down the suspended mirrors of the interferometer without introducing additional noise (such as seismic and acoustic excitation caused by the cryogenic liquid boiling processes, or the vibration of mechanical pumps). Cryo-cooling systems are appealing because they promise an excellent duty cycle to the interferometer, hence cryogenics is one of the most attractive technologies to reduce the thermal noise of the optics suspension in a third generation GW detector. The first problem to be solved in a cryo-interferometer is how to cool down the test masses without introducing additional vibrations that spoil the very low frequency performance. A promising technology is now available, based on cryo-cooling systems, damped to reduce the seismic vibration (Tomaru *et al.*, 2004).

The cryogenic solution introduces a discontinuity in the evolution of the materials adopted in GW interferometers. In fact, the monolithic suspension, developed for the second generation of GW detectors, cannot be used in cryogenics because of the poor thermal conductivity of the fused silica and because of a well-known dissipation peak of that amorphous material at low temperature. To be suitable for making a cryogenic suspension, a material should satisfy the following requirements: a high thermal conductivity at the operation temperature, to permit efficient heat extraction (which is crucial, because of the relatively large heating power deposited in the test masses by the high optical power stored in the interferometer cavities); a low mechanical dissipation angle (to reduce the Brownian thermal noise); a low thermal expansion coefficient (to minimise the thermoelastic noise); and a good breaking strength (to safely support the test masses). Currently there are two candidate materials for this role: sapphire and silicon. Sapphire has been selected to realise the suspension fibres of LCGT both for its dissipation properties (Uchiyama *et al.*, 2000) and for its thermal conductivity (Tomaru *et al.*, 2002). Silicon has also been found to be a suitable material for both suspension fibres (Alshourbagy *et al.*, 2006) and ribbons (Reid *et al.*, 2006). However, currently only sapphire has been used to realise a full cryogenic suspension and the usage of silicon still needs a successful research and development activity.

At intermediate frequency (40–200 Hz) it is the thermal noise, related to the losses in the suspended test masses, that dominates the noise budget of the advanced detectors. To be more precise, the mechanical dissipation located in the high reflectivity dielectric coatings (mainly in the tantalum pentoxide, Ta_2O_5, layers) gives the largest contribution to the noise level in that frequency range. A large research and development effort in the development of the advanced detectors is focused on reducing this effect, both by decreasing the dissipation of the Ta_2O_5 layers through the introduction of a titanium dopant (Harry *et al.*, 2007), and by reducing the total amount of tantalum pentoxide material while keeping the same reflectivity (Agresti *et al.*, 2006). These developments are also important for the third generation of GW observatories, but the low temperature requirement introduces additional problems. It has been shown that at low temperature, the mechanical dissipation in a multilayer tantalum pentoxide coating is rather constant (Yamamoto *et al.*, 2006). More recent measurements (Martin *et al.*, 2008) have even shown a low temperature dissipation peak in a single layer of Ta_2O_5 doped with TiO_2. Furthermore, because of the previously mentioned broad dissipation peak shown by fused silica at low temperature, in cryogenic interferometers it is impossible to employ the low mechanical loss, low optical absorption substrates developed for the advanced detectors. The best candidate materials for the test masses are, once again, sapphire (selected in LCGT) and crystalline silicon, which shows a very low mechanical dissipation angle (about $3\text{--}4 \times 10^{-9}$) at low temperature. The use of a sapphire test masses simplifies the development of all the electro-optical components of a third generation GW observatory, since it is transparent to the *standard* wavelength adopted in the GW detectors (1064 nm). Furthermore, it shows relatively small thermal lensing (Tomaru *et al.*, 2002) owing to its large thermal conductivity at low temperature ($2330\,\mathrm{W\,m^{-1}K^{-1}}$ at 10 K). However its high optical absorption (about 90 ppm cm; Tomaru *et al.*, 2001), as measured in the substrate samples available, constrains the interferometer design and limits the future light power increase in the main Fabry–Perot cavities. Future improvements in the optical properties of sapphire substrates could enhance the possibilities of this solution. Silicon shows a similar thermal conductivity ($1200\,\mathrm{W\,m^{-1}K^{-1}}$ at 12.5 K), but it is transparent only at a longer wavelength ($\lambda \gtrsim 1450\,\mathrm{nm}$), where it shows a very low absorption (about $3 \times 10^{-8}\,\mathrm{cm^{-1}}$ at 1445 nm; Green and Keevers, 1995). Use of silcon would therefore necessitale a reconsideration of all the optical and electro–optical component choices in the interferometer.

A solution to this problem is sought both by investigating possible new high refractive index materials, which must show a low dissipation at cryogenic temperature when used in a dielectric coating, and by seeking a completely new approach. An arbitrary selection of interesting research and development activities focused on a new approach is described below.

The first promising possibility (Brückner *et al.*, 2009) relates to the production of high reflectivity mirrors with just one dissipative layer of dielectric coating material on the substrate, or even without any additional layer, realising the so-called *resonant waveguide grating* (Bunkowski *et al.*, 2006) by nano-structuring the surface of the silicon substrate. Another possible innovative solution could be based upon the so-called 'Khalili cavities' (Khalili, 2005), which allow high-finesse Fabry–Perot cavities to be built that minimise the thickness of the dissipative high reflectivity coating layers, by replacing the end mirrors with additional antiresonant cavities. This option is currently under preliminary investigation in ET (Hild, 2009) and an overall reduction in the coating Brownian thermal noise by a factor of 1.5 has been evaluated; the realisation difficulties (thermal load on the end mirror substrates, short end cavity control issues, etc.) are still to be investigated. A further method, also under investigation in ET, which promises an effective reduction of the thermal noise level, is the use of flatter beams in the Fabry–Perot cavities. It is well known that the effect of the mirror thermal noise fluctuation on the detector sensitivity diminishes with beam size in the cavity but, because of the physical limitations on the mirror substrate size and on the acceptable level of diffraction losses, it is difficult to have beam sizes in ET that are substantially larger than the beams in the advanced detectors. For this reason, the possibility of using a beam with an intensity distribution flatter than the Gaussian one is appealing and the thermal noise reduction using higher-order Laguerre–Gauss modes has been investigated (Chelkowski *et al.*, 2009).

Quantum noise

As seen in the chapters discussing the second generation of GW detectors, i.e. the 'advanced' instruments, quantum noises are important both a the high and the low frequency end of the detection band. At high frequencies shot noise typically dominates the sensivity, whereas the radiation pressure noise becomes increasingly important towards lower frequencies. Both forms of quantum noise need to be lowered in order to reach the sensitivity goals for a third generation observatory.

Shot noise is an apparent test mass displacement noise as it creates a signal but no real mirror motion. Differential phase noise on the laser beams in the interferometer arms gets converted into amplitude fluctuations of the recombined beam in the output port and hence cause measurable noise on the photodetector. There are two ways of improving the shot noise: increasing the signal and reducing the noise.

The easiest way to increase the signal is to increase the light power in the interferometer arms. The signal size obtained from the photodetector at the output port created by a GW of a given amplitude linearly depends on the light power, whereas the shot noise scales as the square root of the light power. Hence the signal-to-noise ratio (SNR) also scales as the square root of the light power.

In the ET a light power in the interferometer arms of a few megawatts may be necessary. This can be achieved with a high power laser of about 1 kW power, arm cavities and power recycling as a straightforward extension of current technologies. The problems arising from increased local heating by absorbed laser power will require sophisticated thermal compensation schemes, similar to those used for the first generation of GW detectors (Waldman for the LIGO Science Collaboration, 2006; Acernese *et al.*, 2008). As in the second generation of detectors, signal recycling (Meers, 1988) can be used to further increase the signal in parts of the frequency band.

The use of high light power in the interferometer arms can lead to another problem: parametric instabilitly (Braginsky *et al.*, 2001). Parametric instabilities arise from a coupling of higher-order modes in resonant optical cavities and acoustic modes of the cavity mirror substrates and can cause serious problems for the stable operation of advanced interferometers (Ju *et al.*, 2006a, b). Studies for LCGT (Yamamoto *et al.*, 2008) have shown that the problem is less severe in this interferometer because of the different geometry and materials of the mirrors, since the radius of curvature of the mirror and the sound velocity at the operating temperature play a key role in determining the number of unstable modes. Investigation of the parametric instabilities for the ET has shown that the number of unstable modes is much greater than in LCGT (Yamamoto *et al.*, 2008) and most likely suppression techniques will be needed. Reducing the mechanical Q of the elastic mirror modes in a way that does not compromise the thermal noise to an intolerable extent shows promise for treating the parametric instabilities. Active feedback with external actuators, e.g. radiation pressure actuators, is also possible, but this requires sophisticated control loops to safely identify and suppress all possible unstable modes.

The sensitivity of the interferometer can also be improved by reducing the quantum noise. Reducing the shot noise requires lowering of the differential phase fluctuations in the interferometer arms. This can be achieved by injecting squeezed vacuum states into the output port of the interferometer. Details are provided in Kimble *et al.* (2001) and Harms *et al.* (2003). Squeezed states trade the fluctuations in one quadrature of the field against fluctuation in the other. Lowering the phase fluctuations will inevitably increase the amplitude fluctuations, as both quadratures are related to each other by Heisenberg's uncertainty relation. Hence if the noise in the squeezed light is lowered in the quadrature that coincides with the carrier amplitude at the output port in case of DC readout (often called phase quadrature, as it causes differential phase noise in the interferometer arms), the shot noise is reduced but the amplitude noise in the interferometer arms is increased. At present squeezing levels of more than 10 dB can be reached (Vahlbruch *et al.*, 2008; Mehmet *et al.*, 2009) and the squeezing can cover the full frequency range aimed for (Vahlbruch *et al.*, 2006). By the time ET is in operation we anticipate the sum of losses in the path from the squeezing source via the interferometer to the photodetector to be low enough (i.e. on the order of 5%) and the squeezing levels to be high enough (about 15 dB) to obtain an effective squeezing of 10 dB, i.e. a factor of 10 decrease in power spectral density, corresponding to the same SNR improvement as a light power increase of a factor of 10.

Noise in the amplitude quadrature of the light in the interferometer arms causes physical motion of the test masses by the momentum transfer of the light being reflected. This noise is called radiation pressure noise and rises with the inverse of the square of the signal frequency, as the mechanical susceptibility due to inertia rises towards lower frequencies. The radiation pressure effect can be lowered by increasing the mirror mass. Mirror masses on the order of 500 kg are considered for the ET. In the same way as the shot noise can be improved with squeezed light in one quadrature, radiation pressure noise can be lowered by squeezing the other quadrature of light being injected into the output port. In order to gain from squeezing in shot noise and radiation pressure noise, different quadratures will have to be used for the respective frequency ranges. This can be achieved by using filtering cavities (Kimble *et al.*, 2001). The optimal way of filtering the light before detection and filtering the squeezed state before injection for different interferometer topologies for the ET is being investigated.

16.4 Scenarios and timeline for the third generation

The realisation of a third generation GW observatory could follow different paths depending on the scientific targets chosen, the technologies selected and the financial investment to the project. In the following sections a possible scenario is depicted, based on the studies performed within the ET design study.

Single detector or multi-detector observatory

As described in the previous sections, a schematic picture of a third generation GW observatory could be based on the following options:

- long arms, probably about 10 km long, to enhance the sensitivity to the dimensionless spacetime strain h;
- underground site, to suppress the seismic and gravity gradient noises;
- longer seismic filtering chains, with respect to advanced detectors, to push down the low frequency limit toward 1 Hz;
- cryogenic test masses, to suppress suspension and mirror thermal noises;
- large and flat beams, to suppress thermal noise and mitigate the mirror thermal lensing;
- high power laser (about 1 kW), high-finesse Fabry–Perot cavities, high power recycling factor, signal recycling and squeezed light state injection, to suppress the quantum (shot) noise;
- heavy test masses and filtered squeezed state injection, to suppress radiation pressure noise.

All these options have advantages, drawbacks and constraints and, in a realistic design of a GW detector, identification of the limitations of each technology and any cross-incompatibilities is crucial. A first evaluation of the potential of a detector implementing all these options, disregarding the cross-incompatibilities between the selected technologies, has been performed within the ET project (Hild *et al.*, 2008). Targeting for a wide-band detector, the sensitivity (named ET-B) of an underground, long suspension, cryogenic, signal and power recycled single Fabry–Perot enhanced Michelson detector has been evaluated (see Table 1 in Hild *et al.*, 2008) and the resulting sensitivity has been plotted in Figure 16.1. In this evaluation, the cross-compatibility between the different technologies has been neglected, but the technological difficulties are evident. For example, the need for high power in the Fabry–Perot cavities conflicts with the requirement of a cryogenic suspension optimised for thermal noise. As an illustration, suppose it has been possible to develop a coating on the suspended mirrors that shows an excellent absorption of about 0.1 ppm; if the power stored in the cavities is about 3 MW, the deposited power is 300 mW. Even though both sapphire and silicon show excellent thermal conductivity at low temperature, to extract that heat from the text mass requires a cross-section of several tens of mm^2 (corresponding to suspending the test masses with rods several mm of radius). Obviously this affects the geometrical dilution factor of the suspension, spoiling the pendulum thermal noise, and increases the coupling between the angular and translation degrees of freedom of the suspension, making the control of the suspended test mass more complex and probably more noisy.

Developing a wide-band third generation observatory offen the opportunity to simplify the technological difficulties by merging the output of two (or more) detectors, specialised on different frequency bands. This attractive option has been evaluated in (Hild *et al.*, 2010). Here the output of a low-frequency-specialised detector is combined with the output

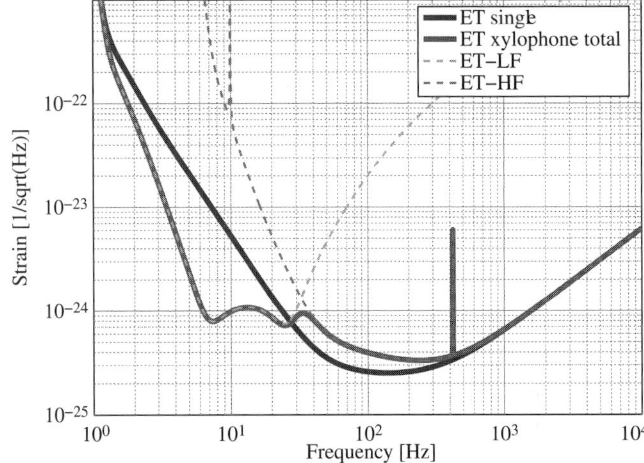

Figure 16.5 Sensitivity of a third generation GW observatory implemented by two frequency specialised detectors low frequency, LF, and high frequency, HF – the so-called 'xylophone' topology (Hild *et al.*, 2010). The curve labelled 'ET single' corresponds to the single wide frequency range ET interferometer implementation ET-B; the 'ET xylophone' curve corrresponds to ET-C.

of a high-frequency interferometer. The former could be a cryogenic interferometer at an underground site, with long suspensions, but moderate optical power, whereas the high-frequency interferometer could essentially be a long arm advanced detector, implementing squeezed light states, a very high power laser and large test masses. The main advantage of this so-called 'xylophone' philosophy (Shoemaker, 2001; Conforto and DeSalvo, 2004) is the fact that it decouples the technological requirements of a high power interferometer from the requirements of a cryogenic detector, simplifying the realisation of an overall wide-band observatory that avoids the cross-incompatibilities of the implemented technologies. As an additional benefit, the capability to tune independently the sensitivities of the two (or more) detectors constituting the observatory allows the sensitivity of such an observatory to be matched better to the GW sources sought. A possible realisation of such a xylophone strategy, evaluated in Hild *et al.* (2010) for the ET design study, is plotted in Figure 16.5.

Detector geometry

Because of the quadrupolar nature of the GW, the L-shaped, geometry with orthogonal arms optimises the sensitivity of a detector in term, of arm length. Consequently, all the currently active GW detectors are 90° L-shaped. However other geometries are possible; in particular, triangular-shaped detectors have been proposed in the past (Winkler *et al.*, 1985) and the LISA geometry is triangular. A detailed analysis of the benefits and drawbacks of a triangular-shaped third generation GW observatory is described in Freise *et al.* (2009) and few highlights are reported here. In a third generation GW observatory, to fully profit from the site infrastructure, which represents by far the dominant part of the costs (underground site excavation, cryogenics, long vacuum pipes), it is likely that more than one wide-band detector will be installed.

Co-located interferometers could be extremely useful for extracting additional information from the GW observation; for example, two L-shaped detectors, forming a 45° angle, could fully resolve the two polarisation amplitudes of the incoming wave. Using virtual interferometry techniques, three co-located interferometers, rotated by an arbitrary angle, could do the same, supplying additional benefits such as null-stream channels and redundancy. The implementation in an underground site of a cluster of similar orthogonal L-shaped detectors presents several drawbacks, mainly because of the large number of tunnels and caverns required to host the detectors and, consequently, the huge cost of the infrastructure. Some optimisation of the costs is considered possible if the angle between the two arms of each detector is reduced to 60° and three detectors can be accommodated in a triangular-shaped underground site. This would minimise the total length of the tunnels, and also the number of caverns, while offering a sensitivity equivalent to two sets of orthogonal L-shaped double detectors, rotated by 45° (see Figure B1 of Freise *et al.*, 2009).

Because of the dominant role of the cost of the infrastructures, the selection of observatory geometry will probably be driven by the selection of the site and not vice versa. If a site that can accommodate a triangular observatory is found, the triple co-located interferometers will be the best choice, otherwise two sets of orthogonal L-shaped double detectors will be more appealing.

Timelines

The evolution to the third generation of GW observatories has been, and will continue to be a long path. Currently the main effort is being made in Europe and only the European scenario will be depicted, but it is crucial to have a network of third generation GW observatories in the world. Indeed, although the enhanced sensitivity of this generation of interferometers allows the directionality of each observatory to be improved, i.e. by using the additional information embedded in the higher harmonics of the full PN modelling of the GW signal, it is recommended to have a network of three distant detectors to reconstruct the position of the GW source.

After a series of preliminary activities supported by the European Commission within the Framework Programme 6 (FP6), a conceptual design study has been funded in the 2008–2011 Framework Programme 7 (FP7): the Einstein Telescope design study (ET design study, 2009). The main goal of the ET project is to deliver a conceptual design for such a facility, investigating the technological feasibility, the scientific targets, the site requirements and prepare a costing draft for the infrastructure and a list of candidates sites.

Other design and preparation phases will follow the end of the conceptual design and it is expected that, after the first detection of GW in the advanced interferometers, it will be possible to move ahead with the realisation of the first third generation GW observatory: the Einstein Telescope observatory. According to current plans, this it will occur approximately in 2018–2019, followed by several years of construction and commissioning of the detectors.

Acknowledgements

The research leading to these results has received funding from the European Community's Seventh Framework Programme (FP7/2007–2013) under grant agreement number 211743. The authors also thank the German Centre for Quantum Engineering and Space-Time Research, QUEST, for support.

References

Abbott, R., *et al.* 2002. *Class. Quant. Grav.*, **19**, 1591–1597.

Acernese, F., *et al.* 2008. *Class. Quant. Grav.*, **25**, 114045.

Agresti, J., *et al.* 2006. *Advances in Thin-Film Coatings for Optical Applications III*, **6286**, 628608.

Ajith, P., *et al.* 2008. *Phys. Rev. D*, **77**, 104017.

Alshourbagy, M., *et al.* 2006. *Review of Scientific Instruments*, **77**, 044502.

Amaro-Seoane, P., *et al.* 2009. *Einstein Telescope Design Study: Vision Document, https://workarea.et–gw.eu/et/WG4–Astrophysics/visdoc/.* ET internal note: ET-031-09.

Amico, P., *et al.* 2002. *Review of Scientific Instruments*, **73**, 3318–3323.

Ando, S., Beacom, J. F. and Yüksel, H. 2005. *Phys. Rev. Lett.*, **95**, 171101.

Arnaud, N., *et al.* 2002. *Physical Review D (Particles, Fields, Gravitation, and Cosmology) Rev. D*, **65**, 042004.

Arun, K. G., *et al.* 2007. *Physical Review D (Particles, Fields, Gravitation, and Cosmology)*, **75**, 124002.

Baker, J. G., *et al.* 2007. *Phys. Rev. Lett.*, **99**, 181101.

Beccaria, M., *et al.* 1998. *Class. Quant. Grav.*, **15**, 3339–3362.

Beker, M. G., *et al.* 2009. *Selection criteria for ET candidate site.* ER internal note: ET-030-09.

Bergh, S. V. and Tammann, G. A. 1991. *Annual Review of Astronomy and Astrophysics*, **29**, 363–407.

Boyle, M., *et al.* 2007. *Phys. Rev. D*, **76**, 124038.

Braccini, S. 2009. *Superattenuator seismic isolation measurements by Virgo interferometer: a comparison with the future generation antenna.* ET internal note ET–025–09.

Braccini, S., *et al.* 2005. *Astroparticle Physics*, **23**, 557–565.

Braginsky, V. B., Strigin, S. E. and Vyatchanin, S. P. 2001. *Physics Letters A*, **287**, 331–338.

Broeck, C. V. D. 2006. *Class. Quant. Grav.*, **23**, L51–L58.

Broeck, C. V. D. and Sengupta, A. S. 2007. *Class. Quant. Grav.*, **24**, 155–176.

Brückner, F., *et al.* 2009. *Opt. Express*, **17**, 163–169.

Brügmann, B., Tichy, W. and Jansen, N. 2004. *Phys. Rev. Lett*, **92**, 211101.

Bunkowski, A., *et al.* 2006. *Journal of Physics: Conference Series*, **32**, 333.

Buonanno, A. and Damour, T. 1999. *Phys.Rev. D*, **59**, 084006.

Buonanno, A. and Damour, T. 2000. *Phys.Rev. D*, **62**, 064015.

Buonanno, A., *et al.*, 2007. *Phys.Rev.D*, **76:104049,2007**.

Cagnoli, G., *et al.* 1996. *Physics Letters A*, **213**, 245–252.

Cagnoli, G., *et al.* 1999. *Physics Letters A*, **255**, 230–235.

Cagnoli, G., *et al.* 2006. *Journal of Physics Conference Series*, **32**, 386.

Callen, H. B. and Welton, T. A. 1951. *Phys. Rev.* **83**, 34–40.

Cella, G. 2009. *Gravity Gradient Noise: estimates & reduction strategies.* Talk at the 2nd ET Annual meeting, http://www.et–gw.eu/2ndgeneralworkshop.

Chelkowski, S., Hild, S. and Freise, A. 2009. *Phys. Rev. D*, **79**, 122002.

Conforto, G. and DeSalvo, R. 2004. *Nuclear Instruments and Methods in Physics Research Section A: Accelerators, Spectrometers, Detectors and Associated Equipment*, **518** , 228–232. Frontier Detectors for Frontier Physics: Proceedin.

Damour, T. and Nagar, A. 2007a. *Phys.Rev.D*, **77:024043,2008**.

Damour, T. and Nagar, A. 2007b. *Phys.Rev.D*, **76:064028,2007**.

Damour, T., Iyer, B. R. and Sathyaprakash, B. S. 2002. *Phys.Rev. D*, **66**, 027502.

Damour, T., *et al.* 2003. *Phys.Rev. D*, **67**, 064028.

ET design study. 2009. The Einstein Telescope design study (FP7–Capacities, Grant Agreement 211743, see http://www.et-gw.eu/.

Freise, A., *et al.* 2009. *Class. Quant. Grav.*, **26**, 085012 (14pp).

Gair, J. R., *et al.* 2009a. *arXiv:0907.5450.*

Gair, J. R., *et al.* 2009b. *Class. Quant. Grav.*, **26**, 204009.

Goßler, S., *et al.* 2004. *Class. Quant. Grav.*, **21**, S923.

Green, M. A. and Keevers, M. J. 1995. *Progress in Photovoltaics: Research and Applications*, **3**, 189–192.

Hannam, M. and Hawke, I. 2009. *arXiv:0908.3139v1 [gr-qc].*

Harms, J., *et al.* 2003. *Phys. Rev. D*, **68**, 042001.

Harry, G. M., *et al.* 2007. *Class. Quant. Grav.*, **24**, 405–415.

Hild, S. 2009. *Khalili-Cavities for ET?* Talk at the ET–WP3 meeting in Glasgow (December 2009).

Hild, S., Chelkowski, S. and Freise, A. 2008. *arXiv:0810.0604v2 [gr-qc].*

Hild, S., *et al.* 2010. *Class. Quant. Grav.*, **27**, 015003 (8pp).

Hughes, S. A. and Thorne, K. S. 1998. *Phys. Rev. D*, **58**, 122002.

Ju, L., *et al.* 2006a. *Physics Letters A*, **355**, 419–426.

Ju, L., *et al.* 2006b. *Physics Letters A*, **354**, 360–365.

Khalili, F. 2005. *Physics Letters A*, **334**, 67–72.

Kimble, H. J., *et al.* 2001. *Phys. Rev. D*, **65**, 022002.

Martin, I., *et al.* 2008. *Class. Quant. Grav.*, **25**, 055005.

Meers, B. J. 1988. *Phys. Rev. D*, **38**, 2317–2326.

Mehmet, M., *et al.* 2009. *http://arxiv.org/pdf/0909.5386.*

Nakar, E. 2007. *Physics Reports*, **442**, 166–236.

Ott, C. D. 2009. *Class. Quant. Grav.*, **26**, 063001.

Pretorius, F. 2005. *Phys. Rev. Lett.*, **95**, 121101.

Punturo, M., *et al.* 2009. *Third Generation of Gravitational Wave Observatories and their Science Reach.* submitted to *Class. Quant. Grav.*

Reid, S., *et al.* 2006. *Physics Letters A*, **351**, 205–211.

Sathyaprakash, B. S., Schutz, B. and Broeck, C. V. D. 2009. *arXiv:0906.4151v1 [astro-ph.CO].*

Sato, S., *et al.* 2004. *Phys. Rev. D*, **69**, 102005.

Saulson, P. R. 1984. *Phys. Rev. D*, **30**, 732–736.

Schutz, B. F. 1986. *Nature*, **323**, 310–311.

Shoemaker, D. 2001. *Future limits to sensitivity.* G010026-00-G.

Tomaru, T., *et al.* 2002. *Physics Letters A*, **301**, 215–219.

Tomaru, T., *et al.* 2001. *Physics Letters A*, **283**, 80–84.

Tomaru, T., *et al.* 2002. *Class. Quant. Grav.*, **19**, 2045.

Tomaru, T., *et al.* 2004. *Class. Quant. Grav.*, **21**, S1005.

Uchiyama, T., *et al.* 2000. *Physics Letters A*, **273**, 310–315.

Vahlbruch, H., *et al.* 2006. *Phys. Rev. Lett.*, **97**, 011101.
Vahlbruch, H., *et al.* 2008. *Phys. Rev. Lett.*, **100**, 033602.
Waldman for the LIGO Science Collaboration, S. J. 2006. *Class. Quant. Grav.*, **23**, S653–S660.
Winkler, W., *et al.* (eds). 1985. Presentation *at the 4th Marcel Grossmann Meeting, Rome, Italy, Jun. 1985.*
Yamamoto, K., *et al.* 2006. *Phys. Rev. D*, **74**, 022002.
Yamamoto, K., *et al.* 2008. *Journal of Physics: Conference Series*, **122**, 012015 (6pp).